Ecology and Conservation of Pinnipeds in Latin America

Gisela Heckel • Yolanda Schramm
Editors

Ecology and Conservation of Pinnipeds in Latin America

Springer

Editors
Gisela Heckel
Departamento de Biología de la Conservación, Centro de Investigación Científica y de Educación Superior de Ensenada, Ensenada, Mexico

Yolanda Schramm
Universidad Autónoma de Baja California
Ensenada, Mexico

ISBN 978-3-030-63176-5 ISBN 978-3-030-63177-2 (eBook)
https://doi.org/10.1007/978-3-030-63177-2

This Springer imprint is published by the registered company Springer Nature Switzerland AG
The registered company address is: Gewerbestrasse 11, 6330 Cham, Switzerland

Acknowledgments

We would like to thank João Pildervasser, editor for Life Sciences in Latin America at Springer, for inviting us to "organize a compendium on the current knowledge of pinnipeds in Latin America," as he wrote in his first e-mail. It has been a great adventure, and we have learned a lot from this experience.

We acknowledge all lead authors and co-authors of the chapters for trusting this project and for writing their complete and excellent contributions, as well as revising them after receiving suggestions from their peers.

We also are grateful to all experts who reviewed one or more chapters; this improved the quality of the book substantially. The reviewers were: Jorge Acevedo, Alejandro Acevedo-Gutiérrez, Sylvia Brunner, Claudio Campagna, Susana Cárdenas-Alayza, Gustavo Daneri, Judith Denkinger, René Durán, Fernando Elorriaga-Verplancken, Jaume Forcada, Valentina Franco-Trecu, Raúl González, Denise Greig, Jana Jeglinski, Mark S. Lowry, Elizabeth McHuron, Sharon Melin, Doris Oliva, Tony Orr, Maritza Sepúlveda, Diana Szteren, Daniel Torres, Stella Villegas-Amtmann, Jochen Wolf, and two anonymous reviewers.

Contents

Contributors

Karina Acevedo-Whitehouse Unit for Basic and Applied Microbiology, School of Natural Sciences, Autonomous University of Queretaro, Queretaro, Mexico

María José Amador-Capitanachi Instituto Politécnico Nacional, Centro Interdisciplinario de Ciencias Marinas, Departamento de Pesquerías y Biología Marina, La Paz, Baja California Sur, Mexico

Enrique Alberto Crespo Laboratorio de Mamíferos Marinos, CESIMAR-CCT CENPAT-CONICET, Puerto Madryn, Chubut, Argentina

Larissa Rosa de Oliveira Mammal Ecology Lab, Universidade do Vale do Rio dos Sinos, São Leopoldo, Brazil

Eugene DeRango Department of Animal Behavior, Bielefeld University, Bielefeld, Germany

Elena Eder Centro para el Estudio de Sistemas Marinos (CESIMAR-CONICET), Puerto Madryn, Chubut, Argentina

Fernando R. Elorriaga-Verplancken Instituto Politécnico Nacional, Centro Interdisciplinario de Ciencias Marinas, Departamento de Pesquerías y Biología Marina, Instituto Politécnico Nacional. Centro Interdisciplinario de Ciencias Marinas, La Paz, Baja California Sur, Mexico

Jaume Forcada British Antarctic Survey, Natural Environment Research Council, Cambridge, UK

Casandra Gálvez Instituto Politécnico Nacional, Centro Interdisciplinario de Ciencias Marinas, Departamento de Pesquerías y Biología Marina, La Paz, Baja California Sur, Mexico

María C. García-Aguilar Departamento de Oceanología Biológica, Centro de Investigación Científica y de Educación Superior de Ensenada, Ensenada, Baja California, Mexico

Gisela Heckel Departamento de Biología de la Conservación, Centro de Investigación Científica y de Educación Superior de Ensenada, Ensenada, Mexico

Claudia J. Hernández-Camacho Laboratorio de Ecología de Pinnípedos "Burney J. Le Boeuf", Centro Interdisciplinario de Ciencias Marinas, Instituto Politécnico Nacional, La Paz, Baja California Sur, Mexico

Stephanie Kalberer Department of Animal Behavior, Bielefeld University, Bielefeld, Germany

Oliver Krüger Department of Animal Behavior, Bielefeld University, Bielefeld, Germany

Mara Landa-Garza Departamento Académico de Biología Marina, Universidad Autónoma de Baja California Sur, La Paz, Baja California Sur, Mexico

Mirtha Lewis Centro para el Estudio de Sistemas Marinos (CESIMAR-CONICET), Puerto Madryn, Chubut, Argentina

Kristine Meise Department of Animal Behavior, Bielefeld University, Bielefeld, Germany

Tenaya Norris The Marine Mammal Center, Sausalito, California, USA

Diego Páez-Rosas Galapagos Science Center, Universidad San Francisco de Quito, San Cristóbal, Islas Galápagos, Ecuador

Lili Pelayo-González Laboratorio de Ecología de Pinnípedos "Burney J. Le Boeuf", Centro Interdisciplinario de Ciencias Marinas, Instituto Politécnico Nacional, La Paz, Baja California Sur, Mexico

Paolo Piedrahita Facultad de Ciencias de La Vida, Escuela Superior Politécnica Del Litoral, Guayaquil, Ecuador

Martha P. Rosas-Hernández Marine Macroecology Department, Centro de Investigación Científica y de Educación Superior de Ensenada, Unidad La Paz, La Paz, Baja California Sur, Mexico

Yolanda Schramm Facultad de Ciencias Marinas, Universidad Autónoma de Baja California, Ensenada, Baja California, Mexico

Jonas Schwarz Department of Animal Behavior, Bielefeld University, Bielefeld, Germany

Maritza Sepúlveda Facultad de Ciencias, Universidad de Valparaíso, Valparaiso, Chile

Fritz Trillmich Department of Animal Behavior, Bielefeld University, Bielefeld, Germany

Chapter 1
Introduction: Pinnipeds in Latin America

Gisela Heckel and Yolanda Schramm

Abstract The purpose of this book is to present reviews on the ecology and conservation of pinnipeds that occur in the waters and on the coasts and islands of Latin America, a vast region extending from Mexico in the north to Argentina and Chile in the south. The aspects covered for each species include the following: general characteristics, Latin American and worldwide (if applicable) distribution and abundance, annual cycle (e.g., pupping/molting phenology), migration or movements, reproductive biology, behavior (social organization, foraging strategies, and breeding strategies), foraging/diet, conservation status (IUCN red list and national lists), and threats (e.g., hunting, fisheries interactions, and climate change, among others). In this introduction, we present basic concepts and generalized information about pinniped species distributed in Latin America (Fig. 1.1).

Keywords Seals · Sea lions · Fur seals · Evolution · Reproduction · Molting · Feeding · Conservation issues · IUCN red list

What Are Pinnipeds? Definition, Families, and Species in Both Latin America and the World

Pinnipeds (from the Latin *pinna*, or fin, and *pes*, or foot) are a group of aquatic mammals whose limbs are modified into fins by a short and wide humerus and the elongation of the first digit of the hand and the side toes of the foot. They are classified within the order Carnivora (Berta 2017; Committee on Taxonomy 2019) and into three families: Otariidae (sea lions and fur seals), Phocidae (seals), and Odobenidae (walruses). Various differences are observed in the external

G. Heckel (✉)
Departamento de Biología de la Conservación, Centro de Investigación Científica y de Educación Superior de Ensenada, Ensenada, Mexico
e-mail: gheckel@cicese.mx

Y. Schramm
Facultad de Ciencias Marinas, Universidad Autónoma de Baja California, Ensenada, Baja California, Mexico
e-mail: yschramm@uabc.edu.mx

G. Heckel, Y. Schramm (eds.), *Ecology and Conservation of Pinnipeds in Latin America*, https://doi.org/10.1007/978-3-030-63177-2_1

Fig. 1.1 Pacific harbor seals (*Phoca vitulina richardii*) hauled out on Todos Santos Island, Mexico, on 2 July 2014. (Photo by Alejandro Arias-Del Razo)

morphology of these families, where while otariids possess ear pinnae, phocids and odobenids do not. Moreover, otariids and odobenids do not have hair on their posterior limbs, while phocids do. Otariids and odobenids are capable of moving their hind limbs forward, enabling them to walk on land, an ability absent in phocids due to the tendon that passes over the astragalar process of the ankle and prevents the foot from moving forward (Berta 2017), forcing them to move on land by undulating their body. Phocids and odobenids swim using their hind limbs for propulsion, while otariids use their forelimbs for the same purpose (Riedman 1990).

There are 33 extant pinniped species in the world (Committee on Taxonomy 2019), with 8 otariid species (South American fur seal, Galápagos fur seal, Antarctic fur seal, Juan Fernández fur seal, Guadalupe fur seal, South American sea lion, California sea lion, and Galápagos sea lion) and 3 phocid species (southern and northern elephant seals and Pacific harbor seal) in Latin America. As the distribution of these species is not limited by human-defined borders, some of them are distributed beyond the territory of Latin American countries. For example, the Pacific harbor seal (*Phoca vitulina richardii*) is widely distributed across the North Pacific rim, from Baja California, Mexico, north along western North America, and across the Pacific rim to Japan (Jefferson et al. 2015; Committee on Taxonomy 2019) (Fig. 1.2).

Fig. 1.2 Pacific harbor seals and male northern elephant seal (*Mirounga angustirostris*) on Todos Santos Island, Mexico, on 25 June 2014. (Photo by Alejandro Arias-Del Razo)

Evolution

Pinnipeds originated from one carnivoran lineage, arctoid carnivorans, which, given that it includes procyonids (e.g., raccoons), mustelids, and ursids (Berta et al. 2015), is monophyletic in origin (Berta et al. 2015; Berta 2017). Pinnipedimorpha (pinnipeds and their ancestors) originated in the eastern North Pacific 30–23 million years ago, adapting to life in the water by developing fins, thick blubber, and fur as insulation against the cold water, as well as the ability to deep dive (Berta 2017).

The earliest well-represented pinnipedimorph is *Enaliarctos*, which was heterodont in dentition, with large blade-like cusps well adapted for shearing. Later diverging *Enaliarctos* had more simplified cheek teeth and, having lost cusps and roots, presented what was to become a characteristic of modern pinnipeds, whose postcanines are similar in shape and size (homodonty) and which use pierce feeding as their most common feeding strategy. *Enaliarctos* swam with its hind limbs, as do modern phocids and walruses (Berta 2017).

The best-known ancestor of the otariids is *Thalassoleon*, which presented pierce feeding and used its forelimbs for locomotion, like modern sea lions and fur seals, although the anatomy of its fore and hind limbs enabled it to walk on land better (Deméré and Berta 2005).

The earliest extant diverging otariids (2–4 million years ago) are the northern fur seal (*Callorhinus*), followed by North Pacific sea lions (*Zalophus*, *Eumetopias*) and southern sea lions and fur seals (*Otaria*, *Phocarctos*, *Arctocephalus*, *Neophoca*) (Churchill et al. 2014).

There are two groups of phocids, Monachinae (southern seals, including elephant and monk seals) and Phocinae (northern seals); both lineages originated 15 million years ago in the North Atlantic (Berta 2017). While the fossil record in South America is dominated by phocids, extant species are mostly otariids, due to a changing sea level that reduced the number of haul-out sites for phocids and an increase in the number of rocky islands surrounded by a deeper-water environment, which favors otariids (Berta 2017).

Reproduction

Pinniped mating systems are diverse, mostly due to evolutionary adaptations in response to their environment. Some pinnipeds that breed on ice at high latitudes are serially monogamous, where a male mates with one female at a time, because females tend to be sparsely distributed in this rapidly changing environment. Another mating system is polygyny, occurring when animals congregate in colonies, with the male mating with several females by means of either defending a territory (otariids) or holding a harem as the dominant alpha male (phocids). In both cases, dominant males keep other males away to prevent access to the females. Land-breeding pinnipeds, such as those living in Latin America, are polygynous and, moreover, with the exception of the harbor seal, are sexually dimorphic, meaning that males and females are different in size and appearance. Those males growing to a larger size than their rivals are more aggressive and are capable of establishing a territory or holding a harem (Riedman 1990) (Fig. 1.3).

Each year, in summer for otariids and winter for phocids, males arrive at the breeding sites earlier than females to establish territories or dominance hierarchies with other males. While located mostly on islands, breeding sites are also found on isolated mainland beaches and sandbars, habitats which offer protection from predators. When females haul out, usually at the same site year on year, they give birth to a single precocious (namely, in an advanced stage of development) pup in a specific territory or harem. A few days (4–14) after parturition, female otariids come into estrus, making them receptive to mating with the territory or harem holder (Berta et al. 2015), while estrus occurs in phocids near the end of lactation or soon after weaning (Riedman 1990). Copulation occurs generally on land, although in warmer climates, such as the Gulf of California, Mexico, copulation occurs most often in the water (García-Rivas 2008). Dominant males usually fast for the entire breeding season, which may last from a few weeks to months depending on the species, after which both males and females return to the sea to forage (Riedman 1990) (Fig. 1.4).

Pregnant females have a gestation period of roughly 10.5–11.75 months, depending on the species (Riedman 1990). The embryo stops growing when it reaches the blastocyst stage, implanting in the uterine wall after 3.5–4 months (Riedman 1990).

Fig. 1.3 California sea lion (*Zalophus californianus*) rookery on San Jerónimo Island, Mexico, on 30 May 2014. (Photo by Alejandro Arias-Del Razo)

Fig. 1.4 Guadalupe fur seal (*Arctocephalus philippii townsendi*) pup on Guadalupe Island, Mexico, on 24 July 2010. (Photo by Alejandro Arias-Del Razo)

Seemingly triggered by changes in day length (Bonner 1994), this physiological condition is known as *delayed implantation*, with fetal development continuing then for 7.75–8.5 months until parturition. This 12-month gestation allows the mothers to give birth when breeding conditions are favorable in spring and summer, with the benefit of warm weather and, for otariids, prey availability in the waters near

Fig. 1.5 Pacific harbor seal mother and pup near Todos Santos Island, Mexico, on 19 February 2014. (Photo by Alejandro Arias-Del Razo)

breeding sites. Females haul out at almost the same time and congregate in large numbers in order to give birth and raise their young (Riedman 1990) and to enable mating a few days or weeks after parturition (Bonner 1994).

Maternal investment is substantial in all pinnipeds, although differences are observed among families and individual species (Riedman 1990). Their milk is highly rich in fat (30–60%) and protein (5–15% or more), with phocid milk presenting higher fat content than otariid milk; the lactation period in the former is shorter, with phocid mothers tending to fast. Pups usually do not eat solid food during lactation and weaning occurs abruptly. Otariid mothers fast for only a few days after giving birth, then entering the sea to forage for 8 to 10 days and returning to the rookery to nurse their young. This process is repeated for several months until weaning, which occurs gradually (except in the Antarctic fur seal), with the pup supplementing its milk diet with solid food later in the lactation period (Riedman 1990). The Pacific harbor seal is an exception and combines both lactation strategies, presenting very short or non-existent fasting (Bowen et al. 1992; Boness et al. 1994), a short nursing period (3–6 weeks), and gradual weaning (Lawson and Renouf 1987) (Fig. 1.5).

Molting

Molting is the process by which the individual replaces and renews its fur coat, occurring, in otariids, over the course of various months without being obviously observable. In phocids, however, molting occurs in a relatively short time (from a

Fig. 1.6 Male northern elephant seal molting on San Benito Centro Island, on 7 August 2010, next to a California sea lion. (Photo by Alejandro Arias-Del Razo)

few weeks to a few months), sometimes involving the shedding of large patches of skin and hair, with the individuals hauling out to molt and staying on land, even fasting, in the case of the elephant seal (Riedman 1990). The molting season varies by age and sex class and is related to reproductive status, where, for example, juvenile harbor seals are the first to molt in the winter, with mothers starting to molt after having weaned their pups, while males molt after the breeding season is completely over and there are no estrous females available for mating (Tapia-Harris et al. 2017). Pups shed their lanugo or natal fur in utero or several months after birth, depending on the species (Riedman 1990) (Fig. 1.6).

Feeding

Unlike other carnivores, pinniped dentition is modified, where while they do have incisors and canines, their premolars and molars tend to have the same size and shape and are, therefore, known as postcanines. Usually swallowing their prey whole, pinnipeds do not need to slice or chew their food, indicating that their dentition has adapted to this way of feeding, with their postcanines presenting only one cusp, which is used for piercing prey (Berta et al. 2015). It has been suggested that their relatively long small intestines (8 and 25 times their body length in the California sea lion and the elephant seal, respectively) help them to digest their prey whole. Another explanation for such long small intestines in elephant seals is related to the frequent, deep diving they undertake for long periods, during which the intestine serves as a storage compartment for food when the animal is at sea for several weeks or months (Krockenberger and Bryden 1994) (Fig. 1.7).

Fig. 1.7 Juvenile Guadalupe fur seals interacting on Guadalupe Island on 22 July 2010. (Photo by Alejandro Arias-Del Razo)

Pinnipeds are top predators, occupying high trophic levels (3.7–4.3), with the composition of their diet presenting varying proportions of benthic invertebrates, large zooplankton, squid, small pelagic fish, mesopelagic fish, and high vertebrates, i.e., marine mammals and birds (Pauly et al. 1998). Foraging on tens of different species, pinnipeds have been deemed to be opportunistic or generalist predators (Riedman 1990); however, in almost all cases, their diet comprises only a few species during any one season in a particular geographical area (Berta et al. 2015). Some species have a limited diet composition, such as elephant seals (*Mirounga leonina* and *M. angustirostris*) which prefer to feed on squid and fish (Pauly et al. 1998).

Pinniped foraging behavior varies according to the type of prey, with non-schooling fish or squid and slow-moving or sessile invertebrates most efficiently captured individually in most phocids, such as elephant and harbor seals. In contrast, schooling fish and squid that occur in a patchy distribution in pelagic waters are more efficiently captured via cooperative foraging strategies, which are more frequently used by otariids when schools are large. Otariids, however, also forage singly, depending on the type of prey (Riedman 1990).

Conservation Issues: Hunting, Fisheries Interactions, and Climate Change

Examination of archaeological sites has revealed that, thousands of years ago, pinnipeds were hunted, with their meat and blubber used for food; their oil used for cooking, light, and heat; and their skins used for clothes, bedding, kayaks, rope, and

Fig. 1.8 Guadalupe fur seal on Guadalupe Island, 24 July 2010. (Photo by Alejandro Arias-Del Razo)

shelter materials (Roberts 2007; Berta et al. 2015). It was not until the early eighteenth century that pinnipeds began to be exploited commercially, mainly for their skins and oil (Busch 1985), with the latter mainly undertaken by whalers, whose whaling expedition had not yielded sufficient oil. However, after the Chinese market for skins emerged in the late eighteenth century, sealing grew in importance. Sealers mostly hunted fur seals because they have denser hair than sea lions, meaning that the pelts of the former were highly valued and while the skin of sea lions had no commercial value, they were hunted for blubber. This was also the case for southern and northern elephant seals, which were hunted to near extinction. During the early 1800s, hundreds of sealing ships reached even the remotest islands in search of profit (Roberts 2007), including the Galápagos Islands and the coasts of Baja California and Guadalupe Island, which yielded approximately 150,000 fur seals during the nineteenth century. By the beginning of the twentieth century, fur seal numbers were very low, and some, such as the Guadalupe fur seal and the Galápagos fur seal, were thought to be extinct (Roberts 2007). In addition, other commercial uses were found for the less-desirable sea lions and the so-called trimmings, referring to the use of dried penises and gall bladders in oriental medicine and whiskers for toothpicks and opium-pipe cleaners (Busch 1985). In recent years, California sea lions have been reportedly killed illegally in the Gulf of California, for the harvesting of their genitalia, presumably for the oriental market, on which they are still highly valued as both a medicine and an aphrodisiac (Martínez 2019, Hernández-Camacho et al., Chap. 7, this volume) (Fig. 1.8).

Due to their feeding habits and strategies, pinnipeds tend to interact with many fisheries (Northridge 2009) and are perceived as direct competition by many fishermen, who observe pinnipeds removing fishing product or becoming entangled in fishing gear. Therefore, fishermen tend to blame pinnipeds for catch reductions and sometimes take retaliatory measures, killing or at least harassing them (Read 2008). As many studies have been carried out to identify or quantify pinniped prey, there is some evidence of their direct competition with fisheries (Szteren et al. 2004). However, this is a controversial subject, with studies showing that catch reductions may also be caused by overfishing (Goñi 1998) or changes in the oceanographic environment that affect the abundance of fish (Sandweiss et al. 2004).

Even more important than illegal killing of seals and sea lions is their entanglement and drowning in fishing nets. Gill nets, mainly discarded or lost nets ("ghost nets"), are the fishing gear in which the highest number of pinnipeds – and other megafauna – are caught accidentally every year (Northridge 2009). Even if the animal manages to survive, small pieces of netting can impede swimming. Other debris, such as packing bands, encircle the neck or shoulders of seals and sea lions and, as they grow, cut into the tissues and, ultimately, may suffocate the animals (Bonner 1994).

Climate change is also a threat for pinniped conservation (Kovacs et al. 2012), with ice-associated pinnipeds, which use the ice for breeding, molting, and resting in the Arctic and Antarctic, at notable risk. For pinnipeds in low and mid-latitude areas, such as those in Latin America, the availability of prey may change, with climate change seemingly increasing the frequency of El Niño and other warming events. It has already been observed that, during El Niño years, biological productivity is drastically reduced in the eastern Pacific Ocean, leading to high pup mortality in Galápagos fur seals (Trillmich, Chap. 3, this volume) and Galápagos sea lions (Krüger et al., Chap. 8, this volume). A northward shift in long-term distribution of northern elephant seals from Mexico has been documented, probably as a consequence of ocean warming (García-Aguilar et al. 2018, García-Aguilar, Chap. 10, this volume).

Conservation Measures: IUCN Red List Status of Latin American Pinnipeds

The International Union for the Conservation of Nature (IUCN) periodically assesses the risk of extinction for thousands of flora and fauna species across the world (IUCN 2020). Pinnipeds in Latin America have been classified in only two categories: *Endangered* (Galápagos fur seal and Galápagos sea lion) and *Least Concern* (South American fur seal, Antarctic fur seal, Juan Fernández fur seal, Guadalupe fur seal, California sea lion, South American sea lion, Pacific harbor seal, southern elephant seal, and northern elephant seal) (IUCN 2020). The Galápagos fur seal and sea lion are still endangered, mainly because their small and declining populations are exposed to repeated El Niño events and other threats (Krüger et al., Chap. 8; Trillmich, Chap. 3, this volume). All other species, notably fur seals and elephant seals, have recovered from near extinction due to the legal protection they have been given in the countries where they occur and, furthermore, because many of their breeding colonies are found within natural protected areas. However, fisheries interactions still pose a major threat for many pinniped species, with law enforcement difficult to achieve due to their marine foraging habits, the remoteness of haul-out sites, and a lack of personnel in the corresponding government agencies.

Governments in Latin America are committed to the conservation and management of pinnipeds, with species and natural protected area management programs implemented in the field, based on the best available scientific information. Non-governmental organizations aid in these efforts with educational programs in the communities living near pinniped colonies. However, conservation strategies have to be designed and implemented at a local, national, regional, and international level in order to achieve goals on a wide geographical and long-term basis.

References

Berta A (2017) The rise of marine mammals: 50 million years of evolution. Johns Hopkins University Press, Baltimore

Berta A, Sumich JL, Kovacs KM (2015) Marine mammals: evolutionary biology, 3rd edn. Academic Press, San Diego

Boness DJ, Bowen WD, Oftedal OT (1994) Evidence of a maternal foraging cycle resembling that of otariid seals in a small phocid, the harbor seal. Behav Ecol Sociobiol 34:95–104

Bonner N (1994) Seals and sea lions of the world. Blandford, London

Bowen WD, Oftedal OT, Boness DJ (1992) Mass and energy transfer during lactation in a small phocid, the harbor seal (*Phoca vitulina*). Physiol Zool 65:884–866

Busch BC (1985) The war against the seals. A history of the North American seal fishery. McGill-Queen's University Press, Montreal

Churchill M, Boessenecker RW, Clementz MT (2014) Colonization of the Southern Hemisphere by fur seals and sea lions (Carnivora: Otariidae) revealed by combined evidence phylogenetic and Bayesian biogeographical analysis. Zool J Linn Soc 172(1):200–225

Committee on Taxonomy (2019) List of marine mammal species and subspecies. Society for Marine Mammalogy. http://www.marinemammalscience.org. Accessed 26 Jun 2020

Deméré TA, Berta A (2005) New skeletal material of *Thalassoleon* (Otariidae: Pinnipedia) from the late Miocene-early Pliocene (Hemphillian) of California. Bull Fla Mus Nat Hist 45(4):379–411

García-Aguilar MC, Turrent-Thompson C, Elorriaga-Verplancken FR et al (2018) Climate change and the northern elephant seal (*Mirounga angustirostris*) population in Baja California, Mexico. PLoS One 13(2):e193211

García-Rivas MC (2008) Sistema de apareamiento del lobo marino de California (*Zalophus californianus californianus*) en la Isla Ángel de la Guarda, BC, México (Mating system of the California sea lion (*Zalophus californianus californianus*) on Ángel de la Guarda Island, BC, Mexico). M.Sc. Thesis, Universidad Nacional Autónoma de México, México, D.F

Goñi R (1998) Ecosystem effects of marine fisheries: an overview. Ocean Coast Manag 40(1):37–64. https://doi.org/10.1016/S0964-5691(98)00037-4

IUCN (2020) The IUCN red list of threatened species. IUCN. https://www.iucnredlist.org/. Accessed 26 Jun 2020

Jefferson TA, Webber MA, Pitman RL et al (2015) Marine mammals of the world. A comprehensive guide to their identification, 2nd edn. Elsevier, London

Kovacs KM, Aguilar A, Aurioles-Gamboa D et al (2012) Global threats to pinnipeds. Mar Mamm Sci 28(2):414–436. https://doi.org/10.1111/j.1748-7692.2011.00479.x

Krockenberger MB, Bryden MM (1994) Rate of passage of digesta through the alimentary tract of southern elephant seals (*Mirounga leonina*) (Carnivora: Phocidae). J Zool 234(2):229–237. https://doi.org/10.1111/j.1469-7998.1994.tb06071.x

Lawson J, Renouf D (1987) Bonding and weaning in harbor seals, *Phoca vitulina*. J Mammal 68(2):445–449. https://doi.org/10.2307/1381496

Martínez S (2019) Matan lobos marinos para tráfico ilegal (Sea lions are killed for illegal trafficking). El Imparcial. https://www.elimparcial.com/mexicali/mexicali/Matan-lobos-marinos-para-trafico-ilegal-20190627-0001.html. Accessed 24 Jun 2020

Northridge S (2009) Effects of fishing industry. In: Perrin WF, Würsig B, Thewissen JGM (eds) Encyclopedia of marine mammals, 2nd edn. Academic Press, San Diego, pp 443–447

Pauly D, Trites AW, Capuli E et al (1998) Diet composition and trophic levels of marine mammals. ICES J Mar Sci 55(3):467–481. https://doi.org/10.1006/jmsc.1997.0280

Read AJ (2008) The looming crisis: interactions between marine mammals and fisheries. J Mammal 89(3):541–548. https://doi.org/10.1644/07-MAMM-S-315R1.1

Riedman M (1990) The pinnipeds: seals, sea lions, and walruses. University of California Press, Berkeley

Roberts CM (2007) The unnatural history of the sea. Island Press, Washington, DC

Sandweiss DH, Maasch KA, Chai F et al (2004) Geoarchaeological evidence for multidecadal natural climatic variability and ancient Peruvian fisheries. Quatern Res 61(3):330–334. https://doi.org/10.1016/j.yqres.2004.02.008

Szteren D, Naya DE, Arim M (2004) Overlap between pinniped summer diet and artisanal fishery catches in Uruguay. LAJAM 3(2):119–125. https://doi.org/10.5597/lajam00057

Tapia-Harris C, Heckel G, Schramm Y et al (2017) Molting phenology of the Pacific harbor seal (*Phoca vitulina richardii*) on two islands off the Baja California Peninsula, Mexico. Mar Mamm Sci 33(3):817–829. https://doi.org/10.1111/mms.12400

Chapter 2
South American Fur Seal (*Arctocephalus australis*, Zimmerman 1783)

Enrique Alberto Crespo and Larissa Rosa de Oliveira

Abstract The South American fur seal is distributed on both sides of South America along the Atlantic and Pacific Oceans mostly in sympatry with the South American sea lion. However, there is gap of distribution between 43°S and 29°S in the Pacific coast of approximately 1600 km. The total population was estimated at 320,000 individuals, and the species is considered as *Least Concern* in the IUCN red list, but the Peruvian population is classified as *Vulnerable*. Genetic and morphological comparisons between the Atlantic and the Peruvian fur seals indicated that they could be distinct evolutionarily significant units. The species is polygynous and sexually dimorphic, with males being larger and heavier than females. Breeding occurs during the austral summer, from December to early January. Longevity is around 20 years for both sexes in wild and captive life. The South American fur seal is an opportunistic species that predates mainly on pelagic species of fish, squids, and crustaceans, some of them of commercial importance. Fur seals were used by aboriginal people as far as 8000 YBP and since the eighteenth century by Europeans, who drove them to the brink of extinction. After the cease of hunting, the stocks were slowly recovering at least on the Atlantic coast. In most of the coastal settlement sites where fur seals occur, they share the places with South American sea lions, although there is some differentiation in the use of space. The fur seals prefer places of difficult access, rocky areas or more steep, and the sea lions occupy sandy or boulder beaches. At present, there is no major threat for South American fur seals, given that the species does not interact with fisheries or salmon farms as South American sea lions do.

Keywords Distribution · South American fur seal · *Arctocephalus australis* · Population dynamics · Exploitation · Genetic structure · Threats

---

E. A. Crespo (✉)
Laboratorio de Mamíferos Marinos, CESIMAR-CCT CENPAT-CONICET, Puerto Madryn, Chubut, Argentina
e-mail: kike@cenpat-conicet.gob.ar; kike@cenpat.edu.ar

L. R. de Oliveira
Mammal Ecology Lab, Universidade do Vale do Rio dos Sinos, São Leopoldo, Brazil
e-mail: larissaro@unisinos.br

G. Heckel, Y. Schramm (eds.), *Ecology and Conservation of Pinnipeds in Latin America*, https://doi.org/10.1007/978-3-030-63177-2_2

Introduction

The South American fur seal *Arctocephalus australis* (Zimmermann 1783) was originally described from a specimen collected from the Falkland (Malvinas) Islands. This was one of three fur seal species described in the late eighteenth century and is still recognized today. The other two are the northern fur seal, *Callorhinus ursinus* (Linnaeus 1758), and the Cape fur seal, *A. pusillus* (Schreber 1776). The holotype of *A. australis* was a specimen from the "Museum of the UK Royal Society sent of late years from the Falkland Islands" (Pennant 1781), but the specimen was lost (Oliveira and Brownell 2014). Some of the specimens from the Royal College of Surgeons of England were transferred to the British Museum (Natural History), but King (1954) did not report any details on the type specimen other than the type location.

King (1954) in the study on "the otariid seals of the Pacific coast of America" examined skulls of fur seals from the Falkland (Malvinas) Islands and South America and concluded that skulls from the Falkland Islands specimens were larger than those of South America, and she suggested that they belong to the subspecies *Arctocephalus australis australis*, while the specimens from the Galapagos Islands must be considered *A. australis galapagoensis*, and finally fur seals from the mainland population of South America must be considered a third subspecies: *A. australis gracilis*. However, Repenning et al. (1971:21) carried out a major revision on the systematics of the fur seal genus *Arctocephalus* and recognized the Galapagos fur seal as a full species.

Oliveira et al. (2008a, b), based on the analysis of skull morphometrics and seven microsatellite loci, found a single Atlantic population (including Falkland (Malvinas) Islands, Brazil, and Uruguay), which was morphologically and genetically different from the Peruvian population, suggesting that both should be considered two evolutionarily significant units (ESUs). However, the authors did not specifically address the *A. australis* taxonomical or subspecies status, as suggested by Berta and Churchill (2012). Moreover, Túnez et al. (2013) and Crespo et al. (2015) presented data that incorporated the Argentine population to the Atlantic ESUs.

Oliveira and Brownell (2014) proposed *A. a. gracilis* as a junior synonym of *A. a. australis*, since the type location of *A. australis gracilis* (Nehring 1887) is in the Atlantic coast (Tramandahy (ca. 29°55′S, Rio Grande do Sul, Brazil)).

More recently, Rodrigues et al. (2018), based on the analyses of mtDNA control region sequences from GenBank and a few samples collected from Guafo Island, suggested a connection between southern Chile and Atlantic populations, because they shared several haplotypes. Additional studies are needed on the relationship between Peruvian and the remaining populations of fur seals, including Falkland (Malvinas) Islands, in order to determine their taxonomic status in relation to other fur seal species. In this context, given the controversial status for the South American fur seals and for simplicity, here we refer to it as a unique species, but when it is necessary we specify the ESU or population.

Geographic Range

The South American fur seal is distributed along the western South Atlantic (southern Brazil, Uruguay, Argentina, and the Falkland (Malvinas) Islands) and eastern South Pacific coast (central Chile) (Cárdenas-Alayza et al. 2016; Cárcamo et al. 2019) (Fig. 2.1). On the Atlantic side, nonbreeding sites can be found along the coasts of Rio Grande do Sul in Brazil (approximately from 29° to 32° S) (Sanfelice et al. 1999; Muelbert and Oliveira 2006; Oliveira 2013), while the limit of breeding colonies is found at Islas del Castillo, Uruguay (34°21′S, 53°44′W). Along the coast of Argentina, there are several breeding and nonbreeding colonies between Mar de Plata and Tierra del Fuego, including Isla de los Estados at the extreme south, and

Fig. 2.1 Distribution of South American fur seals in South America. Striped area: breeding area; shaded area; nonbreeding area

the Falkland (Malvinas) Islands (Crespo et al. 2015). Still on the Atlantic, the breeding colonies have a discontinuous distribution, as a result of a relict pattern of what it was in the past when the species had a continuous distribution. It was argued that the current distribution of colonies is associated with the distance to the continental shelf break and the availability of habitat, such as islands and protected areas along the coast (Túnez et al. 2008). However, archeofaunistic sites documented very extensively the presence of fur seal bones and teeth along a continuous distribution in the Atlantic coast, in places where the species is or was absent in the last few decades and is now recovering (Borella 2014; Crespo et al. 2015).

On the Pacific side, the distribution continues all around the southern tip of South America to the central coast of Chile (Cárcamo et al. 2019). Until recently, there was information on a gap from Isla Guafo to northern Chile. However, Cárcamo et al. (2019), during aerial censuses conducted from 2009 to 2018, found two non-breeding colonies in the gap of distribution in the Pacific coast of Chile, at 38°25′S and 38°26′S, in central Chile. Therefore, the gap today would be around ~1600 km.

The specimens from the Peruvian ESU are distributed along the coast of Peru and northern Chile. In Peru they occur from Isla Mazorca (11°20′S) to Ilo (17°42′S), with a small isolated colony occurring in northern Peru at Isla Foca (5°20′S) (Oliveira et al. 2012b; Cárdenas-Alayza and Oliveira 2016). At the southern limit of Peru, colonies continue to occur on the north coast of Chile to Rocas Abtao (23°05′S) (Torres 1985; Guerra and Torres 1987), although the distribution of this species in Chile is now expanding to the south, reaching to 29°S (Sepúlveda, pers. comm. cited by Cárdenas-Alayza and Oliveira 2016). Thus, the majority of the breeding population of this ESU occurs in Peru from 15° to 17°S.

Migrations or Seasonal Movements

Movements of fur seals have been demonstrated from records of pups marked at birth in the breeding colony at Isla de Lobos, Uruguay, which were seen later at Isla Escondida and other localities of Buenos Aires (Bahía Blanca and Quequén) and Rio Negro Provinces (Islote Lobos; see Fig. 2.2) (Crespo et al. 2015). The increase in numbers of South American fur seals on the Argentine coast occurs from October to February, which coincides with weaning in Uruguay (September–December), and these numbers decrease in the fall. A similar pattern of movements was detected throughout the Beagle Channel where the number of fur seals increases from late February to mid-September reaching a maximum by March (Crespo et al. 2015), with seals probably moving westward to the many islands and islets located in southernmost Chile. More recently Baylis et al. (2018) equipped with satellite tags nine adult females in the Falkland (Malvinas) Islands, and the authors found that the foraging trips included visits to Isla Rasa and Isla de los Estados in the Patagonian shelf (Baylis et al. 2018).

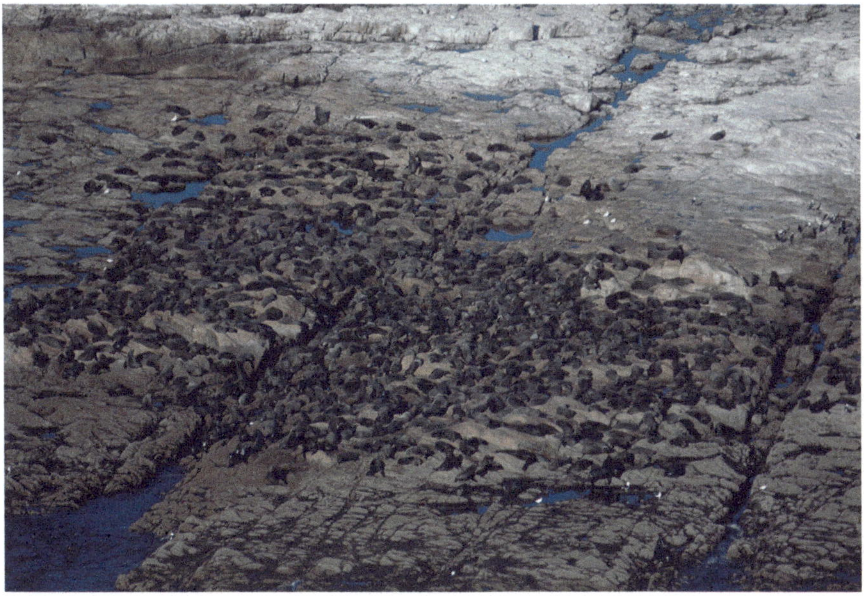

Fig. 2.2 South American fur seal seasonal settlement at Islote Lobos (41°25′S, 65°02′W), Rio Negro, Argentina, in 2008

Population Numbers and Trends at a Global Scale

Total population estimates for the entire distribution of South American fur seals are rather fragmentary, scarce, and outdated. According to Vaz-Ferreira (1982, 1987), there was a total population estimated at 320,000 individuals: 250,000 from Uruguay, less than 3,000 from Argentina, 14,000–16,000 from the Falkland (Malvinas) Islands, 40,000 from Chile, and 12,000 from Peru. Strange (1992) estimated a population of between 18,000 and 20,000 animals in the Falkland (Malvinas) Islands, but there were no surveys or monitoring of colonies for two decades (Otley 2008), until more recently when Baylis et al. (2019) estimated a total of 36,500 pups. If we consider a rough estimation of 4 adults per pup, a total population could be no less than 120,000 individuals (Harwood and Prime 1978).

In Uruguay, although the population size has been estimated during the 1990s at 300,000 individuals, this information comes from simulations based on a previously estimated number of pups (Páez 2000, 2006). More recent information indicates that these figures for Uruguay are overestimated and that the overall population size estimated from aerial surveys would be around half of those figures (Franco-Trecu et al. 2019a). Pup abundance (ca. 31,000 in 2013) was 62% of the previous estimate based on non-exhaustive ground pup counts corrected by the probability of resighting. The finite annual growth rate only at Isla–Islote de Lobos is around 1.5% showing a positive trend between 1956 and 2013 (Franco-Trecu et al. 2019a).

Fig. 2.3 Aerial photograph of Isla Rasa (45°07′S – 65°22′W), Chubut, Argentina, taken in 1995

The Argentine Sea has its main breeding colonies at Isla Escondida, Isla Rasa, and several settlements at Isla de los Estados. During the nineteenth and twentieth centuries, the fur seals almost disappeared. However, during the 1990s, the local population started to recover probably because of the immigration of individuals from Uruguay. In two decades, between the mid-1990s and the mid-2010s, the number of pups born at Isla Rasa increased from 15 to more than 700 individuals. The same happened at Isla de los Estados, where there was an increase in the number of individuals but also in the number of colonies (Milano et al. 2020).

Isla Rasa is an example of expansion and possible change in the social structure of the species. Unpublished information of the Laboratorio de Mamíferos Marinos (CENPAT–CONICET, Puerto Madryn) indicated that in the mid-1990s the island sustained as much as 12,000–13,000 juveniles dispersed all along the surface of the island during spring and summer (Fig. 2.3). Only 15 pups were born per year at that time (Crespo et al. 2015). During the last 2 or 3 years (2017–2019), a disruption seems to have occurred, and the social structure seems to have changed. Now, territorial males, adult females, and pups occupy the border of the island near the water, showing obvious thermoregulation behavior. The majority of the island is unoccupied, with the exception of areas with ponds (Fig. 2.4). These apparent changes in social structure require further monitoring for confirmation. In summary, the total population for the Argentine Sea would be around 25,000 individuals.

Fig. 2.4 Aerial photograph of Isla Rasa (45°07′S – 65°22′W), Chubut, Argentina, taken in 2019

Recent estimations for the coast of Chile indicate that the population is around 32,642 with 90% of the individuals concentrated in the southern zone with the exception of the Magallanes Region. The southern zone is the one which showed the highest increase in numbers (Oliva et al. 2020).

Population estimates for Peru conducted by the *Instituto del Mar del Perú* in 1997 were around 24,500 fur seals. During the severe El Niño in 1997–1998, this population declined by 72% as a result of low food availability, resulting in a remaining population of only 8,223 individuals (Oliveira et al. 2006, 2012a). However, the effective population size (Ne), not the census number, is a main concern for evolutionary aspects. Factors such as variation in the sex ratio of breeding individuals, variation of population size in different generations, and mating system are important. In this context, the estimated Ne after this El Niño was only 2,153 specimens (Oliveira et al. 2012b), which is a critical value of Ne for the Peruvian population, because it is significantly lower than the mean minimum viable population for vertebrates (7,000 breeding age adults). This estimated Ne is critically important, because, combined with the current El Niño events, there is great concern for the survival of the species, and should be taken into account in future management plans to ensure the conservation and protection of the species in the Peruvian coast (Oliveira et al. 2012a). The last census of the Peruvian population was in 2006 and was of approximately 16,000 individuals. However, after the

1997–1998 El Niño, the Ne of the Peruvian population was estimated for the first time, taking into account the effects on the mating system and the variation in population size. Since then, *A. australis* has been classified as *in danger of extinction* in Peru and as *Vulnerable* by IUCN (Cárdenas et al. 1987).

In summary, the total estimates given by Vaz-Ferreira (1982) were 320,000 individuals. The total estimate in present days is 393,500. Nevertheless, the distribution of numbers by country is very different. Uruguay has half of the numbers estimated in the 1980s, and the Falkland (Malvinas) Islands, the Argentine Sea (without Malvinas), and the southern coast of Chile had experienced an important increase in numbers. Present figures have been probably estimated with better techniques and protocols.

Evolutionary Structure of the Species

The genetic diversity and population structure of South American fur seals were analyzed using mitochondrial DNA haplotypes, along with *O. flavescens*, from colonies located along the Atlantic and Pacific coasts of South America (Túnez et al. 2007, 2013). Colonies from these areas did not share haplotypes, and this result, based on a rather low number of samples to allow robust comparisons between oceans, suggested that populations from each ocean correspond to different evolutionarily significant units. Oliveira et al. (2008a, b), based on the analysis of skull morphometrics and seven microsatellite loci, found a single Atlantic population (including Falkland (Malvinas) Islands, Brazil, and Uruguay), which was morphologically and genetically different from the Peruvian population, suggesting that both should be considered two evolutionarily significant units (ESUs). Recently, Rodrigues et al. (2018) suggested a connection between southern Chile and Atlantic populations, because they share several haplotypes. In all these studies, fur seal samples only came from Uruguay, Argentina, and Peru, but not from the Falkland (Malvinas) Islands.

History of Exploitation and Recovery Along the SW Atlantic

The pre-exploitation period lasted for 8000 years until the colonization of America by Europeans. During this time, the sea lions and fur seals were hunted by aboriginal groups for food, leather, and oil with varying degrees of intensity throughout the entire Atlantic coast (Bayón and Politis, 1996, 2014; Bayón et al. 2012; Orquera and Piana, 1999; Borella, 2007, 2014; Gómez Otero, 2007; Castilho and Simões-Lopes 2008; Vales 2015 and references therein). However, hunting pressure on fur seal populations has been considered low in relation to the past abundance of different stocks (Schiavini 1992; Orquera and Piana 1999).

The commercial exploitation period extended from the colonization of America by the Europeans until the beginning of the twentieth century. The first record of commercial fur seal and sea lion hunting dates back to 1515, when the crew of the Juan Díaz de Solís expedition took fur seals on Isla de Lobos (Uruguay), whose skins were sold in the market of Seville (Vaz-Ferreira and Ponce de León 1987). The next mention of commercial exploitation is from 1724, the year in which seal oil was used for illumination of the city of Maldonado. In 1792, exploitation was begun by the Real Compañía Marítima on instructions of the King of Spain and continued until the English invasion of the territory in 1808. After that, harvesting was done by concessionaries under the Uruguayan government's control (Vaz-Ferreira and Ponce de León 1987).

From the sixteenth to the eighteenth centuries, foreign crews (mainly English, French, and North Americans) hunted sea lions and fur seals along the coasts of the Southwestern Atlantic Ocean, primarily for fresh meat and oil and later for skins. The drastic decline of stocks and fashion changes made the fur business an unprofitable activity. According to Weddell (1825), no less than 1,200,000 skins were removed from the southern seas including most of *Arctocephalus* species without any way to discriminate each. Uruguay was the last country to ban the sealing in the South American continent in 1991 (Ximénez and Langguth 2002), while the rest of the countries ceased fur seal commercial activities at the beginning of the twentieth century. After the end of fur seal hunting in Uruguay, there was an increase of sightings of individuals in La Plata River and from there to the south of the country. Six individuals tagged in Uruguay were found in waters of Chubut Province in the following years (Crespo et al. 2015). The number of colonies and individuals continues to increase in the Southwestern Atlantic to date (Crespo et al. 2015; Baylis et al. 2019; Milano et al. 2020).

Abundant archaeological records show that fur seals were exploited by the ancient Peruvians as far back as 4000 years ago (Majluf 1987), but little else is known about the status and exploitation in Peru before 1940, mainly because of failure to distinguish between the species and the sympatric South American sea lion (Majluf 1987). Many authors mentioned large numbers of seals along the Peruvian coast around the beginning of the twentieth century, and commercial exploitation apparently took place indiscriminately until 1946. Between 1925 and 1946, 806,525 seal skins were exported from Peru, averaging 44,252 skins per year (Majluf and Trillmich 1981), but it is not known to which species of otariid seal they belonged. By 1943, only a few small groups of fur seals survived in isolated areas in southern Peru. In 1946, the hunting of both species of seals was prohibited between January and April, but it was not until 1959 that sealing was totally banned. At present, despite this legal protection, seal poaching is a very common practice and is persistently carried out by local fishermen (Cárdenas-Alayza and Oliveira 2016).

Feeding Habits

The South America fur seal is largely a pelagic feeder, with a diet mainly based on small pelagic fish like anchovies and sardines. Squids are also found in the diet and eventually crustaceans. In Patagonian waters, at least 42 different preys have been identified including fish, squids, and crustaceans. However, the most important prey are the shortfin squid *Illex argentinus*, the Argentine anchovy *Engraulis anchoita*, and the common hake *Merluccius hubbsi* (unpublished information of the Marine Mammal Laboratory, Puerto Madryn). Along the coasts of Rio Grande do Sul in Brazil, five species of teleost fishes (cutlassfish *Trichiurus lepturus*, anchovy *Anchoa marinii*, stripped weakfish *Cynoscion guatucupa*, squirrel hake *Urophycis brasiliensis*, and rough scad *Trachurus lathami*), two cephalopods (long-finned squid *Loligo sanpaulensis* and paper nautilus *Argonauta nodosa*), and one crustacean (red shrimp *Pleoticus muelleri*) were found. The most important prey for the species, according to the index of relative importance, were *T. lepturus* and *L. sanpaulensis* (Oliveira et al. 2008a, b). In waters of the Falkland (Malvinas) Islands, species found in fur seal scats included the Falkland herring *Sprattus fugensis*, Patagonian longfin squid *Loligo gahi*, and *Patagonotothen* spp., which occurred frequently (Baylis et al. 2014). In Peru, a preliminary analysis of otoliths suggests that the species forages mainly on sardine *Sardinops sagax*, anchoveta *Engraulis ringens*, and *Trachurus symmetricus*, as well as other top predators like sea lions and seabirds (Majluf 1987).

Predation

Vaz-Ferreira and Ponce de León (1987) mentioned several species of sharks as predators of fur seals in Uruguay, catching pups and attacking adult females; and killer whales *Orcinus orca* are sometimes seen near the islands during the breeding season. Along the Patagonian coast, predation on South American fur seals was documented and included broadnose sevengill sharks *Notorynchus cepedianus* (Crespi Abril et al. 2004). The broadnose shark is known by its preference to prey on marine mammals, which includes, in addition to fur seals, sea lions, elephant seals, dusky and Commerson's dolphins, and southern right whales (Crespi Abril et al. 2004). In Peru, young sea lion males fed on small fur seal pups ashore occasionally, and, very rarely, sharks attacked fur seal adults feeding offshore (Majluf 1987).

Reproductive Biology and Development

As other Otariidae, fur seals are polygynous and sexually dimorphic, with an annual cycle divided into a short reproductive stage and a pelagic phase with frequent and regular visits to the coast. The breeding season occurs from late November through early January (Cárdenas-Alayza et al. 2016).

Territorial males can be found at the colonies in late December, decreasing in numbers in January (Franco-Trecu 2015). In Uruguay, reproductive males defend small territories without resources required for females, which have extensive home ranges that overlap with the territories of many males, indicating that they can move freely in the colony and that males do not monopolize access to females. Under these circumstances, females could assess potential mates and choose among them accordingly. These conditions meet the criteria for a mating system called a "lek." This reproductive system has been documented in Peru (Majluf, 1991; see Figs. 2.5 and 2.6) and Uruguay (Franco-Trecu 2015).

Pupping peaks occur from mid-November to mid-December. However, there are slight variations in time along the distribution range. The males are territorial and remain fasting on land during the entire breeding season. Individual bulls can occupy territories until most of the females have mated, and then they leave their territories to start foraging at sea (Pavés and Schlatter 2008; Cárdenas-Alayza et al. 2016). The females enter in estrus between 5 and 8 days after the pup's birth. After mating, alternating periods of feeding at sea and nursing on land occur for a period

Fig. 2.5 South American fur seals at Punta San Juan (15°22′S, 75°11′W), Peru

Fig. 2.6 South American fur seals at Punta San Juan (15°22′S, 75°11′W), Peru

of between 8 and 12 months. During the first 3 months of maternal care, duration of foraging trips between females is highly variable, which affects the survival of offspring because longer trips increase pup mortality (Franco-Trecu et al. 2010). Most pups are weaned when they are 8 months old, although some continue suckling until they are 12 months old. In the Pacific populations affected by El Niño events, lactation can take more than a year. During the first 7 months, milk is the only stomach content, and after the eighth month, the stomach contents include either milk or milk with items such as sand, small stones, and mollusk shells (Vaz-Ferreira and Ponce de Leon 1987). During the first 3 months of parental care, the duration of foraging trips by females is highly variable, which affects the survival of offspring, since longer trips also increase pup mortality in this area, particularly for the Peruvian population in El Niño years (Majluf 1987, 1991).

Threats, Conservation Status, and Management

South American fur seal populations were exploited for at least 8000 years by aboriginal people along the South American coast (Borella 2014). Modern sealers severely depleted the species during the eighteenth century. They were exploited mainly for their skin and in a lesser degree for oil and genitals of the adult males that were exported to Asian countries for the manufacture of aphrodisiacs. Regarding interactions with fisheries, fur seals do not interact significantly, or the mortality in fishing gear is not considered important (Vaz-Ferreira 1982; Crespo et al. 2015).

Development of large- and small-scale commercial fisheries had a negligible effect on South American fur seals in the Atlantic. Only one adult male was caught in a bottom trawler along the Patagonian coasts. The species does not usually feed close to the shore or in fishing nets; therefore, it is not killed by fishermen as South American sea lions and South African fur seals (Vaz-Ferreira 1987). However, some young animals drown in trammel nets set too close to breeding areas. There is also minimal overlap between fur seal prey items and target commercial species (Crespo pers. comm.). De Maria et al. (2012) reported anecdotal bycatch of fur seals in artisanal fisheries on the coasts of Uruguay. More recently, Franco-Trecu et al. (2019b) recorded in the Uruguayan coastal bottom trawl fleet between January 2009 and April 2012 an annual mortality of 18–38 South American fur seals. This figure is around 0.02% of the local population and one-third the mortality of sea lions in the same fishing gear (Franco-Trecu et al. 2019b). Incidental captures of seals in shark nets have been reported for Uruguay (Scialabba 1989). Very small numbers of fur seals are also caught in the Chilean trawl-fisheries (Sepúlveda pers. comm).

During the 1970s and early 1980s, South American fur seals and other wildlife were hunted illegally in southern Chile and Argentina to bait traps for southern king crab (*Lithodes santolla*). Bait used between 1976 and 1980 was estimated as 200–400 tons per year (Cárdenas et al. 1987). Because this fishery is decreasing due to overexploitation, hunting pressure on the fur seal is being reduced. In spite of the spatial overlap between the colonies of South American fur seals and salmon aquaculture in southern Chile, there is no evidence of interaction with this activity (Durán et al. 2011).

The limited number of large, dense breeding aggregations could make this species particularly sensitive to the effects of oil spills and disease epidemics. Like other fur seals, South American fur seals are vulnerable to oil spills because of their dependence on their thick pelage for thermoregulation. In February 1997, 5000 metric tons of crude oil were spilled from the vessel San Jorge onto the coast of Isla de Lobos in southern Uruguay. Nearly 5000 South American fur seal pups (2–3 months old) were heavily oiled and/or died (Mearns et al. 1999).

In general, the South American fur seal has been interacting with the human population since the very beginning of the colonization of South America by aboriginal people. There is a huge piece of evidence all along the distribution range. At that time, seals were used mainly for food, leather, and oil. When Europeans started conquering the continent, the species was increasingly used with the same purposes but at the industrial level, taking the population to dramatically reduced numbers. When hunting stopped and conservation measures were taken in more recent times during the twentieth century, recovery was recorded in many areas. Since the fur seal does not interact with human activities, particularly fishing and aquaculture, the future is brighter than that for the sea lion. However, some overlapping in the use of target preys exists with the development of fisheries. As well as with the sea lion, the near future will require management measures that allow an agreed balance between fishing interests and population levels.

Acknowledgments The authors want to thank CONICET, ANPCyT, FAPESP, CAPES, Amneville Zoo, CNPq, UNP, GEMARS, and UNISINOS for their support for us and our students for almost 37 years of research and conservation and to our colleagues from LAMAMA and GEMARS for sharing field work and papers. We thank Doris Oliva and René Durán for sharing unpublished information on the present status of sea lion and fur seal populations. We also thank Fernando Lopes for his map and three anonymous reviewers for their help to improve the manuscript.

References

Baylis AMM, Arnould JPY, Staniland IJ (2014) Diet of South American fur seals at the Falkland Islands. Mar Mamm Sci 30:1210–1219

Baylis AMM, Tierney M, Orben RA et al (2018) Geographic variation in the foraging behaviour of South American fur seals. Mar Ecol Prog Ser 596:233–245

Baylis AMM, Orben RA, Arkhipkin AI et al (2019) Re-evaluating the population size of South American fur seals and conservation implications. Aquat Conserv Mar Freshwat Ecosyst. https://doi.org/10.1002/aqc.3194

Bayón C, Politis G (1996) Estado actual de las investigaciones en el sitio Monte Hermoso 1, Provincia de Buenos Aires. Arqueol 6:83–116

Bayón C, Politis G (2014) The inter-tidal zone site of La Olla: early–middle Holocene human adaptation on the Pampean coast of Argentina. In: Evans AM, Flatman JC, Flemming NC (eds) Prehist Archaeol continent shelf: a global review. Springer, New York, p 115130

Bayón C, Frontini R, Vecchi R (2012) Middle Holocene settlements on coastal dunes, Southwest Buenos Aires Province, Argentina. Quat Int 256:54–61

Berta A, Churchill M (2012) Pinniped taxonomy: review of currently recognized species and subspecies, and evidence used for their description. Mammal Rev 42:207–234

Borella F (2007) ¿Dónde están los lobos en la costa atlántica de Norpatagonia? Explorando vías para resolver el registro arqueofaunístico. Werken 9:97–114

Borella F (2014) Zooarchaeological evidence of otariids in continental coast of Patagonia, Argentina: old and new perspectives. In: Muñoz SA, Gotz CM, Roca ER (eds) Neotropical and Caribbean aquatic mammals. Perspectives from archaeology and conservation biology. Nova Science Publishers, Inc, Hauppauge, pp 135–160

Cárcamo D, Pizarro M, Orellana M et al (2019) New non-breeding colonies of the South American fur seal in Central Chile. Is the distribution in the southeastern Pacifc waters extending? Lat Am J Aquat Mamm 14(1):9–17. https://doi.org/10.5597/lajam00249

Cárdenas JC, Oporto J, Stutzin M et al (1987) Impacto de la pesquería de centolla (*Lithodes antarctica*) y centollon (*Paralomis granulosa*) sobre las poblaciones de cetáceos y pinnípedos de Magallanes, Chile. Proposiciones para una política de conservación y manejo. In: Siciliano S, Fiori B (eds) An 2a Reu Trab Esp mam Acuat Amer Sul, pp 32–35

Cárdenas-Alayza S, Oliveira LR (2016) *Arctocephalus australis* (Peruvian/Northern Chilean subpop). IUCN Red List Threat Species 2016:e.T72050476A72050985

Cárdenas-Alayza S, Oliveira LR, Crespo EA (2016) Arctocephalus australis. IUCN Red List Threat Species 2016:e.T2055A45223529. https://doi.org/10.2305/IUCN.UK.2016-1.RLTS.T2055A45223529.en

Castilho PV, Simões-Lopes PC (2008) Sea mammals in archaeological sites on the southern coast of Brazil. Rev Mus Arqueol Etnol S Paulo 18:101–113

Crespi Abril A, García NA, Crespo EA et al (2004) Marine mammals consumption by broadnose sevengill shark *Notorynchus cepedianus* in the northern and central Patagonian shelf. Lat Am J Aquat Mamm 2(2):101–107

Crespo EA, Schiavini A, García N et al (2015) Status, population trend and genetic structure of South American fur seals *Arctocephalus australis* in southwestern Atlantic waters. Mar Mamm Sci 31(3):866–890

De María M, Golluchi M, Szteren D (2012) Registros de *Arctocephalus australis* (Carnivora Mammalia) interaccionando con la pesca artesanal en Uruguay. Bol Soc Zool Urug (2a ép) 21(1–2):50–58

Durán R, Oliva D, Sepúlveda M et al (2011) Interacción entre el lobo marino común y la salmonicultura en Chile: buenas prácticas para su mitigación. Ed Univ Valparaíso, 58 pp

Franco-Trecu V (2015) Tácticas comportamentales de forrajeo y apareamiento y dinámica poblacional de dos especies de otáridos simpátricas con tendencias poblacionales contrastantes. PhD thesis, Univ Rep (UdelaR) Montevideo, Uruguay

Franco-Trecu V, Tassino B, Páez E (2010) Comportamiento maternal de *Arctocephalus australis*, en Isla de Lobos-Uruguay. Revista de Etología 9(1):29–40

Franco-Trecu V, Drago M, Grandi MF et al (2019a) Abundance and population trends of the south American Fur seal (*Arctocephalus australis*) in Uruguay. Aquat Mamm 45(1):48–55

Franco-Trecu V, Szephegyi MN, Doño F et al (2019b) Marine mammal bycatch by the industrial bottom trawl fishery at the Río de la Plata Estuary and the adjacent Atlantic Ocean. Lat Am J Aquat Res 47(1):89–101

Gómez Otero J (2007) Dieta, uso del espacio y evolución en sociedades cazadoras-recolectoras de la costa centro-septentrional de Patagonia durante el Holoceno medio y tardío. Tes Doc Arqueol, University of Buenos Aires

Guerra CC, Torres DN (1987) Presence of South American fur seal, *Arctocephalus australis*, in northern Chile. In: Croxall JP, Gentry RL (eds) Proceedings of an international symposium and workshop: status, biology and ecology of fur seals, Cambridge, UK 23–27 April 1984, pp 169–176

Harwood J, Prime JH (1978) Some factors affecting the size of British grey seal populations. J Appl Ecol 15:401–411

King JE (1954) The otariid seals of the Pacific coast of America. Bull Br Mus (Nat Hist) Zool 2:311–337

Majluf P (1987) South American Fur Seal, *Arctocephalus australis*, in Peru. In: Status, biology, and ecology of fur seals. In: Proceedings of an International Symposium and Workshop, Cambridge, England, 23–27 April 1984, pp 33–36

Majluf P (1991) El Niño effects on Pinnipeds in Peru. In: Trillmich F, Ono KA (eds) Pinnipeds and El Niño: responses to environmental stress. Springer, Berlin, pp 55–65

Majluf P, Trillmich F (1981) Distribution and abundance of sea lions (*Otaria byronia*) and fur seals (*Arctocephalus australis*) in Peru. Z Saugtierkd 46(6):384–393

Mearns A, Levine E, Yender E et al (1999) Protecting Fur Seals during spill response: lessons from the San Jorge (Uruguay) Oil Spill. Int Oil Spill Conf Proc 1999(1):467–470

Milano VN, Grandi MF, Schiavini ACM et al (2020) Recovery of South American fur seals from Fuegian Archipelago (Argentina). Mar Mamm Sci. https://doi.org/10.1111/mms.12686

Muelbert MMC, Oliveira LR (2006) First records of stranded pregnant female South American fur seals, *Arctocephalus australis*, in the southern Brazilian coast. Lat Am J Aquat Mamm 5(1):67–68

Oliva D, Durán LR, Sepúlveda M et al (2020) Estimación poblacional de lobos marinos e impacto de la captura incidental. Inf Fin Proy FIP 2018-54, 184 pp + Anexos

Oliveira LR (2013) Canívoros Marinhos. In: Weber MM, Roman C, Cáceres NC (eds) Mamíferos do Rio Grande do Sul pp 405–427

Oliveira LR, Brownell RL Jr (2014) Taxonomic status of two subspecies of South American fur seals: *Arctocephalus australis australis* vs. *A. a. gracilis*. Mar Mamm Sci 30(3):1258–1263

Oliveira LR, Arias-Schreiber M, Meyer D et al (2006) Effective population size in a bottlenecked fur seal population. Biol Conserv 131:505–509

Oliveira LR, Hoffman JI, Hingst-Zaher E et al (2008a) Morphological and genetic evidence for two evolutionarily significant units (ESUs) in the South American fur seal, *Arctocephalus australis*. Conserv Genet 9:1451–1466
Oliveira LR, Ott PH, Malabarba LR (2008b) Ecologia alimentar dos pinípedes do sul do Brasil e uma avaliação de suas interações com atividades pesqueiras. In: Barros N, Peracchi AL, dos Santos GASD (eds) Ecologia de Mamíferos. Technical Books, Londrina, pp 97–116
Oliveira LR, Darsie Fraga L, Majluf P (2012a) Effective population size for South American sea lions along the Peruvian coast: the survivors of the strongest El Niño event in history. J Mar Biol Assoc UK 92(8):1835–1841
Oliveira LR, Lopes FRV, Almeida R et al (2012b) Status Taxonomico dos lobos-marinhos de Isla Foca (Perú): *Arctocephalus australis*, *A. galapagoensis* (CARNIVORA: OTARIIDAE) ou híbridos? 15 Reu Trab Exp Mam Acuát Am Sur 9 Cong SOLAMAC Pto Mad Argentina
Orquera LA, Piana EL (1999) Arqueología de la región del canal Beagle (Tierra del Fuego, República Argentina). Sociedad Argentina de Antropología, Buenos Aires
Otley H (2008) Falkland Islands Species Action Plan for Seals and Sea Lions 2008–2018. Environment Planning Department (Falkland Is), Stanley
Páez E (2000) Utilización de Boostrap y análisis de poder en estimaciones de abundancia de cachorros de *Arctocephalus australis* [Using Bootstrap and power analysis in abundance estimates of *Arctocephalus australis* pups]. Pag 55–70 in Rey M, Amestoy F (eds). [Synop biol ecol pop fur seals and sea lions Urug] INAPE, Montev, Uruguay
Páez E (2006) Status of the administration of fur seals and sea lions in Uruguay]. Pag 577–583 in R. Menafra L, Rodrıguez-Gallego F et al (eds) (Basis for the Conservation and Management of the Uruguayan Coast) Soc Urug Cons Nat, Montevideo, Uruguay
Pavés HJ, Schlatter RP (2008) Temporada reproductiva del lobo fino austral, *Arctocephalus australis* (Zimmerman, 1783) en la Isla Guafo, Chiloé, Chile. Rev Chil Hist Nat 81:137–149
Pennant T (1781) History of quadrupeds. B. White, London, U.K. 2:285–566
Repenning CA, Peterson RS, Hubbs CL (1971) Contributions to the systematics of the southern fur seals, with particular reference to the Juan Fernández and Guadalupe species. Antarcl Res Ser 18:1–34
Rodrigues P, Seguel M, Gutiérrez J et al (2018) Genetic connectivity of the South American fur seal (*Arctocephalus australis*) across Atlantic and Pacific oceans revealed by mitochondrial genes. Aquat Conserv Mar Freshwat Ecosyst 28(2):315–323
Sanfelice D, Vasques VC, Crespo EA (1999) Ocupação sazonal por Otariidae (Mammalia Pinnipedia) da Reserva Ecológica da Ilha dos Lobos, Torres, Rio Grande do Sul, Brasil. Iher Ser Zool 87:101–110
Schiavini ACM (1992) Hombres y lobos marinos en la Sudamérica Austral. In: Lichter AA (ed) En Huellas en la arena, sombras en el mar. Los mamíferos marinos de la Argentina y la Antártida. Terra Nova, Buenos Aires, pp 250–254
Scialabba N (1989) World review of marine mammal entanglement in fishing gear and plastic marine debris. FAO, Rome
Strange IJ (1992) A field guide to the wildlife of the Falkland Islands and South Georgia. Harper & Collins, London
Torres D (1985) Presencia del lobo fino sudamericano (*Arctocephalus australis*) en el norte de Chile, como consecuencia de El Niño. 1982-1983. Inv Pesq (Chile) 32:225–233
Túnez JI, Centrón D, Cappozzo HL et al (2007) Geographic distribution and diversity of mitochondrial DNA haplotypes in South American sea lions (*Otaria flavescens*) and fur seals (*Arctocephalus australis*). Mamm Biol 72:193–203
Túnez JI, Cappozzo HL, Cassini MH (2008) Regional factors associated with the distribution of South American fur seals along the Atlantic coast of South America. ICES J Mar Sci 65:1733–1738
Túnez JI, Cappozzo HL, Paves H et al (2013) The role of Pleistocene glaciations in shaping the genetic structure of South American fur seals (*Arctocephalus australis*). New Zeal J Mar Freshw Res 47:139–152

Vales D (2015) Uso actual y pasado de los recursos tróficos por parte del lobo marino de dos pelos *Arctocephalus australis* (Zimmermann, 1783) en el Océano Atlántico Sudoccidental. Doctoral thesis, Universidad Nacional de Mar del Plata

Vaz-Ferreira R (1982) *Arctocephalus australis* Zimmermann, South American Fur Seal. Mamm Seas FAO Fish Ser 4(5):497–508

Vaz-Ferreira R (1987) Ecology, behavior, and survival of the South American fur seal in Uruguay. In: Croxall, JP, Gentry, RL (eds) Status, biology, and ecology of fur seals: proceedings of an international symposium and workshop Cambridge, England, 23–27 April 1984, NOAA Technical Report NMFS 51, pp 165–168. Available via Aquatic Commons. http://aquaticcommons.org/2771/. Accessed 13 Aug 2020

Vaz-Ferreira R, Ponce de León A (1987) South American Fur Seal, *Arctocephalus Australis*, in Uruguay. In: Status, Biology, and Ecology of Fur Seals. Proceedings of an international symposium and workshop, Cambridge, UK, 23–27 April 1984, pp 29–32

Weddell J (1825) A Voyage towards the South Pole performed in the years 1822–1824 by James Weddell Master in the Royal Navy, London 1825, Loneman, Hurst, Rees, Orme, Brown y Green

Ximénez IM, Langguth E (2002) Isla de Lobos. Graphis Ed, Montevideo, Uruguay

von Zimmermann EAW (1783) Geographische Geschichte der Menschen, und der allgemein verbreiteten vierfüssigen Thiere. Zweiter Band, vol 3. Wenganschen Buchhandlung, Leipzig, pp 1778–1783

Chapter 3
Galápagos Fur Seal (*Arctocephalus galapagoensis*, Heller 1904)

Fritz Trillmich

Abstract The Galapagos fur seal (GFS; *Arctocephalus galapagoensis*) is the smallest seal species, endemic to the Galapagos archipelago and closely related to the South American fur seal, *Arctocephalus australis*. Females are highly site faithful. Genetic exchange between colonies happens predominantly through male dispersal. The species' distribution overlaps with that of the Galapagos sea lion (*Zalophus wollebaeki*). The largest colonies exist on the western islands where cold, productive waters upwell. GFSs breed when upwelling is strongest, from August to December. They forage pelagically on organisms of the deep scattering layer, mostly cephalopods, myctophids, and bathylagids. These organisms migrate toward the surface when it gets dark, which brings them into the diving range of the GFS. Consequently, GFSs forage at night, and their foraging is strongly influenced by the lunar cycle. Females give birth to a single pup and nurse it usually for 2 years, but under poor environmental conditions may extend the time to weaning. If a female bears another pup during lactation, the newborn's survival is seriously reduced. The extended period of maternal care lowers the reproductive rate. Moreover, in years of strong El Niño, offspring mortality is increased and even adult animals may die of starvation. The small distribution area and low population size make the GFS vulnerable to environmental disturbances, whether anthropogenic or natural like El Niño. Fishery interactions and the potential introduction of diseases are presently the greatest dangers. Tourism should be closely monitored and restricted to protect breeding sites.

Keywords Distribution · Tropical ocean environment · El Niño · Foraging strategy · Social structure

F. Trillmich (✉)
Department of Animal Behavior, Bielefeld University, Bielefeld, Germany
e-mail: fritz.trillmich@uni-bielefeld.de

G. Heckel, Y. Schramm (eds.), *Ecology and Conservation of Pinnipeds in Latin America*, https://doi.org/10.1007/978-3-030-63177-2_3

Introduction: The Environment

The Galapagos fur seal (GFS; *Arctocephalus galapagoensis*; Fig. 3.1) inhabits an area of the eastern tropical Pacific, the Galapagos archipelago that is unusually cold for a tropical ocean. Situated 1000 km off the western coast of Ecuador on the equator, the archipelago consists of 17 islands and many islets of widely varying area, which arose by volcanic activity over the last 4 to 4.5 million years. The archipelago comprises a central rather shallow (ca. 200 m) platform surrounded by deep (>1000 m) oceanic waters. The islands receive cold, productive waters through the Humboldt (or Peru) current, which is driven by SE trade winds. These winds produce strong upwelling of cold-water masses from depth along the coasts of Peru and Chile. In compensation for the massive westward movement of water with the Humboldt current, the cold and nutrient-rich equatorial countercurrent, or Cromwell current, flows at depth from the west and wells up along the western shores of the archipelago (Fiedler and Talley 2006; Kessler 2006). The combined action of the Humboldt current and the local upwelling of the Cromwell current inject huge amounts of cool water into the Galapagos ecosystem and make the archipelago a productive spot amidst an oceanic desert (Fernández-Álamo and Färber-Lorda 2006; Pennington et al. 2006). This creates a rich marine life as the necessary support system for the two local pinniped species, the GFS and the Galapagos sea lion (*Zalophus wollebaeki*; see Chap. 8), as well as many other marine organisms.

When the trade winds weaken – a normal event at the beginning of the southern summer (December to April) – the Humboldt current becomes less intense, and upwelling of the Cromwell current is reduced. Consequently, the surface waters warm by insolation and in addition warm water enters from the Panama Bay. This

Fig. 3.1 Female Galapagos fur seal (left) threatens an approaching territorial male

depresses the thermocline. The situation normally reverts around April, when the southeast trade winds strengthen again (Fahrbach et al. 1991). However, during an El Niño event, warming continues, and the thermocline is depressed, often several tens of meters, making cold-water prey organisms harder to access for the fur seals and sea lions. El Niño occurs approximately every 4–8 years and may last for several months. While El Niño constitutes a recurrent event in the marine ecosystem, which has existed for several thousand years already and therefore constitutes a feature that has influenced the ecology and behavior of GFS and Galapagos sea lions over evolutionary time, it nevertheless often leads to dramatic consequences for their populations because of its influence on a multitude of marine organisms of the Galapagos (Arntz et al. 1991; Robinson and del Pino 1985; Trillmich and Limberger 1985; Trillmich and Dellinger 1991). If the intensity and duration of El Niño events increase due to climate change (as seems likely: Cai et al. 2014, but see Collins et al. 2010), this could endanger the survival of both pinniped species, which have already declined, most likely due to recurrent El Niño events over the last three decades (Alava and Salazar 2006; Trillmich 2015; Trillmich et al. 2016).

On the positive side, El Niño events are followed by La Niña (Timmermann et al. 1999), unusually cold conditions when the thermocline gets close to the surface and marine productivity is increased, making foraging easier and more efficient for the GFS. Indeed, changes in the upwelling regime in the area of the Galapagos may lead to local sea surface cooling in the west of the archipelago (Karnauskas et al. 2015).

Morphology and Systematic Position

The GFS is the smallest and apparently the tamest pinniped species worldwide (Fig. 3.2). The few adult males that were weighed opportunistically during the breeding season weighed between 60 and 68 kg. This is likely an underestimate of the body mass of territorial males at the beginning of the reproductive season, when they come ashore after extended periods at sea carrying maximal fat reserves. Even such maximally fat males most likely weigh less than 100 kg. Adult females have curvilinear lengths of 1.1–1.3 m and an average mass of 28.5 kg (Horning & Trillmich 1997a). Fattening before parturition may lead to maximal female weights of 41 kg (Costa and Trillmich 1988).

All recent phylogenetic analyses agree that the species is most closely related to the South American fur seal (*Arctocephalus australis*) and may indeed be considered a subspecies of the western population of the species, i.e., the animals found in Peru and Chile (Dasmahapatra et al. 2009; Higdon et al. 2007; Wolf et al. 2007; Wynen et al. 2001). The separation date of the two populations was estimated at approximately 700–900 kya ago (Higdon et al. 2007), but a more recent estimate based on whole genome information suggests an even later split at around 400 kya (± 20 kya) (Lopes F, Oliveira L, Bonatto SL, pers. comm.). It appears possible – but

Fig. 3.2 After brief periods of habituation, Galapagos fur seals tend to ignore humans. Note the small size of the fully adult females and the adult male (left, in foreground) in relation to a human

is not proven – that occasional stragglers of the South American population enter the Galapagos and may even reproduce there.

Distribution

The GFS occurs only on the Galapagos archipelago where its distribution overlaps with that of the Galapagos sea lion (GSL; *Zalophus wollebaeki*; see Chap. 8). The GFS is most abundant in the west where upwelling is strongest, particularly on the islands of Fernandina and Isabela (Fig. 3.3). At present, the GSL abundance is low in these areas. The central platform of the archipelago, which is the center of the GSL's distribution, is visited only occasionally by GFSs. This implies that the distribution area of the species is unusually restricted, covering an area smaller than the Galapagos Marine Reserve (less than 140,000 km^2). The species has been observed rarely outside the archipelago on the western coast of South America, east and north of Galapagos (Capella et al. 2002; Felix et al. 2001, 2007) up to the coast of Mexico (Aurioles-Gamboa et al. 2004). In two cases, females were observed to give birth on the mainland coast of Ecuador, but due to disturbance or other reasons, these females abandoned their pups a few days after birth (Felix et al. 2007). Only few of the records have been verified by genetic analysis, which is desirable because the species is difficult to distinguish from its congeneric *A. australis*, particularly if immature are observed.

Fig. 3.3 Map of the Galapagos archipelago, with names of the sites and islands mentioned in the text. The Galapagos National Park office is based on Santa Cruz; the capital of the province is on San Cristobal. The central islands lie on a platform where the sea is only about 200 m deep, whereas toward the west the shelf drops off quickly to depths greater than 1000 m

Females show high natal philopatry, i.e., they tend to stay where they were born. This leads to strong population structuring whereby most colonies, even if separated by as little as 70 km "as the seal swims," present characteristic mixtures of mitochondrial haplotypes (Lopes et al. 2015). Nuclear DNA does not present such population structure, suggesting substantial male dispersal. This agrees with the observation that within the archipelago most fur seals observed on islands without breeding colonies are males. The largest breeding colonies on Fernandina and Isabela harbor about two thirds of the total population. Smaller colonies are situated on Santiago and the northern islands of Pinta, Marchena, and Genovesa.

GFSs prefer to haul out on rocky coasts with large boulders and ledges that provide shade and the opportunity to rest protected from the intense solar radiation in spaces between rocks, in crevices, and in lava tubes. Only a narrow fringe of coast is useful habitat for the species since most animals need access to the water during the hottest time of the day to thermoregulate. During that time, many animals raft in front of the colony (Fig. 3.4).

Population Development

GFSs were slaughtered for their pelts during the nineteenth century, but we have no detailed knowledge of the population size pre-exploitation (Trillmich 1987). Most sealing in the Galapagos was carried out by whalers who called at the islands to

Fig. 3.4 Rafting GFS in front of a colony. To reduce cooling, animals often keep flippers out of the water

collect provisions of fresh water and tortoise meat. In 1816, a single voyage caught 8000 animals, suggesting that the population at that time was probably several 10,000s (Townsend 1934). The harvest continued to the point that it was no longer economically viable (Townsend 1934; Slevin 1959). The species slowly recovered from the 1930s onward (Levêque 1963; Orr 1972) and in the late 1970s was estimated to number around 30,000 animals (Trillmich 1987). Due to a number of strong El Niño events (1983, 1987, 1992, 1998, 2016 following the NOAA classification https://www.esrl.noaa.gov/psd/enso/past_events.html), the population appears to have dropped (Alava and Salazar 2006) but may recover (Trillmich 2015). Unpublished census data suggest the population at present comprises about 11,000 animals (Páez-Rosas 2018 and to be submitted). Despite massive exploitation in the nineteenth century and population reductions due to El Niño events, the population shows no sign of reduced genetic diversity (Lopes et al. 2015).

Foraging Behavior and Diet

Foraging was studied in the early 1980s, again in much more detail in the early 1990s at Cabo Hammond (Kooyman and Trillmich 1986; Horning and Trillmich 1997b, 1999), and 20 years later at Cabo Douglas (Jeglinski et al. 2013; Villegas-Amtmann et al. 2013). At those times, Galapagos sea lions co-occurred in these sites with the GFS, but in recent years, these sea lion colonies have disappeared

(Paez-Rosas, pers. comm.). Both sites are on the western side of Fernandina (Fig. 3.3), one of the most productive marine areas of the Galapagos. In front of this coast, the shelf slopes steeply to depths of >1000 m. GFS forage pelagically above these deep zones, near exclusively at night, 10–70 km off the coast (Fig. 3.5; Jeglinski et al. 2013, Villegas-Amtmann et al. 2013). The average duration of trips to sea by lactating females is short. Mean trips of GFS females last approximately 1.5 days in the cold season (May to November) and around 4 days in the warm season (December to April) (Trillmich and Kooyman 2001; Jeglinski et al. 2013; Villegas-Amtmann et al. 2013). However, foraging absences of females become much longer during El Niño to the extent that pups may die of starvation during maternal absence (Trillmich and Limberger 1985).

The mean depth of foraging dives is 22–32 m, with durations of less than 2 min (Horning and Trillmich 1997b, Jeglinski et al. 2013). The maximum dive depth recorded was 169 m, and the longest duration 6.5 min (Horning and Trillmich 1997b). Adult females swim at a speed of around 1.5 m/s, but in spurts can reach 4 m/s (Horning and Trillmich 1997b; Ponganis et al. 1990).

GFSs feed on organisms of the deep scattering layer which comprise cephalopods and several fish species, most abundantly myctophids (*Lampadena* spec.) and bathylagids (*Leuroglossus* spec.), as determined by analysis of otoliths from scats. Food choice varied among years. Most obvious was the decline in myctophids after the 1982/1983 El Niño and the increase in sardine (*Sardinops sagax*) during the El Niño. In non-El Niño years, sardine contributed very little to the GFSs' diet (Dellinger and Trillmich 1999). During the years of that study (1983–1986), we observed only minimal dietary overlap with Galapagos sea lion prey. *Onychoteuthis banksii* contributed the numerical (96%) and estimated mass (73%) majority of cephalopods in GFS prey. Ommastrephid squids contributed another 25% of the mass (Clarke and Trillmich 1980). The mass of these species varied between 5 and 120 g. The prey spectrum fits the observed diving behavior and foraging space use described above. It was also supported by the analysis of stable isotopes, which showed GFSs to be pelagic foragers using a slightly lower trophic level than Galapagos sea lions (Páez-Rosas et al. 2012).

Foraging at night in pelagic waters exposes the GFS to shark predation. The extent and probability of shark attacks is unknown, but animals with shark bites are seen regularly (Fig. 3.6). Mortality due to shark attacks among adult GFS cannot be very high as of 83 marked females observed for at least 3 months during the reproductive period (i.e., 20 female-years), none was recorded missing (Trillmich 1987). Daytime attacks by cooperatively hunting killer whales (*Orcinus orca*) have been observed near colonies and appear often successful, judging by the feeding of frigate birds on scraps immediately following such attacks. When killer whales approach, fur seals swim toward shore at top speed, searching safety on land. Many, especially inexperienced juvenile animals, may die at sea by unobserved predator attacks, making it impossible to estimate the mortality caused by predation.

Juveniles make their first brief foraging sojourns when older than 6 months and dive more or less regularly to forage for a contribution to their own maintenance when 1-year-old (or even older). At this time, their body mass varies between 10 and

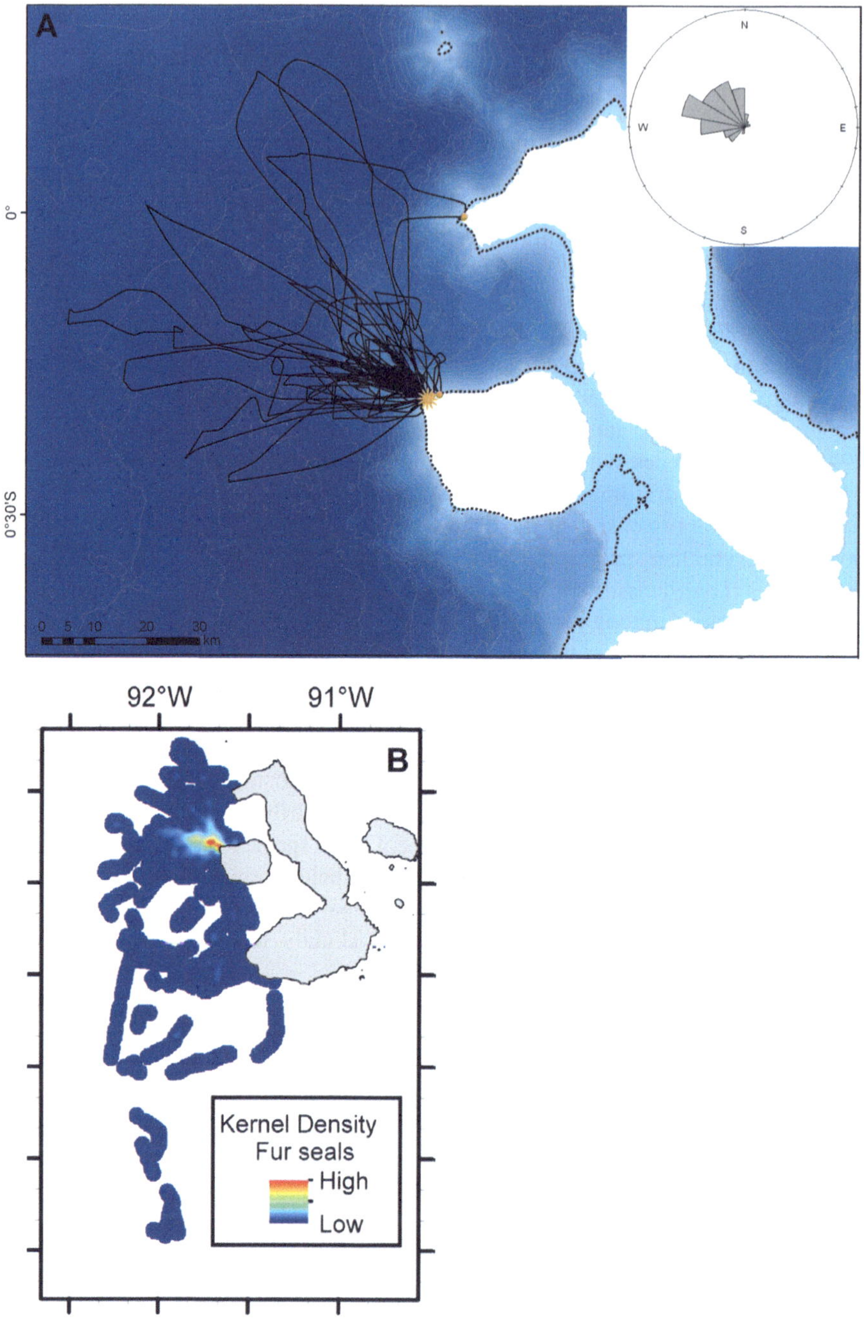

Fig. 3.5 Foraging areas used by adult GFS females in front of Cabo Douglas, Fernandina (upper panel from Jeglinski et al. 2013; lower panel from Villegas-Amtmann et al. 2013)

Fig. 3.6 An adult GFS seriously wounded by a shark attack. The animal is extremely emaciated (note the pelvic bones sticking out) and likely to die because of its reduced swimming ability

15 kg depending on marine productivity (Horning and Trillmich 1997b; Trillmich 1990). At this age, they spend significantly less time at sea (24%) than lactating adult females (50–60%) (Horning and Trillmich 1997b). The late onset of foraging is due to slow development of oxygen stores in blood (Horning and Trillmich 1997a) and muscle, which are needed to permit longer hunting forays under water. By the age of 2 years, diving abilities finally begin to approach those of adult females. This slow development of body mass, oxygen stores in the body, and consequent foraging abilities explains the late weaning in this species. Simultaneously, it buffers young through the continuing energy gain through maternal milk against the natural variability in marine productivity and allows them to gain substantial experience with the marine environment before independence (Trillmich 1990; Horning and Trillmich 1997b, 1999).

Lunar light strongly influences the diving behavior of GFSs. As it suppresses the nightly vertical migration of prey organisms toward the surface, GFSs have to dive deeper under lunar illumination. This further decreases their foraging time window. Indeed, the effect was first detected by the observation that most fur seals were ashore around full moon and numbers were minimal shortly after full moon (Trillmich and Mohren 1981). Even within a single night, diving activity and dive depth vary enormously in relation to lunar light intensity (Fig. 3.7). The effect is strongest for juveniles that are less able divers than adult females and is clearly seen when analyzing the proportion of animals at sea in relation to the lunar cycle. Such an analysis shows that juveniles remain on land for a longer period around full moon than adult females (Fig. 3.8). Whereas females tend to lose weight around full moon, yearlings gain weight during this time, because mothers are in attendance for the longest period (Horning and Trillmich 1999). These observations indicate that cyclic changes in prey accessibility related to lunar light levels modulate foraging

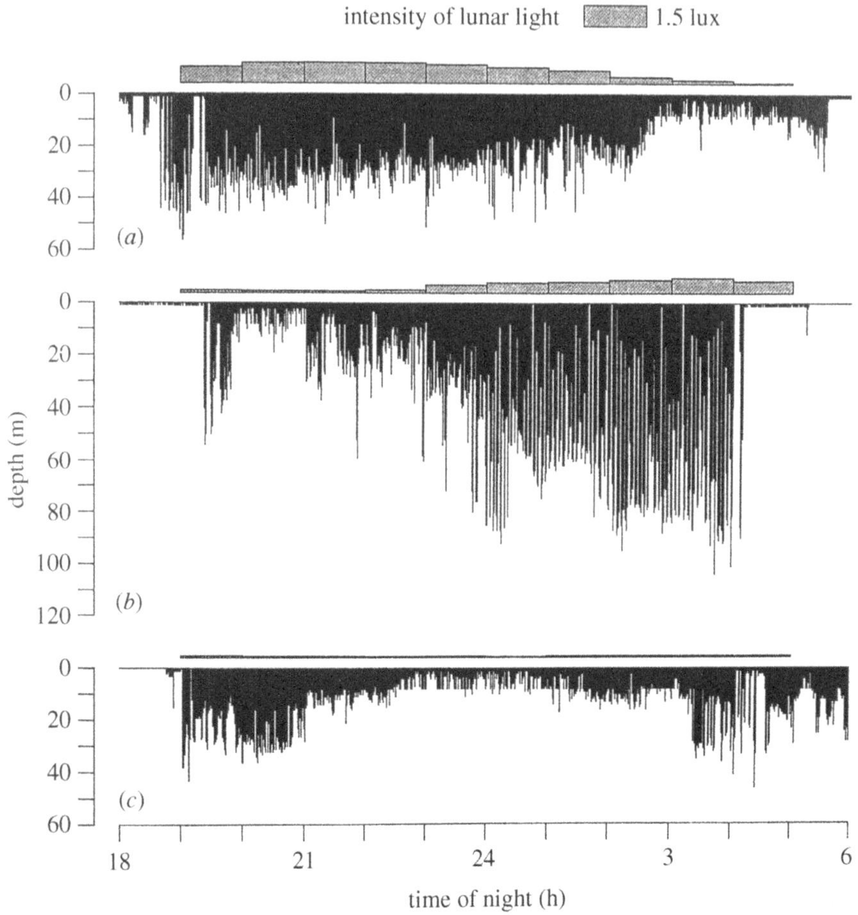

Fig. 3.7 Diving pattern of an adult female GFS in relation to lunar light. (**a**) Fourth night before full moon. (**b**) Fourth night after full moon. (**c**) One night before new moon. (From Horning and Trillmich 1999)

efficiency to the point that hunting becomes inefficient during full moon. The lower diving efficiency of juveniles makes them dependent on the energy input through maternal milk every month around full moon and thereby increases the time to complete weaning. These constraints influence the life history of the GFS: they contribute to reduced juvenile growth rates, to extended dependence on maternal resources, and ultimately to reduced maternal reproductive rates. Ontogenetic constraints thereby prove an influential cause of the species' susceptibility to fluctuations in food availability (Horning and Trillmich 1999).

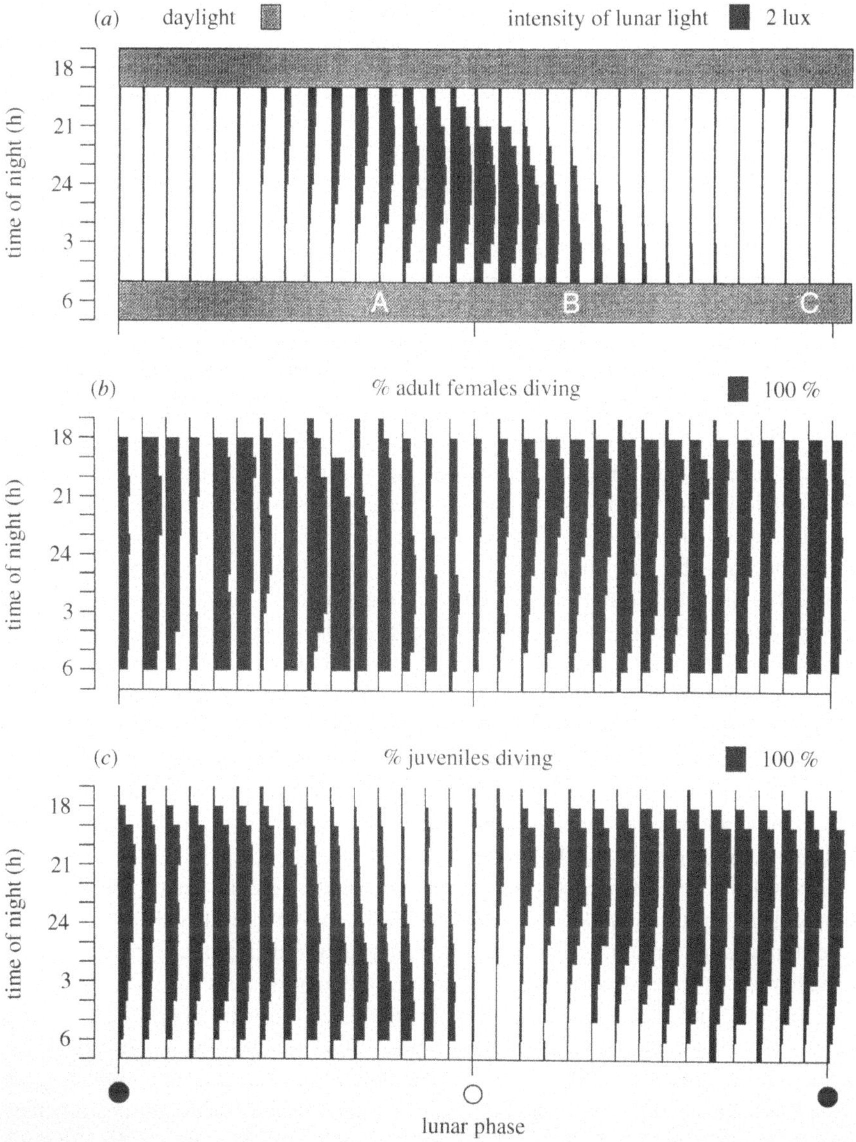

Fig. 3.8 Lunar light conditions influence the foraging activity of GFSs. (**a**) Intensity of lunar light as measured ashore. Gray bars indicate dawn and dusk periods with >10 lux. (**b**, **c**) Percentage of study animals' active foraging. (**b**) Foraging activity of adult females. (**c**) Foraging activity of 1–2-year-old juveniles. A, B, and C refer to the lunar phase of the dive traces in Fig. 6. (From Horning and Trillmich 1999)

Reproductive Biology and Development

Animals on Fernandina island breed between August and December, with most pupping concentrated around the end of September to the beginning of October (Trillmich 1984, 1987; Trillmich and Mohren 1981). This period coincides with the cool *garúa* (light drizzle) season when frequently overcast skies reduce the thermal stress ashore. The season is characterized by the most intense flow of the Humboldt current and intensified upwelling of cold water creating high marine productivity. The reproductive period on other islands may peak in other months. For example, a birth was observed on Genovesa in late March of 1991 (Trillmich pers. obs.), but no systematic data for islands other than Fernandina are available.

Females arrive at the parturition site one or a few days before giving birth to a pup weighing about 3.6 kg (3.8 for male pups, 3.5 for females; Trillmich and Wolf 2008). They remain with the pup for 3–10 days after parturition and fast during that time. Females defend the pup aggressively against the approach of other fur seals. The female frequently drags the pup along against its intense protests as she will initially not separate from it. This conflict originates in different thermoregulatory needs, because the pup often can find a place in the shade whereas its mother is exposed to the sun. Mothers are then forced to move to the zone where waves wash the shoreline. In this way, they travel up and down the shore during the day for thermoregulation, thereby giving the pup the opportunity to explore the area around its birthplace. During this period mother and offspring learn to recognize each other by voice (Trillmich 1981; Pitcher et al. 2010) and presumably by smell. Fasting females lose about 1.7% of their body mass per day (Costa and Trillmich 1988). This low rate of mass loss is due to a lower metabolic rate for this species, compared to the more temperate or subpolar fur seal and sea lion species (Costa and Trillmich 1988; Trillmich and Kooyman 2001).

Toward the end of the peri-parturient period, the female leaves the pup and approaches a male or makes herself obvious to the territorial male by attacking other females or immatures. The male then approaches the female, appears to determine her estrus state by olfaction, and mounts. Copulation lasts for about 8–10 min and is usually ended by the female biting the male wherever she can get him, usually the neck, but sometimes also the flippers where blood may be drawn, attesting to the seriousness of these bites (Fig. 3.9). Immediately after a successful copulation, the female becomes unattractive to males and may even move into a neighbor's territory without being harassed. Females that have lost their offspring during the previous year and virgin females enter estrus spontaneously and also copulate during the reproductive period.

After copulation, females leave the pup to forage at sea. The first foraging excursion often lasts just one night, but her absences increase in duration and, during the cold season, lead to a cycle of about 1.5 days absence and 0.5–1.0 day presence ashore nursing the pup (Trillmich 1986). During the warm season, mothers of then 6-month-old pups are absent for longer periods, often around 4 days, whereas time onshore with the young remains similar (0.5–1.5 days) (Trillmich 1990). Overall,

Fig. 3.9 Female GFS biting a male toward the end of copulation

females of 1-month-old pups spend about 60% of their time away from the pup, but, during the warm season when pups are about 6 months old, they may stay away for 80% of the time. Females also consistently spend time away from their offspring elsewhere on land resting and may occasionally spend the daytime resting at sea. This suggests that females restore their body reserves before returning to their offspring (Trillmich and Kooyman 2001). In effect, mothers visit their pups an estimated 300 times before weaning at an age of 2 years (Trillmich 1986).

Pups become increasingly mobile and, at the age of about 3 weeks, begin to enter the water on their own, initially only in shallow pools and protected inlets that are not exposed to intense wave action. Juveniles begin to swim for longer periods in open water but start to forage only after they molt into adult fur (Horning and Trillmich 1997b), at 4–7 months.

Independence is delayed under low productivity and especially under El Niño conditions (Trillmich and Wolf 2008). Thus, depending on the productivity in a given year or series of years, offspring may be weaned at ages between 1 and 3 (and up to 4) years. In exceptional cases, offspring were still allowed to suck when they were 4–5 years old. Pups born prior to the weaning of an older sibling rarely survive, because they are outcompeted in the contest for maternal milk and, in rare cases, killed by the older sibling. However, due to a lower pregnancy rate of lactating females on the one hand and offspring mortality during the first year of life on the other, this situation arises only in approximately 15% of all births. The reduction in survival of a newborn pup with an older (yearling) sibling varies with marine conditions from normal (i.e., 5% pup mortality during the first month of life) to 100%: The warmer the marine environment, the lower their survival (Trillmich and Wolf 2008). Females usually defend the newborn against the older offspring but often allow the older one to suck which leads to greatly reduced milk intake of the young pup and its eventual starvation. Only in exceptional cases did females successfully suckle two subsequent offspring simultaneously (Trillmich 1986).

Fig. 3.10 The territorial GFS male threateningly approaches a copulating smaller male to chase him off

Little quantitative information is available on male movement patterns. Males haul out in the reproductive colonies throughout the whole year. They appear to spend much time continuously at sea during the non-reproductive period, as evidenced by barnacles growing on their fur. They do not become physically large enough to compete for a territory (Fig. 3.10) until they are perhaps 7–8 years old, considerably older than the average age of maturity of females, but exact data are missing (Trillmich 1987). As the reproductive season approaches, hauling out becomes more frequent and periods spent on land longer until they remain almost continuously on land defending a territory, usually before the first females give birth. Males hold territories that average 200 m^2, which is large compared to the average size of territories held by other otariid males. For example, males of the Steller sea lion (*Eumetopias jubatus*), which weigh up to 900 kg, also hold 200 m^2 territories. This is particularly notable given the GFS's small size and relates to the low density of animals in the colonies. Due to the long reproductive period of approximately 90 days, and their small size, GFS males cannot maintain a territory for the whole reproductive period. The longest documented period of continuous territory tenure was 51 days, but the median time of tenure is only 27 days (Trillmich 1984). Toward the end of long tenure periods, males may occasionally leave to forage but, in doing so, risk losing the territory. Males that arrived early usually disappear from their territory at the middle of the season, but around 30% of the males return toward the end of the season for a second, usually shorter (15 day), tenure period on their former territory (Trillmich 1984). The absence of the largest males during the middle of the season provides smaller males an option to become territorial near the peak of the season, when surprisingly few large males are in attendance. During this second tenure period, males often leave the territory for overnight foraging or even for a day or two. The most successful males may gain 20 or more copulations within a season, but most males are much less successful. Many males survive only one season holding territory, dying from infection of wounds inflicted during territorial fights. Consequently, there is a major turnover of males from year to year. The longest documented reproductive life of a territorial male lasted three seasons (Trillmich 1984). Until we get reliable estimates of male reproductive success through genetic paternity analysis, the best available estimate of average mating success of a territorial male in a given season is 5–7 copulations (Trillmich 1987).

Demography

GFSs develop slowly and begin reproduction late compared to more temperate and subpolar fur seal species. Female GFSs mature as early as their third year, but most are believed to mature when approximately 5 years old (Trillmich 1987). Females can produce one pup a year because they always copulate, likely only once, about a week after parturition. However, as shown for the South American fur seal (*Arctocephalus australis*) in Uruguay (Lima and Paez 1995), pregnancy rates probably vary between 50% and 80% depending on marine productivity and female age. If a female is lactating, she is unlikely to become pregnant or carry the pregnancy through to parturition. Hence, due to the slow development of offspring, most females rear a pup successfully only every second or third year (Trillmich and Wolf 2008). Longevity is unknown.

Males presumably mature at a similar age to females. Territorial males that died during the reproductive season and were aged by counting growth layers in extracted teeth were between 9 and > 11 years old (Trillmich 1987). Territorial males have a higher mortality rate, estimated at 32% per year (Trillmich 1987) than same-age females. In particular, during El Niño years, male mortality may increase significantly (Trillmich and Dellinger 1991), leading to major die-offs of adult males when the animals return to sea after an extended fast on territory, only to find greatly reduced access to food due to poor marine conditions.

Conservation Status and Management

CITES (2008) lists the GFS in Appendix II. In this category, a species can be traded, if the authority of the country of origin certifies that the animals were obtained under permit and that the trade does not affect the species' survival. Nevertheless, the Special Law for Galapagos (Ley Orgánica de Régimen Especial de la Provincia de Galápagos; Oficio No. SAN-2015-0989, Quito, 10 de junio de 2015) regulates that no endemic species of the Galapagos archipelago is subject to any trade, thereby providing complete protection. In addition to the protection offered by the location of colonies within the boundaries of the National Park, the fur seal population forages mostly in an area that has been protected since 1998 as the Galapagos Marine Reserve (GMR), covering an area of 133,000 km^2 and recognized in 2001 by IUCN as a World Heritage Site. The reserve is defined as a 40-mile (64 km) zone around the islands and therefore covers the main foraging area of the GFS. This greatly reduces, but does not exclude, the danger of direct interaction with fishing gear because some GFS may forage outside the Galapagos Marine Reserve boundaries. Ghost nets and drifting long lines endanger marine life within the reserve as international fishing activity around the perimeter of the GMR is intense and sometimes illegally encroaches into the reserve.

Legal fishing by licensed local fishing boats probably interacts little with GFSs because major sections of the coastline where fur seal colonies exist are designated as zones of "Conservation and Non-extractive Use" (Subzona 2.2. Conservación y Uso No Extractivo; Dirección del Parque Nacional Galápagos 2014). Special "Conservation and Non-extractive Use" zones overlap with the habitat used by foraging GFSs (see also Ventura et al. 2019; special conservation zones plotted by these authors are not approved by the Galapagos National Park). In recent years, the catch per unit effort of the local legal fishery has declined. This implies a risk that in the future it may target lower trophic level species, but it appears unlikely that it will target species of the deep scattering layer that are the most important fur seal prey. In addition, growing tourism and lost or drifting fishing gear led to a substantial increase in drifting and beached plastic pollution (Alava et al. 2014). In particular, drifting nets have been documented to cause substantial mortality through entanglement (Poeta et al. 2017). Connected to the increase in local human population and tourism, marine traffic has increased substantially over the last 20 years, and ships have become much faster with the introduction of high-powered fiber glass boats. This increases the risk of collision with pinnipeds as well as with whales and dolphins, but no systematic monitoring data are available.

Domestic animals, in particular rats, dogs, and cats in the settlements on Galapagos, pose a potential threat to the health of the GFS. All of these species must be considered a reservoir of infectious diseases such as toxoplasmosis, leptospirosis, and even distemper virus (Levy et al. 2008; Alava et al. 2017). Although dogs are unlikely to enter fur seal colonies, cats and rats on Isabela certainly do. Moreover, the sympatric sea lions are in much more intense contact with the domestic animals as they often haul out in human settlements, most obvious in San Cristobal where they often rest on the dock area (Denkinger et al. 2015). Sea lions share many haul outs and colonies with fur seals and could easily transmit diseases to them. Pesticides may also affect the GFS by reducing fertility, which may lead to premature birth or infertility as is shown for California sea lions (DeLong et al. 1973). Alava et al. (2017) found total DDT concentrations between 277 and 1970 ng/g lipid in fur seal samples, giving reason for concern.

Finally, as pointed out above, the potential increase in the frequency and intensity of El Niño events due to greenhouse warming (Cai et al. 2014) may endanger the persistence of the population in the future. As the detailed local effects are difficult to predict (Karnauskas et al. 2015), the concerns described in this chapter make further regular population monitoring as undertaken over the last years by Páez-Rosas (2018) and the Galapagos National Park Service an urgent need.

Acknowledgments I would like to thank the Galapagos National Park for granting the permits that allowed our work on the Galapagos fur seal and the Charles Darwin Research Station for their continued support. This work would have been impossible without the steady support by the Max-Planck Society through its director Wolfgang Wickler at the Institute for Behavioral Physiology, Seewiesen. The German Research Foundation financed our work at Cabo Douglas through grant TR 105/19-1 and the National Geographic Society through grant No 8682.09. Gerald Kooyman and Paul Ponganis established the experimental dive work with the fur seals. Dan P. Costa helped through his kind offer of instruments, and he and Stella Villegas-Amtmann and field assistants too numerous to name them all provided enormous, much appreciated help. Diego Páez-Rosas, Paolo Piedrahita, and Alex Hearn helped in getting information about the National Park zoning scheme. Diego Páez-Rosas provided much appreciated information about his and the Galapagos National Park's recent census results. Susana Cardenas, Jana Jeglinski, and an anonymous reviewer provided much constructive input. Meinolf Ottensmann kindly introduced me to the use of QGIS.

References

Alava JJ, Salazar S (2006) Status and conservation of Otariids in Ecuador and the Galapagos Islands. In: Trites AW, Atkinson SK, DeMaster DP, Fritz LW, Gelatt TS, Rea LD, Wynne KM (eds) Sea Lions of the World. Alaska Sea Grant College Program, Alaska, USA, Fairbanks, pp 495–520

Alava JJ, Palomera C, Bendell L et al (2014) Pollution as an emerging threat for the conservation of the Galapagos marine reserve: environmental impacts and management perspectives. In: Denkinger J, Vinueza L (eds) The Galapagos marine reserve. Springer Science +Business Media, New York, pp 247–283. https://doi.org/10.1007/978-3-319-02769-2_12

Alava JJ, Denkinger J, Jimenez P et al (2017) Chapter 7: Population status, anthropogenic stressors and conservation of the Galapagos fur seal (*Arctocephalus galapagoensis*). In: Alava JJ (ed) Tropical pinnipeds: bio-ecology, threats and conservation. CRC Press, Boca Raton, pp 120–131

Arntz W, Pearcy WG, Trillmich F (1991) Biological consequences of the 1982-83 El Niño in the eastern Pacific. In: Trillmich F, Ono KA (eds) Pinnipeds and El Niño. Responses to environmental stress. Springer Verlag, Berlin, pp 22–42

Aurioles-Gamboa D, Schramm Y, Mesnick S (2004) Galapagos fur seals, *Arctocephalus galapagoensis*, in Mexico. Latin Am J Aquat Mamm 9:77–90

Cai W, Borlace S, Lengaigne M et al (2014) Increasing frequency of extreme El Niño events due to greenhouse warming. Nat Clim Chang 4:111. https://doi.org/10.1038/NCLIMATE2100

Capella JJ, Flórez-González L, Falk-Fernández P et al (2002) Regular appearance of otariid pinnipeds along the Colombian coast. Aquat Mamm 28:67–72

Clarke MR, Trillmich F (1980) Cephalopods in the diet of fur seals of the Galapagos islands. J Zool Lond 190:211–215

Collins M, An S-I, Cai W et al (2010) The impact of global warming on the tropical Pacific Ocean and El Niño. Nat Geosci 3:391–397. https://doi.org/10.1038/NGEO868

Costa DP, Trillmich F (1988) Mass changes and metabolism during the perinatal fast: a comparison between Antarctic (*Arctocephalus gazella*) and Galapagos fur seals (*Arctocephalus galapagoensis*). Physiol Zool 61:160–169

Dasmahapatra KK, Hoffman JI, Amos W (2009) Pinniped phylogenetic relationships inferred using AFLP markers. Heredity 103:168. https://doi.org/10.1038/hdy.2009.25
Dellinger T, Trillmich F (1999) Fish prey of the sympatric Galápagos fur seals and sea lions: seasonal variation and niche separation. Can J Zool 77:1204–1216
DeLong RL, Gilmartin WG, Simpson JG (1973) Premature births in California Sea lions: association with high organochlorine pollutant residue levels. Science 181:1168–1170
Denkinger J, Gordillo L, Montero-Serra I et al (2015) Urban life of Galapagos Sea lions (*Zalophus wollebaeki*) on San Cristobal island, Ecuador: colony trends and threats. J Sea Res 105:10–14. https://doi.org/10.1016/j.seares.2015.07.004
Dirección del Parque Nacional Galápagos (2014) Plan de Manejo de las Áreas Protegidas de Galápagos para el Buen Vivir. Puerto Ayora, Galápagos
Fahrbach E, Trillmich F, Arntz W (1991) The time sequence and magnitude of physical effects of El Niño in the eastern Pacific. In: Trillmich F, Ono KA (eds) Pinnipeds and El Niño. Responses to environmental stress. Springer Verlag, Berlin, pp 8–21
Félix F, Lento G, Davis J et al (2001) El lobo fino de Galápagos *Arctocephalus galapagoensis* en la costa continental de Ecuador, primeros registros confirmados a través de análisis morfológicos y genéticos. Estudios Oceanológicos 20:63–68
Félix F, Jiménez P, Falconi J et al (2007) New records and first births of the Galapagos fur seal, *Arctocephalus galapagoensis* (Heller, 1904) from the mainland of Ecuador. Rev Biol Mar Oceanogr 42:77–82
Fernández-Álamo MA, Färber-Lorda J (2006) Zooplankton and the oceanography of the eastern tropical Pacific: a review. Prog Oceanogr 69:318–359. https://doi.org/10.1016/j.pocean.2006.03.003
Fiedler PC, Talley LD (2006) Hydrography of the eastern tropical Pacific: a review. Prog Oceanogr 69:143–180. https://doi.org/10.1016/j.pocean.2006.03.008
Heller E (1904) Mammals of the Galapagos archipelago, exclusive of the Cetacea. Proc Calif Acad Sci 3:233–251
Higdon JW, Bininda-Emonds ORF, Beck RMD et al (2007) Phylogeny and divergence of the pinnipeds (Carnivora: Mammalia) assessed using a multigene dataset. BMC Evol Biol 7:216
Horning M, Trillmich F (1997a) Development of hemoglobin, hematocrit, and erythrocyte values in Galapagos fur seals. Mar Mamm Sci 13:100–113
Horning M, Trillmich F (1997b) Ontogeny of diving behaviour in the Galapagos fur seal. Behaviour 134:1211–1257
Horning M, Trillmich F (1999) Lunar cycles in diel prey migrations exert a stronger effect on the diving of juveniles than adult Galápagos fur seals. Proc R Soc Lond B Biol Sci 266:1127–1132
Jeglinski JWE, Goetz KT, Werner C et al (2013) Same size – same niche? Foraging niche separation between sympatric juvenile Galapagos Sea lions and adult Galapagos fur seals. J Anim Ecol 82:694–706. https://doi.org/10.1111/1365-2656.12019
Karnauskas KB, Jenouvrier S, Brown CW et al (2015) Strong sea surface cooling in the eastern tropical Pacific and implications for Galapagos penguin conservation. Geophys Res Lett 42:6432–6437. https://doi.org/10.1002/2015GL064456
Kessler WS (2006) The circulation of the eastern tropical Pacific: a review. Prog Oceanogr 69:181–217. https://doi.org/10.1016/j.pocean.2006.03.009
Kooyman GL, Trillmich F (1986) Diving behavior of Galapagos fur seals. In: Gentry RL, Kooyman GL (eds) Fur seals: maternal strategies on land and at sea. Princeton University Press, Princeton, pp 186–195
Levêque R (1963) Le statut actuel des vertebres rares et menacés de'l archipel des Galápagos. Terre et Vie 110:397–430
Levy JK, Crawford PC, Lappin MR et al (2008) Infectious diseases of dogs and cats on Isabela island, Galapagos. J Vet Intern Med 22:60–65
Lima M, Paez E (1995) Growth and reproductive patterns in the South American fur seal. J Mammal 76:1249–1255

Lopes F, Hoffman JI, Valiati VH et al (2015) Finescale matrilineal population structure in the Galapagos fur seal and its implications for conservation management. Conserv Genet 16:1099–1113. https://doi.org/10.1007/s10592-015-0725-1

Orr RT (1972) Galápagos fur seal (*Arctocephalus galapagoensis*). In: Seals. Proceedings of a working meeting of seal specialists on threatened and depleted seals of the world. IUCN, Morges, pp 124–128

Páez-Rosas D (2018) Lobo fino de Galápagos o lobo de dos pelos. In: Fundación Charles Darwin (FCD) y WWF-Ecuador. Especies Nativas e Invasoras. Quito, FCD y WWF-Ecuador, Atlas de Galápagos, pp 151–153

Páez-Rosas D, Aurioles-Gamboa D, Alava JJ et al (2012) Stable isotopes indicate differing foraging strategies in two sympatric otariids of the Galapagos islands. J Exp Mar Biol Ecol 424:44–52. https://doi.org/10.1016/j.jembe.2012.05.001

Pennington JT, Mahoney KL, Kuwahara VS et al (2006) Primary production in the eastern tropical Pacific: a review. Prog Oceanogr 69:285–317. https://doi.org/10.1016/j.pocean.2006.03.012

Pitcher BJ, Harcourt RG, Charrier I (2010) Rapid onset of maternal vocal recognition in a colonially breeding mammal, the Australian sea lion. PLoS One 5:e12195. https://doi.org/10.1371/journal.pone.0012195

Poeta G, Staffieri E, Acosta ATR et al (2017) Ecological effects of anthropogenic litter on marine mammals: a global review with a "black list" of impacted taxa. Hystric, Italian. J Mammal 28:253–264. https://doi.org/10.4404/hystrix-00003-2017

Ponganis PJ, Ponganis EP, Ponganis KV et al (1990) Swimming velocities in otariids. Can J Zool 68:2105–2112

Robinson G, del Pino EM (1985) El Niño en las Islas Galápagos. El evento de 1982–1983. Fundación Charles Darwin para las Islas Galápagos. Fundación Charles Darwin, Quito

Slevin JR (1959) The Galapagos islands. In: A history of their exploration. Occasional papers of the California Academy of Sciences, San Francisco

Timmermann A, Oberhuber J, Bacher A et al (1999) Increased El Niño frequency in a climate model forced by future greenhouse warming. Nature 398:694–697

Townsend CH (1934) The fur seal of the Galapagos Islands. Zool N Y 18:43–56

Trillmich F (1981) Mutual mother-pup recognition in Galápagos fur seals and sea lions: cues used and functional significance. Behaviour 78:21–42

Trillmich F (1984) The natural history of the Galapagos fur seal (*Arctocephalus galapagoensis*, Heller 1904). In: Perry R (ed) Key environments – Galapagos. Pergamon Press, Oxford, pp 215–223

Trillmich F (1986) Attendance behaviour of Galapagos fur seals. In: Gentry RL, Kooyman GL (eds) Fur seals: maternal strategies on land and at sea. Princeton University Press, Princeton, pp 168–185

Trillmich F (1987) Galapagos fur seal (*Arctocephalus galapagoensis*). In: Croxall JP, Gentry RL (eds) Status, biology and ecology of fur seals. NOAA technical reports NMFS 51. Seattle, Washington, US, pp 23–28

Trillmich F (1990) The behavioural ecology of maternal effort in fur seals and sea lions. Behaviour 114:1–20

Trillmich F (2015) *Arctocephalus galapagoensis*. The IUCN red list of threatened species 2015: e.T2057A45223722. https://doi.org/10.2305/IUCN.UK.2015-2.RLTS.T2057A45223722.en

Trillmich F, Dellinger T (1991) The effects of El Niño on Galapagos pinnipeds. In: Trillmich F, Ono KA (eds) Pinnipeds and El Niño. Springer, Berlin, pp 66–74

Trillmich F, Kooyman GL (2001) Field metabolic rate of lactating female Galapagos fur seals (*Arctocephalus galapagoensis*): the influence of offspring age and environment. Comp Biochem Physiol 129(A):741–749

Trillmich F, Limberger D (1985) Drastic effects of El Niño on Galapagos pinnipeds. Oecologia 67:19–22

Trillmich F, Mohren W (1981) Effects of the lunar cycle on the Galapagos fur seal, *Arctocephalus galapagoensis*. Oecologia 48:85–92

Trillmich F, Wolf JBW (2008) Parent-offspring and sibling conflict in Galápagos fur seals and sea lions. Behav Ecol Sociobiol 62:363–375. https://doi.org/10.1007/s00265-007-0423-1

Trillmich F, Meise K, Kalberer S et al (2016) On the challenge of interpreting census data: insights from a study of an endangered pinniped. PLoS One 11:e0154588. https://doi.org/10.1371/journal.pone.0154588ne

Ventura F, Matthiopoulos J, Jeglinski JEW (2019) Minimal overlap between areas of high conservation priority for endangered Galápagos pinnipeds and the conservation zone of the Galápagos Marine Reserve. Aquat Conserv 29:115–126

Villegas-Amtmann S, Jeglinski JWE, Costa DP et al (2013) Individual foraging strategies reveal niche overlap between endangered Galapagos pinnipeds. PLoS One 8:e70748. https://doi.org/10.1371/journal.pone.0070748

Wolf JBW, Tautz D, Trillmich F (2007) Galápagos and Californian sea lions are separate species: genetic analysis of the genus *Zalophus* and its implications for conservation management. Front Zool 4:20. https://doi.org/10.1186/1742-9994-4-20

Wynen LP, Goldsworthy SD, Insley SJ et al (2001) Phylogenetic relationships within the eared seals (Otariidae: Carnivora): implications for the historical biogeography of the family. Mol Phylogenet Evol 21:270–284. https://doi.org/10.1006/mpev.2001.1012

Chapter 4
Antarctic Fur Seal (*Arctocephalus gazella*, Peters 1875)

Jaume Forcada

Abstract The Antarctic fur seal lives south of the Antarctic Polar Front and has never been reported breeding in Latin America. Individuals encountered occasionally, from Peru, northern Chile, and the Juan Fernández Archipelago to South Brazil, are likely to be vagrants from the populations of the sub-Antarctic archipelago of South Georgia and the South Shetland Islands, Antarctic Peninsula. The species is one of the smaller otariids, has a short lactation of less than 4 months, is highly sexually dimorphic, and has a high polygyny. This determines its reproductive behavior and different foraging strategies, distribution, and diet between sexes. Breeding females alternate 2- to 11-day foraging trips at sea with 1- to 3-day nursing periods on land during lactation, which constraints their foraging range and breeding success. Bulls start holding breeding territories at ages 8–9, but most will be unsuccessful, especially in densely populated areas. In the Southwest Atlantic, Antarctic fur seals are important secondary consumers and highly dependent on Antarctic krill, which is a very abundant but variable resource. Most of the world population occurs at South Georgia, where some colonies are declining because of climate effects on the food supply. Currently of least concern, its conservation status requires re-assessment with up-to-date population evaluations.

Keywords Antarctic fur seal · *Arctocephalus gazella* · Antarctic krill · Heterozygosity · Scotia Sea · Threats · Vagrant

The Antarctic fur seal is one of the smaller otariids. Newborn pups have a standard length between 58 and 66 cm, and males are marginally heavier at birth (4.9–6.6 kg) than females (4.8–5.9 kg). As adults, they are highly sexually dimorphic (Fig. 4.1); bulls are almost 1.5 times longer (170–200 cm) and four times heavier (90–197 kg) than females (117–140 cm and 20–51 kg, respectively) (Bonner 1968; Payne 1979).

Antarctic fur seals are part of the subfamily Arctocephalinae and evolved to their present form in the past 2–3 million years (Yonezawa et al. 2009). They are very

J. Forcada (✉)
British Antarctic Survey, Natural Environment Research Council, Cambridge, UK
e-mail: jfor@bas.ac.uk

G. Heckel, Y. Schramm (eds.), *Ecology and Conservation of Pinnipeds in Latin America*, https://doi.org/10.1007/978-3-030-63177-2_4

Fig. 4.1 Antarctic fur seal female with newborn pup and territorial male behind. (Photo by Jaume Forcada)

similar to other southern fur seals, particularly the females. Hybridization between Antarctic fur seals, sub-Antarctic fur seals (*Arctocephalus tropicalis*), and New Zealand fur seals (*Arctocephalus forsteri*) has been reported at Macquarie Island (Southwestern Pacific sub-Antarctic), with some degree of hybrid reproductive success (Lancaster et al. 2006).

Antarctic fur seals have nine pairs of teeth in the upper jaw and eight pairs in the lower jaw. The tooth rows diverge posteriorly and have small unicuspid postcanines with distinctive abrasion on their lingual surface and wide diastemas between posterior premolars and molars, which distinguish the species (Repenning et al. 1971). The lower canines in fur seal bulls, laterally compressed and large, are used in territorial fights with other bulls. The size of the upper canines positively correlates with skull size and individual quality (heterozygosity) (Hoffman et al. 2010). Bulls have heavy manes and larger body foreparts than females, and both sexes can have very long facial vibrissae. These can extend beyond the pinnae, and in bulls, these can grow up to 48 cm, longer than in any other pinniped.

Pelage coloration is grizzled dark brown, shading paler below, because of differences in structure and length of three different types of hair: an underlayer of fine fur, which provides thermal insulation, and two types of guard hairs (Bonner 1968). The pelage of pups is black until first molt, but one in approximately 600 (1–2‰) is born with a hypopigmented (blond) coloration in the guard hairs and paler underfur and skin (Fig. 4.2). They are not albino but have a mutation in the melanocortin 1 receptor (MCR1) that regulates pigment production, with a substitution of a serine with phenylalanine. Blond seals are homozygous for the serine alleles through simple Mendelian inheritance, and dark coloration is dominant (Peters et al. 2016). Surveys at Bird Island, South Georgia, where the species shows significant individual variation in inbreeding levels, suggest no differences between blond and

Fig. 4.2 Antarctic fur seal pups with wild and hypopigmented colorations blond and brindled. (Photo by Jaume Forcada)

Fig. 4.3 Antarctic fur seal territorial males including a blond individual. (Photo by Jaume Forcada)

wild-type individuals and therefore no selective disadvantage for blonds. The presence of hypopigmented territorial males (ages 8–9 or more) (Fig. 4.3) supports a similar fitness between blond and wild-type seals. Additional reported unusual pigmentations in pups from Cape Shirreff, Livingston Island (Acevedo et al. 2009), include "tiger like," "brown morph," "tricolor morph," and piebald, which are likely to be variants of the mutation. At Bird Island, South Georgia, such "brindled" pups tend to be born to just a few females which carry the mutation and are approximately 1 in less than 5000 (Forcada and Staniland 2009). Blond individuals, probably originating at South Georgia, have also been reported at Marion Island, Prince Edward Islands, South Africa (de Bruyn et al. 2007).

Abundance, Distribution, and Movements

Antarctic fur seals have a circumpolar distribution, but the majority of the world population is in the Southwest Atlantic sector of the Southern Ocean and largely in the South Georgia archipelago, with ca 95–97% of the global pup production. Available, although not current (see Forcada and Staniland (2009)), population estimates based on complete surveys for South Georgia are 2,700,000, with pup production of 269,000, for the austral summer of 1990–1991 (from Boyd (1993)). For the South Shetland Islands, the population is 21,190, with a pup production of 10,100, for season 2000–2001, with Cape Shirreff as the most important breeding site (Goebel et al. 2003). Additional, smaller colonies in the Scotia Sea, Southwest Atlantic region, are found in the South Orkney Islands and the South Sandwich Islands. By proximity, most of the individuals found in South America are likely to be vagrants from South Georgia and the South Shetland Islands, in the Antarctic Peninsula.

During the breeding season (late mid-November to early April), the foraging range of breeding females is restricted by their need to provision their land-based offspring, mainly within 100 km from the pupping location (Boyd et al. 1998; Guinet et al. 2001; Staniland and Boyd 2003). In the Southwest Atlantic region, the densities of fur seals at sea during the summer are correlated with bathymetry (water depth) and distance to the continental shelf break (Hunt et al. 1992; Staniland et al. 2012; Santora 2013). Through the summer, densities at sea increase (Santora 2013), showing positive correlations with Antarctic krill (*Euphausia superba*) density across a wide range of spatial scales (Hunt et al. 1992).

After the breeding season, adult females from South Georgia forage in much greater ranges, within approximately 500 km of the pupping locations, and may travel as far south as the pack ice edge and to the northwest, around the Falkland Islands and the Patagonian Shelf, up to Río de la Plata, Argentina, and Montevideo, Uruguay. The winter distribution of females breeding at Cape Shirreff, South Shetlands, includes the southern coast of Chile, near the shelf break and pelagic waters further west of this region (Arthur et al. 2017).

Movements between islands also occur, mainly from high-density breeding areas to areas of lower population density. Antarctic fur seals tagged at South Georgia and Seal Island (north Elephant Island, South Shetland Islands) have been recorded at Cape Shirreff over different summer seasons (Torres 1991). Other incidental records report "land-crossing" of seals from east to west across the tip of the Antarctic Peninsula (Torres et al. 2006).

Antarctic fur seal males stay mostly within waters near South Georgia during the breeding season (Staniland and Robinson 2008) but may travel in winter to the ice edge, in the West Antarctic Peninsula, or areas in the Weddell Sea, including the South Orkney Islands (Boyd et al. 1998). Tagged juveniles (0- and 1-year-olds) of both sexes born at South Georgia stayed under the Antarctic Polar Front, although some travelled 500 km or more from the natal colony (Warren et al. 2006).

In South America, records of Antarctic fur seals range from northern Chile (Coquimbo Region, 29°54'S) to southern Brazil (Rio Grande do Sul, 30°S), although a malnourished adult female, stranded on Regatas Sur beach in Lima, Peru, on December 28, 2018, is the most northern confirmed record for the species (Stepchew and Yaipen-LLanos 2019). For the rest of the range, most records are from stranded dead or alive individuals, or live encounters of a single individual. In Chile, records include one juvenile in August 2014 at Coquimbo Region (Miranda-Urbina et al. 2015); one individual in September 1982 in Alejandro Selkirk Island and one in November 1983 in Robinson Crusoe Island, Juan Fernández Archipelago (33°38′S, 78°50′W) (Torres et al. 1984); one juvenile ringed at Bird Island, South Georgia, found in September 1973 at Hoste Island (55°15′S, 69°0′W), near Cape Horn (Texera 1974); between October and December 2009, three seals – one juvenile male, one adult male, and one unidentified stage – in Punta Dungeness (52°23'S, 69°25'W), Magellan Strait; and one juvenile male in December 2009 in Azopardo Bay (54°27'S, 68°58'W), Almirantazgo Sound (Acevedo et al. 2011). In Argentina, at least five have been observed in Tierra del Fuego, four in the North coast, and one in the South coast, since 1992 (Goodall et al. 2005); over 23 individuals have been recorded at various locations of central and northern Argentina, including the north of Buenos Aires province, 18 in Mar del Plata (38°S) (Rodríguez et al. 1994; Fernández et al. 1998). In Uruguay, one seal was recorded in July 2000 (La Floresta, Canelones) and one female in Punta Colorada (Maldonado) in September 2002 (Naya and Achaval 2006). In Brazil, one female and a non-sexed seal were found in September 1984 in Rio Grande do Sul coast (Pinedo and Marmontel-Rosas 1987); one adult female (in August 1994, at 30°30'S, 50°20'W) and two juvenile males (in September 1994, at 30°42'S, 50°29'W and 29°52'S, 50°04'W) were collected dead in Rio Grande do Sul (Oliveira et al. 2001). In Brazil, strandings of this species corresponded to only 0.03% of all marine mammal strandings reported from 1976 to 2013 for the southern Rio Grande do Sul province (Prado et al. 2016).

Population Structure

The global population structure was initially divided into two groups (Wynen et al. 2000): a western group, including populations at South Georgia, South Shetlands, and Bouvetøya (southwest to central south Atlantic), and an eastern group with populations on Kerguelen Islands, Heard Island, and Macquarie Island (Indian Ocean to southwest Pacific). A reassembly of the Antarctic fur seal genome (Humble et al. 2018a) still identifies the eastern group as a single population but separates South Georgia, South Shetlands, and Bouvetøya as different populations (Paijmans et al. 2020).

The Antarctic fur seal underwent a severe reduction in population size because of commercial exploitation in the eighteenth and nineteenth centuries, mostly for their pelts. At least 1.2 million seal skins were harvested (Weddell 1825), and the species was considered practically extinct in all of its range. After the sealing operations collapsed, the species recovered rapidly at South Georgia (Bonner 1968; Payne 1977) and also South Shetlands (Hucke-Gaete et al. 2004), with speculation that recovery and recolonization would have started from the west of South Georgia (Laws 1973; McCann and Doidge 1987). However, genome sequencing (Humble et al. 2018b) and the analysis of 2000 individuals sampled from 8 circumpolar breeding locations suggest that these populations experienced severe bottlenecks down to effective population sizes of around 150–200, although the declines appear to not have depleted allelic richness by more than around 15%, retaining comparably high levels of neutral variability (Paijmans et al. 2020). This is corroborated by the species being one of the most genetically diverse pinnipeds. The current population structure thus supports the hypothesis that relict fur seal populations probably survived sealing at multiple locations, although with some degree of connection through gene flow (Hoffman et al. 2018).

Annual Cycle

The breeding season lasts from late October to early November until late March to early April (Bonner 1968; Duck 1990). Fur seal bulls begin to arrive to the breeding beaches in numbers during early November and fight to establish territories; they start leaving their territories and return to the sea by late December to early January. Pregnant females start arriving a few weeks later (late mid to late November) and give birth within a few days of hauling out, although births may occur on the day of arrival. There is high breeding synchrony across the species' range, with 90% of the pups born in a 10-day window, around December 7–10 (Duck 1990; Forcada et al. 2005; Hofmeyr et al. 2007). The perinatal period lasts for 5–8 days, after which they experience an estrus and mate. They then start several cycles of foraging trips at sea (2–11 days) followed by suckling bouts on land (1–3 days) for the approximately 4-month lactation (mean 117 days), which is one of the shortest in otariids. Most pups wean by the end of March, although they molt their natal coat around early February at about 1–2 months of age. The molt in adults is incomplete or partial, unlike in phocids. Females molt in February and March while they are still suckling their pups, and they can continue foraging at sea throughout the molt.

Life History and Vital Rates

Females breed annually, having a gestation that lasts between 8 and 9 months and a diapause period of 3–4 months, before implantation of the blastocyst, similar to most other pinnipeds; this closely approaches the birth of a pup with the conception of the next pup (Boyd 1996). They give birth to a single pup, although cases of twins, with single and multiple paternity, are confirmed for South Georgia and Cape Shirreff South Shetlands (Hoffman and Forcada 2009; Bonin et al. 2012). Cases of adoption/foster nursing of unrelated pups and milk stealing are frequently observed at high-density colonies in South Georgia (Lunn 1992). At Bird Island, allonursing is common in females that lose their pup early in the season, and in inexperienced (first-time) mothers that do not seem to effectively bond with their own pups, and these are often rejected. Instead, they attempt kidnapping and nursing non-offspring pups, sometimes with subsequent sharing of a pup. This has also been observed at Cape Shirreff, where densities of seals are much lower, and it resulted in an increased growth rate for the pup (Acevedo et al. 2016). Allonursing has also been observed exceptionally between adult females (Fig. 4.4), although it is probably exceptional.

Mean age at first conception is 3–4 years (range 2–7) (Payne 1977; McCann and Doidge 1987) and depends on long-term environmental conditions and female heterozygosity (Forcada and Hoffman 2014). By ages six to seven, most females have attained full adult size (Lunn et al. 1994), and longevity is around 20–21 years (maximum 24). Fecundity increases rapidly from age two, peaks at 0.80 at age eight, and remains high (mean 0.75, range 0.68–0.77) until ages 11–13, with the onset of a senescent decline. Females mostly defer annual breeding with reduced food availability (Lunn et al. 1994) from adverse environmental conditions (Forcada et al. 2005; Forcada et al. 2008). Pup production responds to climate anomalies cascading through the local physical and biological environments that reduce the food supply.

Fig. 4.4 Fur seal mother allonursing an adult fur seal female. (Photo by Jaume Forcada)

Weaning success increases with age (Doidge and Croxall 1989) and varies with an individual's heterozygosity (Forcada and Hoffman 2014). At South Georgia, pup survival primarily depends on food availability although trauma inflicted by adults is common in dense colonies (Doidge et al. 1984a). At the South Shetlands, predation by leopard seals (*Hydrurga leptonyx*) can be a main cause of mortality among first-year individuals, potentially exerting top-down control at colonies on Seal Island, near Elephant Island (Boveng et al. 1998), and at Cape Shirreff (Schwarz et al. 2013), where pup predation could be as much as 25–30% of the total production (Vera et al. 2006). At both archipelagos (South Georgia and South Shetlands), environmental and climate forcing are main determinants of survival of older individuals and of reproductive rates (Forcada et al. 2008; Schwarz et al. 2013).

Males reach sexual maturity at ages three to four, although they mostly start breeding at ages eight to nine when they reach maximum body size and can establish breeding territories (Payne 1979). Their survival rate after age eight is below 0.5, which is much lower than the average survival for adult females (0.87–0.93).

At South Georgia, variation in female survival is mainly caused by fluctuation of the biological environment. In years with severe climate anomalies (Southern Annular Mode, El Niño-Southern Oscillation), it can be reduced by as much as 15–20% (Forcada et al. 2008). Female fitness, measured as the asymptotic population growth rate, is most sensitive to changes in the survival of breeders and their propensity to breed – probability to breed between years. With adverse environments, females' body condition may decrease, and they defer or alter breeding rather than put their life at risk.

Parental Investment

Female fur seals are income breeders (foraging cycle strategy). They nurse their pup postpartum using body reserves for approximately 7–10 days and then alternate periods of suckling with foraging trips during the lactation period (Doidge et al. 1986; Costa et al. 1989). They can reproduce annually, although they incur higher costs of breeding in years with reduced food supply (Lunn et al. 1994; Forcada et al. 2008). Females' body condition may decrease, and breeding can be altered by not implanting or reabsorbing the blastocyst, by abortion, or by pup abandonment.

Lactation is the highest maternal investment, even though at less than 4 months is the shortest among otariids. Milk consumption increases with pup age before decreasing in the last 30–40 days, depends on the duration of the maternal foraging trip and subsequent attendance bout, and is related to pup mass. At Bird Island, South Georgia, milk consumption is similar for both sexes and is 2.5–3.2 kg (42–53 MJ) during the 6-day perinatal period and 2.9–3.6 kg (49–68 MJ) during each subsequent attendance bout (Arnould et al. 1996a). Female pups invest more energy into fat reserves, while male pups invest more on lean tissue (Arnould et al. 1996a), but there seems to be an equal maternal investment in pups regardless of sex (Lunn and Arnould 1997). At Cape Shirreff, however, those differences were only observed during the perinatal period, and allocation of energy to growth was more dependent on pup mass than sex (McDonald et al. 2012a).

Male pups tend to have significantly higher growth rate than female pups at Bird Island (Doidge et al. (1984b); but see Lunn and Arnould (1997)), Macquarie Island (Guinet et al. 1999), and Cape Shirreff (Vargas et al. 2009). Fasting female pups lose mass at a significantly higher daily rate than male pups of the same mass (Guinet et al. 1999), which would be consistent with a higher proportion of fat reserves, because those are mobilized before lean tissue during fasting (Arnould et al. 2001). This would also explain why females grow at a slower daily rate than males (Guinet et al. 1999). However, under food stress conditions, growth rates can be substantially altered by food availability and foraging trip duration (equivalent to pup fasting time) (Lunn and Arnould 1997; Vargas et al. 2009). Therefore, geographic differences in results need to be put into context, taking study duration and method, population density, and environmental conditions into account.

Maternal mass and pup traits such as pup mass and energy demands affect energy allocation and therefore determine maternal care and diving behavior. Larger females tend to give birth to larger pups and invest more in pups during the perinatal and molt stages (McDonald et al. 2012b).

The body mass gain of mothers during a foraging trip is related to the time and total energy expended at sea, and there is no apparent energetic advantage for females in undertaking foraging trips of any particular length (Arnould et al. 1996b). However, limited food availability prolongs the foraging trip duration and is at detriment of the pup's ability to avoid starvation (Costa et al. 1989). When foraging conditions are poor, longer foraging trips may result in greater food delivery per trip, although pups from mothers that make shorter trips still display faster growth rates over longer periods of consecutive foraging trips (Lunn et al. 1993; Guinet et al. 2000). With less food availability, only more efficient mothers may be able to spend less time diving and less time overall at sea, which allows them a higher-quality milk delivery to their pups, more frequent suckling, or both (Jeanniard-du-Dot et al. 2017).

Colonial breeding in Antarctic fur seals is likely to determine mother-offspring effects in response to density of individuals. Maternal stress hormones (cortisol and testosterone) can be elevated under crowded conditions. Offspring cortisol levels are usually high, uncorrelated to those in mothers, while testosterone levels correlate with maternal cortisol levels, depending on the density of the colonies (Meise et al. 2016). This could reflect a degree of fetal programming under the more stressful conditions typical of high densities. Additionally, social stress may contribute to depress bacterial diversity, and seal-associated microbial diversity is found to be significantly lower in high-density colonies (Grosser et al. 2019).

At locations of high population density, early pup survival can be highly density dependent (Doidge et al. 1984a; Reid and Forcada 2005) but is uncorrelated with pup heterozygosity, a measure of individual quality (Hoffman et al. 2006; Litzke et al. 2019). However, higher heterozygosity in mothers can determine a higher pup survival under environmental stress, regardless of pup heterozygosity levels (Forcada and Hoffman 2014). Among other advantages, individual heterozygosity negatively correlates with bacterial diversity, suggesting that higher-quality mothers and pups would be more effective at suppressing detrimental microbes (Grosser et al. 2019).

Communication

Communication between mothers and pups is through both smell and vocalization, establishing a bond immediately after birth that is constantly re-enforced subsequently. Mothers returning from foraging trips or separated from their pups ashore locate each other by call and response, important for initial long-range recognition (Dobson and Jouventin 2003; Aubin et al. 2015). In close proximity, identity is confirmed by smell. Chemical profiles, predefined by pheromones, microbiome, and colony-/individual-specific substances are highly similar between mothers and pups and reflect a combination of genetic and environmental influences (Stoffel et al. 2015). Mothers and their offspring share similar microbial communities, supporting the notion that microbes may also facilitate mother-offspring recognition (Grosser et al. 2019).

Fur seal males perform two main vocalizations: a threatening roar, in response to specific threats, but often used in aggression directed against other males, and a "huff-chuff" sound used when interacting with females within a territory or in close proximity to other seals. Females and pups may occasionally use this vocalization during either play or aggression toward other seals.

Mating System

Antarctic fur seals are highly polygynous, and males compete for females through direct aggression toward other males (Fig. 4.3); successful males hold territories with up to ten females on average (Bonner 1968). At South Georgia and with high density of seals, males can hold territories for 34 days on average (McCann 1980), median of 14 and maximum of 75 days (Arnould and Duck 1997). Males can start holding territories at ages seven to eight (Payne 1977), when they are physically mature and able to challenge other males. Successful males can hold territories for several years, up to ages 11 or 12, and there are very few observations of territorial males aged 15 or more. The probability to return as territorial is correlated with having held long territory tenures previously (Arnould and Duck 1997). By ages eight to nine, male apparent survival (confounded with permanent emigration from study sites) is approximately 0.47, suggesting that an average individual will hold territories in less than 1.5 seasons in a lifetime. During territory tenure, males mainly fast but may also go to the sea and back for short trips and attempt to recover their territories (McCann 1980). For a fasting male of an average starting mass of 188 kg, the cost of territory tenure was estimated at approximately 1.5–2 kg of daily weight loss (Boyd and Duck 1991).

Polygyny and high breeding philopatry in both sexes (Lunn and Boyd 1991; Arnould and Duck 1997; Hoffman and Forcada 2012) lead to multiple mating among individuals over time and may lead to increased relatedness, negatively affecting genetic diversity. In large density areas such as northwestern South Georgia, a few territorial males can father up to 60% of the pups in a single beach (Hoffman et al. 2003), with most paternities assigned to a limited number of top individuals. Moreover, males able to return in subsequent years can potentially increase breeding success (Arnould and Duck 1997; Hoffman et al. 2003).

In densely populated areas, there appear to be mechanisms in place for females to avoid males that are genetically related (Hoffman et al. 2007), and chemical fingerprints could play an important part in mate choice (Stoffel et al. 2015). Despite high natal philopatry (Hoffman and Forcada 2012), females exhibit mate choice by moving from their pupping territory, with a given male, through a number of neighboring territories in order to mate with a more heterozygous and less related male (Hoffman et al. 2007). This choice is also likely to play a role in male mating success. Nonetheless, a recent analysis of the same population confirms that inbreeding is present with evidence of multiple second- and third-order relatives among a sample of ostensibly unrelated individuals (Humble et al. 2020). This "cryptic relatedness" suggests increased consanguineous mating, with potential implications for fitness variation and mate choice. In lower density areas such as Cape Shirreff, South Shetlands, with less competition for space and resources, re-mating frequency and its effects on pairwise relatedness are much smaller (Bonin et al. 2016).

Foraging Strategies

Foraging activity significantly relates to oceanographic conditions, prey distribution and availability, and distance from the colony (McCafferty et al. 1998b; Guinet et al. 2001; Lea et al. 2006; Staniland and Robinson 2008). This is determined by central place foraging, which together with sexual dimorphism leads to sexual segregation in foraging strategies (Staniland and Robinson 2008; Jones et al. 2020).

As income breeders, females can only spend 2–11 days foraging at sea during the breeding season before returning to nurse their pups on land, which limits the foraging range and activity. At South Georgia, females dive to shallow depths (0–40 m), exploiting diurnally migrating prey such as Antarctic krill and lantern fish (Myctophidae) within the mixed surface layer, and predominately during the night. For this, they travel to the shelf break and beyond, to deeper oceanic waters. In contrast, males dive mostly on the continental shelf, but in deeper areas than females, and benthic dives are frequent. Maximum dive durations and depths are around 5 min and 210 m in females and 10 min and 350 m in males.

When Antarctic krill is locally very abundant at South Georgia, females can make shorter foraging trips and fewer dives and spend more time ashore, feeding almost exclusively on krill; frequent and shallow dives in daytime indicate that the krill is close to the surface and accessible during the day. In poorer krill years, deeper and longer duration of daytime dives are associated with a higher proportion of fish in the diet. Females increase the time spent foraging and also the activity during foraging (Boyd et al. 1994; McCafferty et al. 1998a).

Foraging trip durations in females tend to increase as the pup-rearing period progresses to meet increased lactation demands. Some individuals travel to more distant oceanic waters, where higher-quality (energy-dense) prey is found (Staniland et al. 2007), potentially resulting in greater energetic gain. However, there is a high individual variation in foraging trip types during the breeding season, which appears to be dictated by prey availability (Staniland et al. 2004) as driven by the climate (Lea et al. 2006). Less is known about the post-breeding winter period, but there appears to be some fidelity to foraging sites with lower annual variance in the physical environment (Arthur et al. 2015).

Diet and Trophic Interactions

Antarctic fur seals have a generalist diet, which highly depends on local prey availability and regional differences in food web composition. The same constraints that determine foraging strategies (central place foraging, income breeding, and high sexual dimorphism) determine prey preferences locally. Breeding females predate on epipelagic and diurnally migrating mesopelagic prey, whereas bulls may also undertake benthic dives and forage on benthic prey, which may not be consumed by females (Staniland and Robinson 2008). Studies in the Indian Ocean indicate that female yearlings show a total segregation in isotopic niche (i.e., diet and foraging distribution) from breeding females until age two, whereas juvenile males show a progressive change in niche throughout their development until age five, very different from bulls (Kernaleguen et al. 2016). At Bird Island, South Georgia, stable isotope analysis (Jones et al. 2020) shows a high sexual segregation in foraging habitats, with males spending more time foraging south of the Polar Front in maritime Antarctica than females and spending more time south as they age.

In the Southwest Atlantic region, the main prey is Antarctic krill (*Euphausia superba*) (Doidge and Croxall 1985; Reid and Arnould 1996). This highly productive region has over half of the estimated global biomass of this crustacean (Atkinson et al. 2009), which influences the population dynamics of many seabirds and seals that depend on it (Croxall et al. 1988). At South Georgia, the krill accounts for 75–88% of the fur seal combined diet samples of males and females in the summer and is also important in winter although in a lower proportion (Reid 1995). Other than krill, fur seals mainly eat mackerel icefish (*Champsocephalus gunnari*), *Pseudochaenichtys georgianus*, notothenids (inter alia *Lepidonotothen larseni*, *Notothenia* spp., *Gobionotothen gibberifions*), and myctophids (inter alia *Protomyctophum choriodon*, *Gymnoscopelus* spp., *Electrona* spp.) (North et al. 1983; North 1996). In years of low krill availability, notothenids and myctophids occur in a greater proportion of fur seal diet samples (Croxall et al. 1999; Waluda et al. 2010).

In the South Shetlands, krill is also the main prey, particularly during the breeding season. Fecal analysis from five consecutive years (1997–2001) at Cape Shirreff showed that Antarctic krill was the most frequent prey item, followed by myctophid species (*Gymnoscopelus nicholsi*, *Electrona antarctica*, and *Electrona carlsbergi*), squid, and penguin remains. There was an increase in modal krill size consumed from 1998 to 2001, supporting the dependence of fur seals on strong krill cohorts. Analysis of krill size distribution and selectivity suggests a preference for larger krill (>34 mm), despite the broader size range of prey items available (Osman et al. 2004). This is comparable to the krill consumed at Bird Island, South Georgia (Croxall and Pilcher 1984), although variation is very high (Reid et al. 1999).

In non-breeding males from the South Shetlands, krill is also important, followed by myctophid fish (including *Electrona antarctica* and *Gymnoscopelus nicholsi*), Antarctic silverfish (*Pleuragramma antarcticum*), and *Notolepis coatsi* (Daneri and Carlini 1999; Daneri et al. 2005). The presence of squid in diet samples in the Southwest Atlantic region is marginal (Reid and Arnould 1996; Daneri et al. 1999).

Penguins are predated by juvenile and adult fur seal males in some areas. Individuals may specialize in this prey where they are abundant (Bonner and Hunter 1982; Hofmeyr and Bester 1993), although the predatory impact on prey population is likely to be marginal. In the southwest Atlantic, macaroni (*Eudyptes chrysolophus*) and gentoo penguins (*Pygoscelis papua*) which breed in large numbers are taken occasionally at South Georgia (Fig. 4.5), but carcasses are often discarded before being consumed (Bonner and Hunter 1982).

Limited diet data are available for vagrant seals found in South America. In dead specimens collected in Rio Grande do Sul, Brazil, the stomach contents contained fish, including *Menticirrhus littoralis*, *Micropogonias furnieri*, *Cynoscion*

Fig. 4.5 Antarctic fur seal male catching a gentoo penguin. (Photo by Jaume Forcada)

guatucupa, *Macrodon ancylodon*, *Paralonchurus brasiliensis*, *Pomatomus saltatrix*, and several catfish from the family Ariidae (c.f. Oliveira et al. (2001)) and cutlass fish (*Trichiurus lepturus*), and cephalopods, including Antarctic neosquid (*Alluroteuthis antarcticus*) and long-finned squid (*Loligo sanpaulensis*) (Oliveira et al. 2001).

Given their abundance, Antarctic fur seals can have a significant trophic impact in the Southwest Atlantic region. They live in very high densities, compared, for example, with terrestrial carnivores of similar body size, have high metabolic rates, and have highly localized seasonal prey demands. They are probably one of the most important secondary consumers in the South Georgia food chain, judging from an estimated consumption of 3.84 million tons of krill (Boyd 2002), which amounts to 6.4% of the best krill biomass estimate available for the Scotia Sea and northwest Antarctic peninsula region (Siegel and Watkins 2016). Similarly, they could consume significant proportions of the stocks of other krill predators, such as mackerel icefish (Reid et al. 2005), with cascading effects on the food web.

The main predators of the Antarctic fur seal in the Southwest Atlantic and the rest of its distribution range are probably killer whales (*Orcinus orca*), although there are no published records of predation. However, pups and 1-year-olds are predated by leopard seals, which could be a main cause of mortality among first-year individuals at the South Shetlands (Boveng et al. 1998; Schwarz et al. 2013). By contrast, the impact of leopard seal predation at South Georgia is likely to be marginal, judging from Bird Island data (Forcada et al. 2009).

Trauma, Diseases, and Parasites

Trauma and infections are commonly found in Antarctic fur seals, particularly during the breeding season. At Bird Island, South Georgia, trauma mostly affects pups, is usually caused by crashing by territorial bulls, and is one of the main causes of early mortality (Doidge et al. 1984a; Reid and Forcada 2005; Baker and Doidge 1984). Trauma is also common in territorial males and their competitors, which inflict each other deep wounds during fights. Bull fur seal mortality fluctuates highly between years, with the main causes being infections from such wounds, broken teeth and jaws and other internal damage (Baker and McCann 1989; Erb et al. 1996), and pneumonia. In females, trauma is also related with birthing. Death after prolonged labor/dystocia or prolapsed uterus is occasionally observed leading to a limited number of deaths every season. An additional source of injury is related to entanglement in man-made debris, mostly from the fishing industry. Up to 30% of the early entanglements at Bird Island caused physical injury (Croxall et al. 1990) (see also section on conservation). Seals introduce their heads through collar bands from discarded box packaging or broken fishing nets. As the bands move down the neck and the body grows, they break the skin and penetrate deep in the blubber layer causing severe injuries in some cases.

Not many studies have addressed presence of pathogens from human origin in Antarctic seals, and increasing temperatures and anthropogenic activity are likely to introduce those in Antarctica and sub-Antarctic locations. True Antarctic seals are known to carry antibodies against herpesviruses (Stenvers et al. 1992) and have been found more recently in Antarctic fur seals (Tryland et al. 2012). Antarctic seals also carry antibodies against phocine distemper virus (PDV) and canine distemper virus (CDV) (Bengtson et al. 1991), although with low prevalence (Smeele et al. 2018). These viruses have not been examined in Antarctic fur seals, as this was mostly a concern for the early widespread use of sled dogs in early Antarctic expeditions.

Brucellosis is a worldwide infectious zoonotic disease, which sometimes leads to termination of pregnancies by starvation of oxygen in placentas from terrestrial and marine mammals (Ewalt et al. 1994). It has been detected in a limited number (5 positives in 86 samples in 2000) of Antarctic fur seals at Cape Shirreff (Blank et al. 2001; Retamal et al. 2000), suggesting the presence of the infection, but with unknown prevalence. At Bird Island, South Georgia, anti-*Brucella* antibodies were not found in serum of 21 dead Antarctic fur seal bulls (Jensen et al. 2013) and 20 aborted fetuses, although low trites led to inconclusive results. A few female fur seals at Bird Island are observed aborting their fetuses in some winters (McCafferty 1999), with a maximum in a single winter of 20 between July and October of 2013. The apparent absence of *Brucella* and evidence of poor food availability provide circumstantial evidence that abortions are likely to result from poor body condition, but pathogens have not been ruled out.

Campylobacter insulaenigrae and *C. lari* have been found in fur seal fecal swabs at Deception Island. The genus is also found in penguins and polar skuas, but the source of infection is unknown, the prevalence in seals, and whether it is the result of microbial pollution associated with human activity (García-Peña et al. 2010). Another potential anthropogenic pathogen, Enteropathogenic *Escherichia coli*, was found in 2 out of 33 fur seal pups at Cape Shirreff, in colonies close to human research stations (Hernandez et al. 2007). The origin of the bacteria is also unknown.

At Bird Island, reported bacterial infections not necessarily from human origin include sepsis and opportunistic pathogens, such as various strains of streptococci and *Staphylococcus* sp., *Bordetella* sp., and *Corynebacterium phocae*, mostly from dead pups and territorial bulls (Baker and McCann 1989).

Toxoplasma gondii is an intracellular protozoan parasite with a worldwide distribution, commonly transmitted by domestic cats, also in sub-Antarctic locations where those were introduced and are still present. Tests for antibodies against *T. gondii* were carried out at several locations. In Antarctica, positive test for antibodies was found in 4 out of 165 fur seal samples from Deception Island and Aitcho Islands, South Shetlands, and Avian Island, Marguerite Bay (West Antarctic Peninsula); however, the possible origin of *T. gondii* oocysts leading to infection is unknown (Rengifo-Herrera et al. 2012). At Bird Island, South Georgia, in blood extractions of freshly dead fur seal bulls, the results suggested absence of infection (Jensen et al. 2012).

Nematodes in the lung are observed in dead bulls, and hookworms (*Uncinaria* sp.) are very common in fur seal pups at Bird Island, both in the intestines of dead and feces of live individuals. The occurrence, consequences, and transmission of *Uncinaria* disease, potentially lethal (Keyes 1965), have not been explored. Other gastric nematodes and tapeworms are commonly observed in fecal analyses (Forcada and Staniland 2009). Acanthocephalans, usually ingested as cystacanths in fish, and in particular, *Corynosoma arctocephali* and *C. pseudohamanni*, have been reported for fur seals at South Georgia and the South Shetlands, respectively (Zdzitowiecki 1984). These thorny-headed worms are known to cause local inflammation (McFarlane et al. 2009).

Attachment to the fur by barnacles *Lepas australis*, notably on pregnant females returning to give birth after long periods at sea, has been reported at Bird Island, South Georgia (Bonner 1968). The barnacles die and drop off within a few days of the seal being ashore. A study found an average infestation of ten barnacles on 4% of returning females (Barnes et al. 2004). Occasionally, seals with more than half of the fur colonized by the barnacle were observed (Forcada and Staniland 2009).

Conservation

Antarctic fur seals were almost exterminated in the nineteenth century, and a brief summary of the exploitation (Forcada and Staniland 2009), mostly from Bonner and Laws (1964), follows. When Captain Cook discovered South Georgia in 1774, he found that the beaches were "swarmed" with seals, and soon after (1786) there appear the first records of fur seals taken from the area; large-scale sealing began in the Southern Hemisphere in 1775. Sealing peaked at South Georgia in 1800–1801, when 17 British and American vessels took 112,000 skins, including a single ship with a recorded catch of 57,000 seals for that season. Near extirpation was recorded by James Weddell by 1821, who estimated a total capture of 1.2 million seals. Sealing continued in 1838–1839, and then a few more seals where taken in the 1870s and in 1908. At South Georgia, the last seal found was killed in 1915. In the South Shetland Islands, seal catches began as soon as they were discovered in 1819, and the species was nearly exterminated in just three seasons. James Weddell calculated that 320,000 were taken during 1821 and 1822.

After the collapse of sealing, subsequent large-scale whaling and possibly thanks to the removal of these major Antarctic krill consumers, the species recovered rapidly and successfully (see Population Structure section) to become one of the most abundant pinnipeds. The most recent IUCN assessment for the species is of *Least Concern* (Red List Category & Criteria), although current population estimates are lacking for all of its range. Potential major threats to the species include fisheries interactions and regional warming.

Antarctic fur seal entanglements in plastic packaging bands, synthetic lines (Fig. 4.6), and fishing nets are reported throughout their distributional range and continue to be a main cause for concern at South Georgia and Cape Shirreff, South

Fig. 4.6 Antarctic fur seal juvenile male entangled in a synthetic line. (Photo by Jaume Forcada)

Shetlands, despite mitigation measures put in place by CCAMLR (Commission for the Conservation of Antarctic Marine Living Resources) (Hucke-Gaete et al. 1997; Waluda and Staniland 2013), which regulates Southern Ocean fisheries. There are no good observer data for more direct types of interactions, although the species distribution overlaps with major krill fishery areas of the South Shetlands (Hinke et al. 2017).

Regional warming is also a cause of concern, particularly at the northwest end of South Georgia, where recent data suggests that the species is responding to long-term fluctuations of the food supply, notably the Antarctic krill. The krill range appears to be contracting in the Southwest Atlantic region, where it is most abundant (Atkinson et al. 2019), and the fur seal population at Bird Island (South Georgia) is showing a population decline and phenotypic plasticity correlated with the Southern Annular Mode (Forcada and Hoffman 2014), a main climate variability of high latitudes. Fur seal pups are born approximately 10% lighter than three decades ago, whereas first-time mothers recruit later and are bigger on average and are more heterozygous (Forcada and Hoffman 2014). The effects for the wider South Georgia and other colonies of the Southwest Atlantic region are unknown, although there is evidence of a declining pup production in colonies of the South Shetlands (Goebel et al. 2014), which could be due to multiple causes.

Trace metals are present in fur seals breeding at South Georgia and the South Shetlands, particularly mercury and cadmium, with concentrations thought to be in the toxic range (Malcolm et al. 1994; Cipro et al. 2017). Similarly, persistent organic pollutants, including PCBs and DDTs, have been found in seals from the South Shetlands (Cipro et al. 2012; Brault et al. 2013) and also brominated flame retardants such as polybrominated diphenyl ethers (PBDEs) (Schiavone et al. 2009b). Halogenated flame retardants are mostly used in household appliances, office electronics, textiles, and furniture, and their presence in tissues of Antarctic fur seals are proof of long-range transport of these substances. While detected levels are low, these compounds are persistent in the environment and biomagnify in the food webs, and their toxicity may affect hormonal regulation as well as hepatic, neuronal, and thyroid activities. Emerging compounds such as dechloranes may cross the blood-brain barrier and are found in fur seals of the South Shetlands (Aznar-Alemany et al. 2019). Similarly, concentrations of perfluorochemicals (PFCs), which are also emerging anthropogenic global pollutants, including perfluorooctanesulfonic acid (PFOS) and perfluorooctanoic acid (PFOA) in juveniles from Cape Shirreff, were marginal and far below the levels of toxicity in laboratory species but are present nonetheless (Schiavone et al. 2009a).

The presence of microplastics in the diet of the species has long been documented (Eriksson and Burton 2003). It is speculated that plastics could be incorporated in the diet of fish prey, such as *Electrona* spp., which would be consumed by fur seals. However, the potential impact on the species remains unevaluated.

References

Acevedo J, Torres D, Aguayo-Lobo A (2009) Rare piebald and partially leucistic Antarctic fur seals, *Arctocephalus gazella*, at Cape Shirreff, Livingston Island, Antarctica. Polar Biol 32(1):41–45. https://doi.org/10.1007/s00300-008-0500-6

Acevedo J, Matus R, Droguett D et al (2011) Vagrant Antarctic fur seals, *Arctocephalus gazella*, in southern Chile. Polar Biol 34(6):939–943. https://doi.org/10.1007/s00300-010-0951-4

Acevedo J, Torres D, Aguayo-Lobo A (2016) Offspring kidnapping with subsequent shared nursing in Antarctic fur seals. Polar Biol 39(7):1225–1232. https://doi.org/10.1007/s00300-015-1841-6

Arnould JPY, Duck CD (1997) The cost and benefits of territorial tenure, and factors affecting mating success in male Antarctic fur seals. J Zool 241:649–664. https://doi.org/10.1111/j.1469-7998.1997.tb05739.x

Arnould JPY, Boyd IL, Socha DG (1996a) Milk consumption and growth efficiency in Antarctic fur seal (*Arctocephalus gazella*) pups. Can J Zool Rev Can Zool 74(2):254–266. https://doi.org/10.1139/z96-032

Arnould JPY, Boyd IL, Speakman JR (1996b) The relationship between foraging behaviour and energy expenditure in Antarctic fur seals. J Zool 239:769–782. https://doi.org/10.1111/j.1469-7998.1996.tb05477.x

Arnould JPY, Green JA, Rawlins DR (2001) Fasting metabolism in Antarctic fur seal (*Arctocephalus gazella*) pups. Comp Biochem Physiol A Mol Integr Physiol 129(4):829–841. https://doi.org/10.1016/S1095-6433(01)00339-7

Arthur B, Hindell M, Bester M et al (2015) Return customers: foraging site fidelity and the effect of environmental variability in wide-ranging Antarctic fur seals. PLoS One 10(3):e0120888. https://doi.org/10.1371/journal.pone.0120888

Arthur B, Hindell M, Bester M et al (2017) Winter habitat predictions of a key Southern Ocean predator, the Antarctic fur seal (*Arctocephalus gazella*). Deep Sea Res Pt II 140:171–181. https://doi.org/10.1016/j.dsr2.2016.10.009

Atkinson A, Siegel V, Pakhomov EA et al (2009) A re-appraisal of the total biomass and annual production of Antarctic krill. Deep-Sea Res Pt I 56(5):727–740. https://doi.org/10.1016/j.dsr.2008.12.007

Atkinson A, Hill SL, Pakhomov EA et al (2019) Krill (*Euphausia superba*) distribution contracts southward during rapid regional warming. Nat Clim Chang 9:142. https://doi.org/10.1038/s41558-018-0370-z

Aubin T, Jouventin P, Charrier I (2015) Mother vocal recognition in Antarctic fur seal *Arctocephalus gazella* pups: a two-step process. PLoS One 10(9):13. https://doi.org/10.1371/journal.pone.0134513

Aznar-Alemany O, Yang XF, Alonso MB et al (2019) Preliminary study of long-range transport of halogenated flame retardants using Antarctic marine mammals. Sci Total Environ 650:1889–1897. https://doi.org/10.1016/j.scitotenv.2018.09.287

Baker JR, Doidge DW (1984) Pathology of the Antarctic fur-seal (*Arctocephalus gazelle*) in South Georgia. Br Vet J 140(2):210–219

Baker JR, McCann TS (1989) Pathology and bacteriology of adult male Antarctic fur seals, *Arctocephalus gazella*, dying at Bird Island, South Georgia. Br Vet J 145(3):263–275. https://doi.org/10.1016/0007-1935(89)90079-1

Barnes DKA, Warren NL, Webb K, Phalan B, Reid K (2004) Polar pedunculate barnacles piggy-back on pycnogona, penguins, pinniped seals and plastics. Mar Ecol Prog Ser 284:305–310. https://doi.org/10.3354/meps284305

Bengtson JL, Boveng P, Franzén U et al (1991) Antibodies to canine distemper virus in Antarctic seals. Mar Mamm Sci 7(1):85–87. https://doi.org/10.1111/j.1748-7692.1991.tb00553.x

Blank O, Retamal P, Abalos P et al (2001) New data on anti-Brucella antibodies in *Arctocephalus gazella* from Cape Shirreff, Livingston Island, Antarctica. Ccamlr Sci 8:147–154

Bonin CA, Goebel ME, O'Corry-Crowe GM et al (2012) Twins or not? Genetic analysis of putative twins in Antarctic fur seals, *Arctocephalus gazella*, on the South Shetland Islands. J Exp Mar Biol Ecol 412:13–19. https://doi.org/10.1016/j.jembe.2011.10.010

Bonin CA, Goebel ME, O'Corry-Crowe GM et al (2016) Impacts of ecology and behavior on Antarctic fur seal remating and relatedness. J Exp Mar Biol Ecol 476:72–77. https://doi.org/10.1016/j.jembe.2015.12.008

Bonner NW (1968) The fur seal of South Georgia. Br Antarctic Surv Sci Rep, vol 56. London

Bonner WN, Hunter S (1982) Predatory interactions between Antarctic fur seals, macaroni penguins and giant petrels. Br Antarct Surv B 56:75–79

Bonner WN, Laws RM (1964) Seals and sealing. In: Priestley R, Adie RJ, Robin GDQ (eds) Antarctic research. Butterworths, London, pp 163–190

Boveng PL, Hiruki LM, Schwartz MK et al (1998) Population growth of Antarctic fur seals: limitation by a top predator, the leopard seal? Ecology 79(8):2863–2877. https://doi.org/10.1890/0012-9658(1998)079[2863:pgoafs]2.0.co;2

Boyd IL (1993) Pup production and distribution of breeding Antarctic fur seals (*Arctocephalus gazella*) at South Georgia. Antarct Sci 5(1):17–24

Boyd IL (1996) Individual variation in the duration of pregnancy and birth date in Antarctic fur seals: the role of environment, age, and sex of fetus. J Mammal 77(1):124–133. https://doi.org/10.2307/1382714

Boyd IL (2002) Estimating food consumption of marine predators: Antarctic fur seals and macaroni penguins. J Appl Ecol 39(1):103–119. https://doi.org/10.1046/j.1365-2664.2002.00697.x

Boyd IL, Duck CD (1991) Mass changes and metabolism in territorial male Antarctic fur seals (*Arctocephalus gazella*). Physiol Zool 64(1):375–392. https://doi.org/10.1086/physzool.64.1.30158530

Boyd IL, Arnould JPY, Barton T et al (1994) Foraging behavior of Antarctic fur seals during periods of contrasting prey abundance. J Anim Ecol 63(3):703–713. https://doi.org/10.2307/5235

Boyd IL, McCafferty DJ, Reid K et al (1998) Dispersal of male and female Antarctic fur seals (*Arctocephalus gazella*). Can J Fish Aquat Sci 55(4):845–852. https://doi.org/10.1139/cjfas-55-4-845

Brault EK, Goebel ME, Geisz HN et al (2013) Inter-annual variation of persistent organic pollutants (POPS) in an Antarctic top predator *Arctocephalus gazella*. Environ Sci Technol 47(22):12744–12752. https://doi.org/10.1021/es403577f

Cipro CVZ, Bustamante P, Taniguchi S et al (2012) Persistent organic pollutants and stable isotopes in pinnipeds from King George Island, Antarctica. Mar Pollut Bull 64(12):2650–2655. https://doi.org/10.1016/j.marpolbul.2012.10.012

Cipro CVZ, Montone RC, Bustamante P (2017) Mercury in the ecosystem of Admiralty Bay, King George Island, Antarctica: occurrence and trophic distribution. Mar Pollut Bull 114(1):564–570. https://doi.org/10.1016/j.marpolbul.2016.09.024

Costa DP, Croxall JP, Duck CD (1989) Foraging energetics of Antarctic fur seals in relation to changes in prey availability. Ecology 70(3):596–606. https://doi.org/10.2307/1940211

Croxall JP, Pilcher MN (1984) Characteristics of krill *Euphausia superba* eaten by Antarctic fur seals *Arctocephalus gazella* at South Georgia. Brit Antarct Surv B (63):117–125

Croxall JP, McCann TS, Prince PA et al (1988) Reproductive performance of seabirds and seals at South Georgia and Signy Island, South Orkney Islands, 1976–1987: implications for Southern Ocean monitoring studies. In: Sahrhage D (ed) Antarctic Ocean and resources variability. Springer, Berlin/Heidelberg, pp 261–285

Croxall JP, Rodwell S, Boyd IL (1990) Entanglement in man-made debris of Antarctic fur seals at Bird Island, South Georgia. Mar Mamm Sci 6(3):221–233. https://doi.org/10.1111/j.1748-7692.1990.tb00246.x

Croxall JP, Reid K, Prince PA (1999) Diet, provisioning and productivity responses of marine predators to differences in availability of Antarctic krill. Mar Ecol Prog Ser 177:115–131. https://doi.org/10.3354/meps177115

Daneri GA, Carlini AR (1999) Spring and summer predation on fish by the Antarctic fur seal, *Arctocephalus gazella*, at King George Island, South Shetland Islands. Can J Zool Rev Can Zool 77(7):1157–1160. https://doi.org/10.1139/cjz-77-7-1157

Daneri GA, Piatkowski U, Coria NR et al (1999) Predation on cephalopods by Antarctic fur seals, *Arctocephalus gazella* at two localities of the Scotia Arc, Antarctica. Polar Biol 21(1):59–63. https://doi.org/10.1007/s003000050333

Daneri GA, Carlini AR, Hernandez CM et al (2005) The diet of Antarctic fur seals, *Arctocephalus gazella*, at King George Island, during the summer-autumn period. Polar Biol 28(4):329–333. https://doi.org/10.1007/s00300-004-0693-2

de Bruyn PJN, Pistorius PA, Tosh CA et al (2007) Leucistic Antarctic fur seal *Arctocephalus gazella* at Marion island. Polar Biol 30(10):1355–1358. https://doi.org/10.1007/s00300-007-0320-0

Dobson FS, Jouventin P (2003) How mothers find their pups in a colony of Antarctic fur seals. Behav Process 61(1–2):77–85. https://doi.org/10.1016/s0376-6357(02)00164-x

Doidge DW, Croxall JP (1985) Diet and energy budget of the Antarctic fur seal, *Arctocephalus gazella*, at South Georgia. In: Siegfried WR, Condy PR, Laws RM (eds) Antarctic nutrient cycles and food webs. Springer, Berlin/Heidelberg, pp 543–550

Doidge DW, Croxall JP (1989) Factors affecting weaning weight in Antarctic fur seals *Arctocephalus gazella* at South Georgia. Polar Biol 9(3):155–160. https://doi.org/10.1007/bf00297170

Doidge DW, Croxall JP, Baker JR (1984a) Density-dependent pup mortality in the Antarctic fur-seal *Arctocephalus gazella* at South Georgia. J Zool 202(Mar):449–460. https://doi.org/10.1111/j.1469-7998.1984.tb05095.x

Doidge DW, Croxall JP, Ricketts C (1984b) Growth rates of Antarctic fur-seal, *Arctocephalus gazella,* pups at South Georgia. J Zool 203:87–93

Doidge DW, McCann TS, Croxall JP (1986) Attendance behaviour of Antarctic fur seals. In: Gentry RL, Kooyman GL (eds) Fur seals: maternal strategies on land and at sea. Princeton University Press, Princeton, pp 102–114

Duck CD (1990) Annual variation in the timing of reproduction in Antarctic Fur seals, *Arctocephalus gazella*, at Bird Island, South-Georgia. J Zool 222:103–116. https://doi.org/10.1111/j.1469-7998.1990.tb04032.x

Erb E, Shaughnessy PD, Norman R (1996) Dental and mandibular injury in an Antarctic fur seal, *Arctocephalus gazella*, at Heard Island, Southern Ocean. J Wildl Dis 32(2):376–380. https://doi.org/10.7589/0090-3558-32.2.376

Eriksson C, Burton H (2003) Origins and biological accumulation of small plastic particles in fur seals from Macquarie Island. Ambio 32(6):380–384. https://doi.org/10.1639/0044-7447(2003)032[0380:Oabaos]2.0.Co;2

Ewalt DR, Payeur JB, Martin BM et al (1994) Characteristics of a Brucella species from a bottlenose dolphin (*Tursiops truncatus*). J Vet Diagn Investig 6(4):448–452. https://doi.org/10.1177/104063879400600408

Fernández C, Faiella A, Saubidet A (1998) Frecuencia de aparición de ejemplares de pinípedos en las costas de la ciudad del Mar del Plata, Argentina. In: Anais da 8a Reunião de Trabalho de Especialistas em Mamíferos Aquaticos da América do Sul, Olinda, Brazil, 1998

Forcada J, Hoffman JI (2014) Climate change selects for heterozygosity in a declining fur seal population. Nature 511(7510):462–465. https://doi.org/10.1038/nature13542

Forcada J, Staniland IJ (2009) Antarctic fur seal: *Arctocephalus gazella*. In: Perrin WF, Würsig B, Thewissen JGM (eds) Encyclopedia of marine mammals, 2nd edn. Academic, London, pp 36–42. https://doi.org/10.1016/B978-0-12-373553-9.00010-9

Forcada J, Trathan PN, Reid K (2005) The effects of global climate variability in pup production of Antarctic fur seals. Ecology 86(9):2408–2417. https://doi.org/10.1890/04-1153

Forcada J, Trathan PN, Murphy EJ (2008) Life history buffering in Antarctic mammals and birds against changing patterns of climate and environmental variation. Glob Chang Biol 14(11):2473–2488. https://doi.org/10.1111/j.1365-2486.2008.01678.x

Forcada J, Malone D, Royle JA et al (2009) Modelling predation by transient leopard seals for an ecosystem-based management of Southern Ocean fisheries. Ecol Model 220(12):1513–1521. https://doi.org/10.1016/j.ecolmodel.2009.03.020

García-Peña FJ, Pérez-Boto D, Jiménez C et al (2010) Isolation and characterization of *Campylobacter* spp. from Antarctic Fur seals (*Arctocephalus gazella*) at Deception Island, Antarctica. Appl Environ Microb 76(17):6013–6016. https://doi.org/10.1128/aem.00316-10

Goebel M, Vallejos VI, Trivelpiece WZ et al (2003) Antarctic fur seal pup production in the South Shetland Islands. In: Lipsky J (ed) AMLR 2001/2002 field season report, NOAA-Technical Memorandum-NMFS-SWFSC-350. National Marine Fisheries Service, Southwest Fisheries Science Center, La Jolla, pp 139–152

Goebel ME, Pussini N, Buchheit R et al (2014) Pinniped research at Cape Shirreff, Livingston Island, Antarctica, AMLR 2010-2011 Field Season Report. NOAA Technical Memorandum NMFS-SWFSC-524. National Marine Fisheries Service, Southwest Fisheries Science Center, La Jolla, pp 50–57

Goodall RNP, Boy CC, Benegas LG (2005) Antarctic seals on the coasts of Tierra del Fuego, Argentina – review and update. In: Thatje S, Calcagno JA, Arntz WE (eds) Evolution of Antarctic fauna: extended abstracts of the IBMANT/ANDEEP International Symposium and Workshop in 2003, Berichte zur Polar-und Meeresforschung, vol 507. Bremerhaven, Germany, pp 56–57

Grosser S, Sauer J, Paijmans AJ et al (2019) Fur seal microbiota are shaped by the social and physical environment, show mother–offspring similarities and are associated with host genetic quality. Mol Ecol 28(9):2406–2422. https://doi.org/10.1111/mec.15070

Guinet C, Goldsworthy SD, Robinson S (1999) Sex differences in mass loss rate and growth efficiency in Antarctic fur seal (*Arctocephalus gazella*) pups at Macquarie Island. Behav Ecol Sociobiol 46(3):157–163. https://doi.org/10.1007/s002650050605

Guinet C, Lea MA, Goldsworthy SD (2000) Mass change in Antarctic fur seal (*Arctocephalus gazella*) pups in relation to maternal characteristics at the Kerguelen Islands. Can J Zool-Rev Can Zool 78(3):476–483. https://doi.org/10.1139/cjz-78-3-476

Guinet C, Dubroca L, Lea MA et al (2001) Spatial distribution of foraging in female Antarctic fur seals *Arctocephalus gazella* in relation to oceanographic variables: a scale-dependent approach using geographic information systems. Mar Ecol Prog Ser 219:251–264. https://doi.org/10.3354/meps219251

Hernandez J, Prado V, Torres D et al (2007) Enteropathogenic *Escherichia coli* (EPEC) in Antarctic fur seals *Arctocephalus gazella*. Polar Biol 30(10):1227–1229. https://doi.org/10.1007/s00300-007-0282-2

Hinke JT, Cossio AM, Goebel ME et al (2017) Identifying risk: concurrent overlap of the Antarctic krill fishery with krill-dependent predators in the Scotia Sea. PLoS One 12(1):24. https://doi.org/10.1371/journal.pone.0170132

Hoffman JI, Forcada J (2009) Genetic analysis of twinning in Antarctic fur seals (*Arctocephalus gazella*). J Mammal 90(3):621–628. https://doi.org/10.1644/08-Mamm-a-264r1.1

Hoffman JI, Forcada J (2012) Extreme natal philopatry in female Antarctic fur seals (*Arctocephalus gazella*). Mamm Biol 77(1):71–73. https://doi.org/10.1016/j.mambio.2011.09.002

Hoffman JI, Boyd IL, Amos W (2003) Male reproductive strategy and the importance of maternal status in the Antarctic fur seal *Arctocephalus gazella*. Evolution 57(8):1917–1930

Hoffman JI, Forcada J, Amos W (2006) No relationship between microsatellite variation and neonatal fitness in Antarctic fur seals, *Arctocephalus gazella*. Mol Ecol 15(7):1995–2005. https://doi.org/10.1111/j.1365-294X.2006.02894.x

Hoffman JI, Forcada J, Trathan PN et al (2007) Female fur seals show active choice for males that are heterozygous and unrelated. Nature 445(7130):912–914. https://doi.org/10.1038/nature05558

Hoffman JI, Forcada J, Amos W (2010) Exploring the mechanisms underlying a heterozygosity–fitness correlation for canine size in the Antarctic fur seal *Arctocephalus gazella*. J Hered 101(5):539–552. https://doi.org/10.1093/jhered/esq046

Hoffman JI, Bauer E, Paijmans AJ et al (2018) A global cline in a colour polymorphism suggests a limited contribution of gene flow towards the recovery of a heavily exploited marine mammal. R Soc Open Sci 5(10):181227. https://doi.org/10.1098/rsos.181227

Hofmeyr GJG, Bester MN (1993) Predation on king penguins by Antarctic fur seals. S Afr J Antarct Res 23(1–2):71–74

Hofmeyr GJG, Bester MN, Pistorius PA et al (2007) Median pupping date, pup mortality and sex ratio of fur seals at Marion Island. S Afr J Wildl Res 37(1):1–8. https://doi.org/10.3957/03794369-37.1.1

Hucke-Gaete R, Torres D, Vallejos V (1997) Entanglement of Antarctic fur seals, *Arctocephalus gazella*, in marine debris at Cape Shirreff and San Telmo Islets, Livingston Island, Antarctica: 1988-1997. Serie Científica INACH 47:123–135

Hucke-Gaete R, Osman LP, Moreno CA et al (2004) Examining natural population growth from near extinction: the case of the Antarctic fur seal at the South Shetlands, Antarctica. Polar Biol 27(5):304–311. https://doi.org/10.1007/s00300-003-0597-8

Humble E, Dasmahapatra KK, Martinez-Barrio A et al (2018a) RAD sequencing and a hybrid Antarctic fur seal genome assembly reveal rapidly decaying linkage disequilibrium, global population structure and evidence for inbreeding. G3 Genes Genom Genet 8(8):2709–2722. https://doi.org/10.1534/g3.118.200171

Humble E, Stoffel MA, Forcada J et al (2018b) Inferring recent demographic history in the Antarctic fur seal (*Arctocephalus gazella*). In: Developing a genomic toolkit for the Antarctic fur seal. New insights into old questions, PhD thesis. Bielefeld, Germany, pp 137–149

Humble E, Paijmans AJ, Forcada J et al (2020) A 90K SNP array uncovers inbreeding and cryptic relatedness in an Antarctic fur seal breeding colony. bioRxiv 2020.2004.2001.020123. https://doi.org/10.1101/2020.04.01.020123

Hunt GL, Heinemann D, Everson I (1992) Distributions and predator-prey interactions of Macaroni penguins, Antarctic fur seals, and Antarctic krill near Bird Island, South Georgia. Mar Ecol Prog Ser 86(1):15–30. https://doi.org/10.3354/meps086015

Jeanniard-du-Dot T, Trites AW, Arnould JPY et al (2017) Reproductive success is energetically linked to foraging efficiency in Antarctic fur seals. PLoS One 12(4):19. https://doi.org/10.1371/journal.pone.0174001

Jensen S-K, Nymo IH, Forcada J, Godfroid J, Hall A (2012) Prevalence of *Toxoplasma gondii* antibodies in pinnipeds from Antarctica. Vet Rec 171(10):249–249. https://doi.org/10.1136/vr.100848

Jensen SK, Nymo IH, Forcada J et al (2013) Brucella antibody seroprevalence in Antarctic seals (*Arctocephalus gazella*, *Leptonychotes weddellii* and *Mirounga leonina*). Dis Aquat Org 105(3):175–181

Jones KA, Ratcliffe N, Votier SC et al (2020) Intra-specific niche partitioning in Antarctic fur seals, *Arctocephalus gazella*. Sci Rep UK 10(1):3238. https://doi.org/10.1038/s41598-020-59992-3

Kernaleguen L, Arnould JPY, Guinet C, Cazelles B, Richard P, Cherel Y (2016) Early-life sexual segregation: ontogeny of isotopic niche differentiation in the Antarctic fur seal. Sci Rep UK 6(1):33211. https://doi.org/10.1038/srep33211

Keyes MC (1965) Pathology of the northern fur seal. J Am Vet Med Assoc 157(10):1090–1095

Lancaster ML, Gemmell NJ, Negro S (2006) Menage a trois on Macquarie Island: hybridization among three species of fur seal (*Arctocephalus* spp.) following historical population extinction. Mol Ecol 15(12):3681–3692. https://doi.org/10.1111/j.1365-294X.2006.03041.x

Laws RM (1973) Population increase of fur Seals at South Georgia. Polar Rec 16(105):856–858. https://doi.org/10.1017/S003224740006397X

Lea MA, Guinet C, Cherel Y et al (2006) Impacts of climatic anomalies on provisioning strategies of a Southern Ocean predator. Mar Ecol Prog Ser 310:77–94. https://doi.org/10.3354/meps310077

Litzke V, Ottensmann M, Forcada J et al (2019) Heterozygosity at neutral and immune loci is not associated with neonatal mortality due to microbial infection in Antarctic fur seals. Ecol Evol 9(14):7985–7996. https://doi.org/10.1002/ece3.5317

Lunn NJ (1992) Fostering behavior and milk stealing in Antarctic fur seals. Can J Zool-Rev Can Zool 70(4):837–839. https://doi.org/10.1139/z92-119

Lunn NJ, Arnould JPY (1997) Maternal investment in Antarctic fur seals: evidence for equality in the sexes? Behav Ecol Sociobiol 40(6):351–362. https://doi.org/10.1007/s002650050351

Lunn NJ, Boyd IL (1991) Pupping-site fidelity of Antarctic fur seals at Bird Island, South Georgia. J Mammal 72(1):202–206. https://doi.org/10.2307/1381999

Lunn NJ, Boyd IL, Barton T et al (1993) Factors affecting the growth rate and mass at weaning of Antarctic fur seals at Bird Island, South Georgia. J Mammal 74(4):908–919. https://doi.org/10.2307/1382429

Lunn NJ, Boyd IL, Croxall JP (1994) Reproductive performance of female Antarctic fur seals – the influence of age, breeding experience, environmental variation and individual quality. J Anim Ecol 63(4):827–840. https://doi.org/10.2307/5260

Malcolm HM, Boyd IL, Osborn D et al (1994) Trace metals in Antarctic fur seal (*Arctocephalus gazella*) livers from Bird Island, South Georgia. Mar Pollut Bull 28(6):375–380. https://doi.org/10.1016/0025-326x(94)90275-5

McCafferty DJ (1999) Premature pupping in Antarctic fur seals (*Arctocephalus gazella*). Mar Mamm Sci 15(3):882–885. https://doi.org/10.1111/j.1748-7692.1999.tb00853.x

McCafferty DJ, Boyd IL, Taylor RI (1998a) Diving behaviour of Antarctic fur seal (*Arctocephalus gazella*) pups. Can J Zool-Rev Can Zool 76(3):513–520. https://doi.org/10.1139/cjz-76-3-513

McCafferty DJ, Boyd IL, Walker TR et al (1998b) Foraging responses of Antarctic fur seals to changes in the marine environment. Mar Ecol Prog Ser 166:285–299. https://doi.org/10.3354/meps166285

McCann TS (1980) Territoriality and breeding-behavior of adult male Antarctic fur seal, *Arctocephalus gazella*. J Zool 192:295–310

McCann TS, Doidge DW (1987) Antarctic fur seal, *Arctocephalus gazella*. In: Croxall JP, Gentry RL (eds) Status, biology, and ecology of fur seals. Proceedings of an international symposium and workshop, Cambridge, England, 23-27 April 1984. NOAA technical report NMFSNOAA technical report NMFS, Springfield, USA, p 5–8

McDonald BI, Goebel ME, Crocker DE et al (2012a) Biological and environmental drivers of energy allocation in a dependent mammal, the Antarctic fur seal pup. Physiol Biochem Zool 85(2):134–147. https://doi.org/10.1086/664948

McDonald BI, Goebel ME, Crocker DE et al (2012b) Dynamic influence of maternal and pup traits on maternal care during lactation in an income breeder, the Antarctic fur seal. Physiol Biochem Zool 85(3):243–254. https://doi.org/10.1086/665407

McFarlane RA, de B Norman RJ, Jones HI (2009) Diseases and parasites of Antarctic and sub-Antarctic seals. In: Knowles RK, Riddle M (eds) Health of Antarctic wildlife: a challenge for science and policy. Springer, Berlin/Heidelberg, pp 57–93. https://doi.org/10.1007/978-3-540-93923-8_4

Meise K, von Engelhardt N, Forcada J et al (2016) Offspring hormones reflect the maternal prenatal social environment: potential for foetal programming? PLoS One 11(1):e0145352. https://doi.org/10.1371/journal.pone.0145352

Miranda-Urbina D, Portflitt-Toro M, Serratosa J et al (2015) Vagrant Antarctic fur seal, *Arctocephalus gazella*, in northern Chile. Polar Biol 38(8):1311–1314. https://doi.org/10.1007/s00300-015-1693-0

Naya DE, Achaval F (2006) Nuevos registros de especies poco comunes de Pinnipedia y primer registro de *Actocephalus gazella* (Peters, 1875) (Pinnipedia: Otariidae) para el Uruguay. Bol Soc Zool Urug 15:23–27

North AW (1996) Fish in the diet of Antarctic fur seals (*Arctocephalus gazella*) at South Georgia during winter and spring. Antarct Sci 8(2):155–160. https://doi.org/10.1017/s0954102096000223

North AW, Croxall JP, Doidge DW (1983) Fish prey of the Antarctic fur seal *Arctocephalus gazella* at South Georgia. Br Antarct Surv B (61):27–37

Oliveira LR, Danilewicz DS, Martins MB et al (2001) New records of the Antarctic fur seal, *Arctocephalus gazella* (Peters, 1875) (Carnivora: Otariidae) for the southern Brazilian coast. Comunicacões do Museu de Ciencia y Tecnologia PUCRS Ser Zool (Porto Alegre) 14:201–207

Osman LP, Hucke-Gaete R, Moreno CA (2004) Feeding ecology of Antarctic fur seals at cape Shirreff, south Shetlands, Antarctica. Polar Biol 27(2):92–98. https://doi.org/10.1007/s00300-003-0555-3

Paijmans AJ, Stoffel MA, Bester MN et al (2020) The genetic legacy of extreme exploitation in a polar vertebrate. Sci Rep UK 10(1):5089. https://doi.org/10.1038/s41598-020-61560-8

Payne MR (1977) Growth of a fur seal population. Philos T R Soc B 279(963):67–79. https://doi.org/10.1098/rstb.1977.0072

Payne MR (1979) Growth in the Antarctic fur seal *Arctocephalus gazella*. J Zool 187(Jan):1–20

Peters WCH (1875 Über eine neue Art von Seebären, *Arctophoca gazella*, von den Kerguelen-Inseln. Monatsb K Preuss Akad Wiss Berlin (Monatsberichte der Königlichen Preussischen Akademie der Wissenschaften zu Berlin) 1875) n.d.:393–399

Peters L, Humble E, Kröcker N et al (2016) Born blonde: a recessive loss-of-function mutation in the melanocortin 1 receptor is associated with cream coat coloration in Antarctic fur seals. Ecol Evol 6(16):5705–5717. https://doi.org/10.1002/ece3.2290

Pinedo MC, Marmontel-Rosas M (1987) Primeiros registros de lobo marinho antártico, *Arctocephalus gazella* e novos registros de *A. tropicalis* para o Rio Grande do Sul, Brasil. II Reunião de Trabalhos de Especialistas em Mamíferos Aquáticos da América do Sul (Rio de Janeiro), Anais: 109

Prado JHF, Mattos PH, Silva KG et al (2016) Long-term seasonal and interannual patterns of marine mammal strandings in subtropical western South Atlantic. PLoS One 11(1):23. https://doi.org/10.1371/journal.pone.0146339

Reid K (1995) The diet of Antarctic fur seals (*Arctocephalus gazella* Peters 1875) during Winter at South Georgia. Antarct Sci 7(3):241–249

Reid K, Arnould JPY (1996) The diet of Antarctic fur seals *Arctocephalus gazella* during the breeding season at South Georgia. Polar Biol 16(2):105–114. https://doi.org/10.1007/Bf02390431

Reid K, Forcada J (2005) Causes of offspring mortality in the Antarctic fur seal, *Arctocephalus gazella*: the interaction of density dependence and ecosystem variability. Can J Zool 83(4):604–609. https://doi.org/10.1139/Z05-045

Reid K, Watkins JL, Croxall JP et al (1999) Krill population dynamics at South Georgia 1991–1997, based on data from predators and nets. Mar Ecol Prog Ser 177:103–114. https://doi.org/10.3354/meps177103

Reid K, Hill SL, Diniz TCD et al (2005) Mackerel icefish *Champsocephalus gunnari* in the diet of upper trophic level predators at South Georgia: implications for fisheries management. Mar Ecol Prog Ser 305:153–161

Rengifo-Herrera C, Ortega-Mora LM, Álvarez-García G et al (2012) Detection of *Toxoplasma gondii* antibodies in Antarctic pinnipeds. Vet Parasitol 190(1):259–262. https://doi.org/10.1016/j.vetpar.2012.05.020

Repenning CA, Peterson RS, Hubbs CL (1971) Contributions to the systematics of the southern fur seals, with particular reference to the Juan Fernández and Guadalupe species. In: Burt WH (ed) Antarctic Pinnipedia, Antarctic research series, vol 18. AGU, Washington, DC, pp 1–34. https://doi.org/10.1029/AR018p0001

Retamal P, Abalos P, Blank O et al (2000) Detection of anti-Brucella antibodies in pinnipeds from the Antarctic territory. Vet Rec 146(6):166–167. https://doi.org/10.1136/vr.146.6.166

Rodríguez D, Bastida R, Moron S et al (1994) *Arctocephalus gazella* y *A. tropicalis* en la Argentina. In: Anais da 6a Reunião de Trabalho de Especialistas em Mamíferos Aquaticos da América do Sul, Florianópolis, Brazil, vol 1994, p 127

Santora JA (2013) Dynamic intra-seasonal habitat use by Antarctic fur seals suggests migratory hotspots near the Antarctic Peninsula. Mar Biol 160(6):1383–1393. https://doi.org/10.1007/s00227-013-2190-z

Schiavone A, Corsolini S, Kannan K et al (2009a) Perfluorinated contaminants in fur seal pups and penguin eggs from South Shetland, Antarctica. Sci Total Environ 407(12):3899–3904. https://doi.org/10.1016/j.scitotenv.2008.12.058

Schiavone A, Kannan K, Horii Y et al (2009b) Occurrence of brominated flame retardants, polycyclic musks, and chlorinated naphthalenes in seal blubber from Antarctica: comparison to organochlorines. Mar Pollut Bull 58(9):1415–1419. https://doi.org/10.1016/j.marpolbul.2009.05.019

Schwarz LK, Goebel ME, Costa DP et al (2013) Top-down and bottom-up influences on demographic rates of Antarctic fur seals *Arctocephalus gazella*. J Anim Ecol 82(4):903–911. https://doi.org/10.1111/1365-2656.12059

Siegel V, Watkins JL (2016) Distribution, biomass and demography of Antarctic krill, *Euphausia superba*. In: Siegel V (ed) Biology and ecology of Antarctic krill. Springer, Cham, pp 21–100. https://doi.org/10.1007/978-3-319-29279-3_2

Smeele ZE, Ainley DG, Varsani A (2018) Viruses associated with Antarctic wildlife: from serology based detection to identification of genomes using high throughput sequencing. Virus Res 243:91–105. https://doi.org/10.1016/j.virusres.2017.10.017

Staniland IJ, Boyd IL (2003) Variation in the foraging location of Antarctic fur seals (*Arctocephalus gazella*) and the effects on diving behavior. Mar Mamm Sci 19(2):331–343. https://doi.org/10.1111/j.1748-7692.2003.tb01112.x

Staniland IJ, Robinson SL (2008) Segregation between the sexes: Antarctic fur seals, *Arctocephalus gazella*, foraging at South Georgia. Anim Behav 75:1581–1590. https://doi.org/10.1016/j.anbehav.2007.10.012

Staniland IJ, Reid K, Boyd IL (2004) Comparing individual and spatial influences on foraging behaviour in Antarctic fur seals *Arctocephalus gazella*. Mar Ecol Prog Ser 275:263–274. https://doi.org/10.3354/meps275263

Staniland IJ, Boyd IL, Reid K (2007) An energy-distance trade-off in a central-place forager, the Antarctic fur seal (*Arctocephalus gazella*). Mar Biol 152(2):233–241. https://doi.org/10.1007/s00227-007-0698-9

Staniland IJ, Robinson SL, Silk JRD et al (2012) Winter distribution and haul-out behaviour of female Antarctic fur seals from South Georgia. Mar Biol 159(2):291–301. https://doi.org/10.1007/s00227-011-1807-3

Stenvers O, Plötz J, Ludwig H (1992) Antarctic seals carry antibodies against seal herpesvirus. Arch Virol 123(3):421–424. https://doi.org/10.1007/BF01317275

Stepchew CL, Yaipen-LLanos C (2019) Off-range Southern elephant seal (*Mirounga leonina*) and Antarctic fur seal (*Arctocephalus gazella*) stranded in the coast of Peru: A La Niña side effect? In: Abstracts of the World Marine Mammal Conference, Barcelona, Spain. Available at https://www.wmmconference.org/wp-content/uploads/2020/02/WMMC-Book-of-Abstracts-3.pdf

Stoffel MA, Caspers BA, Forcada J et al (2015) Chemical fingerprints encode mother–offspring similarity, colony membership, relatedness, and genetic quality in fur seals. PNAS 112(36):E5005–E5012. https://doi.org/10.1073/pnas.1506076112

Texera WA (1974) Nuevos antecedentes sobre mamíferos de Magallanes. II Hallazgo de *Arctocephalus gazella* (Mammalia: Otariidae) en Isla Hoste, de la región de Magallanes, anillado en Isla Bird, Georgia del Sur. Anales del Instituto de la Patagonia 5:189–198

Torres D (1991) Pinnípedos con marcas de distinta procedencia avistados en las islas Shetland del Sur, Antártica. Boletín Antártico Chileno 10(2):16–19

Torres D, Guerra C, Cárdenas JC (1984) Primeros registros de *Arctocephalus gazella* y nuevos hallazgos de *Arctocephalus tropicalis* y *Leptonychotes weddelli* en el archipiélago de Juan Fernández. Serie Científica INACH 31:115–148

Torres D, Ulloa J, Torres DE (2006) Ejemplares de lobo fino antártico cruzan la península Antártica. Boletín Antártico Chileno. 25(2):17–19

Tryland M, Nymo IH, Nielsen O et al (2012) Serum chemistry and antibodies against pathogens in Antarctic fur seals, Weddell seals, crabeater seals, and Ross seals. J Wildl Dis 48(3):632–645. https://doi.org/10.7589/0090-3558-48.3.632

Vargas R, Osman LP, Torres D (2009) Inter-sexual differences in Antarctic fur seal pup growth rates: evidence of environmental regulation? Polar Biol 32(8):1177–1186. https://doi.org/10.1007/s00300-009-0615-4

Vera C, Vargas R, Torres D (2006) Focas leopardo en cabo Shirreff: presencia, depredación e interrogantes. In: Leppe M, Barticevic E (eds) II Simposo Lationamericano sobre Investigaciones Antárticas y VI Reunión Chilena de Investigación Antártica, Concepción, Chile, 2006, p 369–372

Waluda CM, Staniland IJ (2013) Entanglement of Antarctic fur seals at Bird Island, South Georgia. Mar Pollut Bull 74(1):244–252. https://doi.org/10.1016/j.marpolbul.2013.06.050

Waluda CM, Collins MA, Black AD et al (2010) Linking predator and prey behaviour: contrasts between Antarctic fur seals and macaroni penguins at South Georgia. Mar Biol 157(1):99–112. https://doi.org/10.1007/s00227-009-1299-6

Warren NL, Trathan PN, Forcada J et al (2006) Distribution of post-weaning Antarctic fur seal *Arctocephalus gazella* pups at South Georgia. Polar Biol 29(3):179–188. https://doi.org/10.1007/s00300-005-0037-x

Weddell J (1825) A voyage towards the south pole, performed in the years 1822–1824. London

Wynen LP, Goldsworthy SD, Guinet C et al (2000) Postsealing genetic variation and population structure of two species of fur seal (*Arctocephalus gazella* and *A. tropicalis*). Mol Ecol 9(3):299–314. https://doi.org/10.1046/j.1365-294x.2000.00856.x

Yonezawa T, Kohno N, Hasegawa M (2009) The monophyletic origin of sea lions and fur seals (Carnivora; Otariidae) in the Southern Hemisphere. Gene 441(1):89–99. https://doi.org/10.1016/j.gene.2009.01.022

Zdzitowiecki K (1984) Some Antarctic acanthocephalans of the genus *Corynosoma* parasitizing Pinnipedia, with description of three new species. Acta Parasitologica Polonica 29(30):359–377

Chapter 5
Guadalupe Fur Seal (*Arctocephalus philippii townsendi*, Brunner 2004)

Fernando R. Elorriaga-Verplancken, Tenaya Norris, Karina Acevedo-Whitehouse, Casandra Gálvez, María José Amador-Capitanachi, and Mara Landa-Garza

Abstract The Guadalupe fur seal (*Arctocephalus philippii townsendi* or *A. townsendi*) was at the brink of extinction due to overhunting during the eighteenth and nineteenth centuries. Today, the Guadalupe fur seal remains classified as an *Endangered* species by Mexican law and as *Threatened* under the US Endangered Species Act of 1973. Currently, the entire population is estimated at around 40,000 individuals, with a single well-established breeding site on Guadalupe Island. Around 20 years ago, it was discovered that Guadalupe fur seals had recolonized San Benito Archipelago, which is 260 km southeast of Guadalupe Island. However, breeding activity at this site continues to be negligible with <1% of pups born here each year. It is likely that recovery of the Guadalupe fur seal depends on prey availability, which is influenced by oceanographic factors in the California Current Ecosystem (CCE). In this regard, differences in foraging habits have been found between the two colonies (Guadalupe and San Benito) in recent years and within the Guadalupe colony across decades. The population recovery rate has been lower than that observed for other pinnipeds that inhabit the CCE. Additionally, there are recent records of unusual mortality events along the US west coast; higher neonatal mortality due to starvation at Guadalupe Island; and marked reduction of the colony size at San Benito, which are all related to anomalous warm water events. These threats

F. R. Elorriaga-Verplancken (✉) · C. Gálvez · M. J. Amador-Capitanachi
Instituto Politécnico Nacional, Centro Interdisciplinario de Ciencias Marinas, Departamento de Pesquerías y Biología Marina, La Paz, Baja California Sur, Mexico
e-mail: felorriaga@ipn.mx

T. Norris
The Marine Mammal Center, Sausalito, California, USA

K. Acevedo-Whitehouse
Unit for Basic and Applied Microbiology, School of Natural Sciences, Autonomous University of Queretaro, Queretaro, Mexico

M. Landa-Garza
Universidad Autónoma de Baja California Sur, Departamento Académico de Biología Marina, La Paz, Baja California Sur, Mexico

G. Heckel, Y. Schramm (eds.), *Ecology and Conservation of Pinnipeds in Latin America*, https://doi.org/10.1007/978-3-030-63177-2_5

are intensified and more frequent because of climate change, which is the most significant threat to the species, especially given that over 99% of Guadalupe fur seals breed at one island, making it more vulnerable to this impact.

Keywords Fur seals · Recovery · Health · Climate change

General Features

The Guadalupe fur seal (*Arctocephalus philippii townsendi*, Brunner 2004, or *Arctocephalus townsendi*, Merriam 1897) has thick pelage with dense underfur. It has a long muzzle, fleshy nose, and long and conspicuous ear pinnae. There is marked sexual dimorphism; adult males are significantly larger than adult females (average of 219 cm length and over 170 kg vs. 148.2 cm length and 49.1 kg, Gallo-Reynoso and Figueroa-Carranza 1996; Fig. 5.1). Large males are dark grayish-brown to grayish-black, whereas adult females, subadults, and juveniles are dark brown to grayish-black above and paler below (Gallo-Reynoso and Figueroa-Carranza 1996, Jefferson et al. 2015; Fig. 5.1). Pups are born with a black coat (Jefferson et al. 2015; Fig. 5.2) and a birth weight of 4.7-6.3 kg (Gálvez et al. 2020).

Historical and Current Distribution

The Guadalupe fur seal population was under severe pressure due to overexploitation (hunting for their pelts) during the eighteenth and nineteenth centuries, owing to the commercial value of its pelt. They were nearly extirpated by the end of the

Fig. 5.1 Adult male (left) and adult female with nursing pup (right) Guadalupe fur seal

Fig. 5.2 Neonate Guadalupe fur seal with a typical black coat

nineteenth century (Townsend 1931). Prior to exploitation, the Guadalupe fur seal population was estimated to be between 20,000 (Fleisher 1978a, b) and 200,000 individuals (Hubbs 1979) and was distributed from Socorro Island at the Revillagigedo Islands, Mexico, to northern Washington, USA (Hamilton 1951; Hubbs 1956; Repenning et al. 1971; Etnier 2002). In addition to Guadalupe Island, reproductive colonies were once located on the Channel Islands in the United States and at Cedros Island and San Benito Archipelago in Mexico (Berdegué 1956; Repenning et al. 1971; Walter and Craig 1979).

Hunters from Russia, Europe, and the United States placed heavy pressure on the Guadalupe fur seal population, often killing not only adult males and pups but also reproductive females, causing a negative impact on the number of births in subsequent years. More than 8300 fur seals were killed on Baja California islands in a 3-week period in 1805, and another hunting bout in 1808 resulted in 3000 fur seals being killed at Socorro Island over 2 weeks (Hamilton 1951). Between 1812 and 1840, hunters harvested up to 1500 fur seals annually in California. Overall, the species was declared commercially extinct by the end of the 1800s, although some individuals were sighted in 1928 and subsequently killed (Townsend 1931). The Mexican government issued official protection for Guadalupe Island in 1927. In 1954, the species was rediscovered when at least 14 individuals were counted on Guadalupe Island (Hubbs 1956). In 1975, the island was declared a sanctuary for wildlife, a decision that was critical to help prevent Guadalupe fur seal extinction. In 1993, 7400 individuals were counted on Guadalupe Island (Gallo-Reynoso 1994). In 2005, the island was declared a Biosphere Reserve (Reserva de la Biósfera Isla Guadalupe), regulated by the National Commission for Natural Protected Areas (Comisión Nacional de Áreas Naturales Protegidas, CONANP), a protection status that is maintained today.

The Guadalupe fur seal is the only member of the *Arctocephalus* genus in the northern hemisphere, and its current population is estimated between 34,000 and

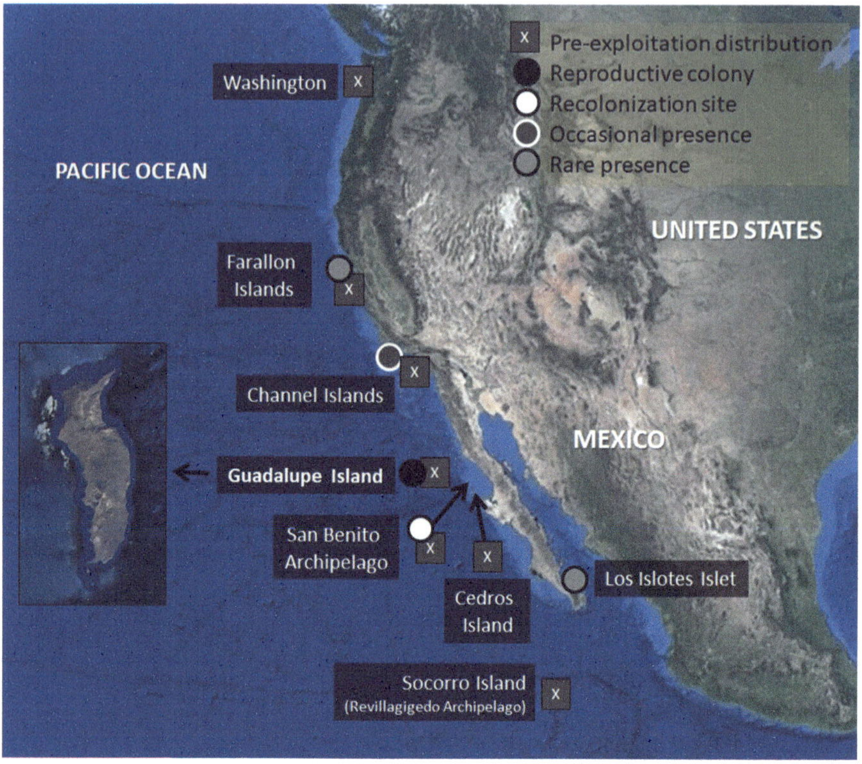

Fig. 5.3 Historical and current terrestrial distribution of Guadalupe fur seals (Hamilton 1951, Hubbs 1956, Hanni et al. 1997; Etnier 2002; Elorriaga-Verplancken et al. 2016a, b)

44,000 individuals. Its increase showed an average annual growth rate of 5.9% (range 4.1–7.7%) from 1984 to 2013 (García-Aguilar et al. 2018); however, there is a single, well-established reproductive colony on Guadalupe Island (Gallo-Reynoso 1994; García-Aguilar et al. 2018). A recolonization site of 2500–3500 individuals exists on San Benito Archipelago, located 260 km southeast of Guadalupe Island. It is composed mainly of juveniles and subadult males; however, a few breeding females and pups are present during the summer (Elorriaga-Verplancken et al. 2016a; García-Capitanachi et al. 2017). There are occasional sightings of animals in the Gulf of California (Aurioles-Gamboa et al. 1999; Elorriaga-Verplancken et al. 2016b) and on the Channel Islands, especially San Miguel Island (Hanni et al. 1997, Melin and DeLong 2006; Fig. 5.3). Between 2015 and 2019, nine live Guadalupe fur seals (eight juveniles and one subadult male, Fig. 5.4) were recorded during monthly surveys at the California sea lion (*Zalophus californianus*) rookery at Los Islotes rookery in Espíritu Santo Archipelago (La Paz Bay, southern Gulf of California, Mexico).

Guadalupe fur seal recovery has been relatively successful, but not as rapid as that of other pinniped species, such as the northern elephant seal (*Mirounga*

Fig. 5.4 Some of the Guadalupe fur seals observed at the California sea lion rookery at Los Islotes in La Paz Bay in the southern Gulf of California, Mexico, during monthly surveys conducted between 2015 and 2019

angustirostris) that was almost extinguished at the same time as the Guadalupe fur seal. Northern elephant seals recovered from only a few animals at Guadalupe Island in the early 1900s to more than 200,000 individuals, also expanding out to different islands (Lowry et al. 2014). The Guadalupe fur seal is still listed as *Endangered* under Mexican law (SEMARNAT-059-2010), *Threatened* under the US Endangered Species Act of 1973, and as *Least Concern* by the International Union for Conservation of Nature (Aurioles-Gamboa 2015). Despite these status inconsistencies, the most important aspect regarding the species protection should focus on the fact that there is only one breeding island, Guadalupe Island, which renders the viability of the entire species dependent on a single location.

Reproductive Biology

Terrestrial activities of Guadalupe fur seals, including pupping, mating, and resting, primarily take place on rocky substrate and among boulders of various sizes that create caves and crevices (García-Aguilar et al. 2013). At sites like San Benito Archipelago, where breeding is negligible, it is possible to observe Guadalupe fur seals 30–40 m inland from the shoreline (Fig. 5.5), which is a long distance compared to what is observed at Guadalupe Island.

Fig. 5.5 Guadalupe fur seal at San Benito Archipelago, Mexico, resting 30–40 m from the shoreline

The Guadalupe fur seal has a polygynous reproductive strategy, like other otariid species (sea lions and fur seals), which consists of a territorial adult male or "bull" that defends (mainly with vocalizations and displays) a territory with an average of 16 adult females or "cows" (Gallo-Reynoso 1994; Fig. 5.6). These territories are concentrated on the east and south sides of Guadalupe Island, in areas with less exposure to weather from the northwest. Adult males usually arrive at Guadalupe Island in June and depart by early August, whereas adult females begin arriving by late May (Seagars 1984; Gallo-Reynoso 1994). The breeding season extends from June to August, when adult females give birth to one pup, which is born relatively soon (4 h to 8 days) after they arrive to the rookery (Gallo-Reynoso 1994). The lactation period is around 9 months (Gallo-Reynoso and Figueroa-Carranza 2010), following a 9-month active gestation period, after an embryonic diapause of around 3 months, in order to synchronize breeding in the summer each year. Females are ready to copulate and get pregnant again only 5–10 days after parturition; the earliest age recorded for a female that gave birth was 4 years (Gallo-Reynoso 1994). Maternal care is characterized by alternating periods of nursing and foraging trips at sea while their pups stay on land (Gallo-Reynoso 1994; Boness and Bowen 1996). Mean length of maternal attendance on land is 5.03 days (Gallo-Reynoso and Figueroa-Carranza 2010). Pups double their mass during their first 70 days due to consumption of the high fat content (43.2%) of maternal milk (Figueroa-Carranza 1994; Gallo-Reynoso and Figueroa-Carranza 2010), and their black neonatal coat is molted at approximately 4–5 months of age.

Fig. 5.6 Territorial Guadalupe fur seal bull and three adult females during the breeding season at Guadalupe Island, Mexico

Births are uncommon at the recolonization location on San Benito Archipelago, with less than 30 pups sighted annually (Elorriaga-Verplancken et al. 2016a). These pups are generally sighted alone, on the west and east island of the archipelago, while they wait for their mothers to return from foraging. There are small aggregations of up to three adult mother-pup pairs, occasionally next to a subadult male or a relatively small adult male, forming a very small territory compared to those found on Guadalupe Island.

If the current recovery of the San Benito Archipelago continues, it is plausible that this location will become a viable breeding colony once again (Aurioles-Gamboa et al. 2010).

Foraging Ecology

It is argued that prey availability, and its effect on maternal provisioning of pups, and therefore pup survival, has been the primary limiting factor for the recovery of Guadalupe fur seals, rather than reduced genetic variability (Weber et al. 2004). If true, it is important to recognize that adult female Guadalupe fur seal foraging trips are among the longest for any otariid, lasting up to 3 weeks and traveling an average total distance of 2375 ± 1389 km, with feeding grounds located at 444 ± 151 km from Guadalupe Island (Gallo-Reynoso et al. 2008). Satellite-tagged individuals typically travel north of Guadalupe Island, although some travel south of the island, and remain offshore (Gallo-Reynoso et al. 2008, Norris et al. 2017,

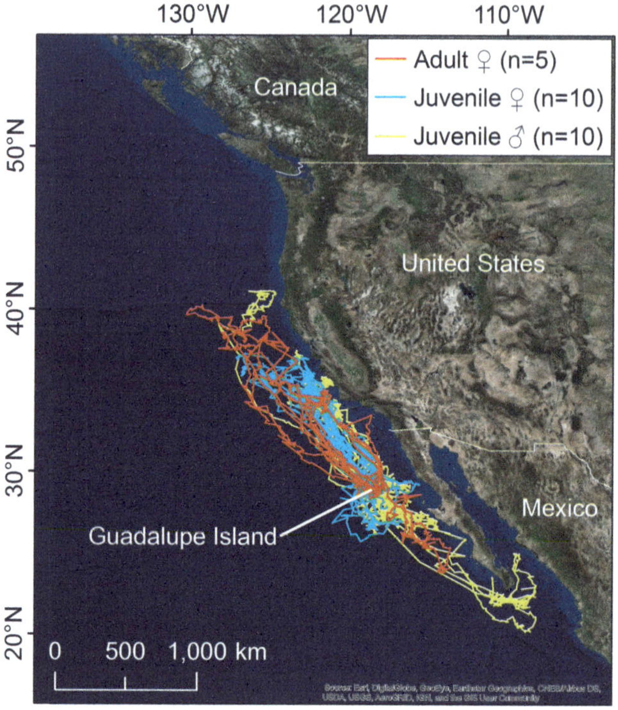

Fig. 5.7 Satellite tracks from April to September 2017 for 25 juvenile and adult female Guadalupe fur seals tagged at Guadalupe Island, Mexico (Norris et al. 2017; Amador-Capitanachi 2018)

Amador-Capitanachi 2018; Fig. 5.7). Additionally, records suggest the presence of foraging areas in the Gulf of Ulloa, especially for males, along the Pacific coast of Baja California (Aurioles-Gamboa et al. 2017), and as far south as central Mexico (Ortega-Ortiz et al. 2019). Guadalupe fur seals forage hundreds of kilometers offshore, which is explained by the distribution of their most commonly consumed prey. These prey are mostly squids [e.g., jumbo (*Dosidicus gigas*), opalescent (*Doryteuthis opalescens*), hooked (*Onychoteuthis* spp.), enope (*Abraliopsis felis*), and luminous flying (*Eucleoteuthis luminosa*) squid], as well as some fishes (e.g., myctophids of the *Symbolophorus*, *Ceratoscopelus*, and *Diaphus* genera; Amador-Capitanachi et al. 2017, Juárez-Ruiz et al. 2018). The Guadalupe fur seal does not dive as deep as other otariids (Aurioles-Gamboa and Camacho-Ríos 2007); its typical vertical dive depth is around 17 ± 10 m (Gallo-Reynoso et al. 2008). Therefore, they take advantage of the diel vertical migration of their mesopelagic prey, foraging primarily from sunset to dawn (Gallo-Reynoso 1994).

There seems to be trophic and geographic segregation between fur seals from Guadalupe Island and San Benito Archipelago. Findings of a study conducted in 2013 using scat and stable isotope analyses (used as a proxy for foraging grounds) indicated that individuals at Guadalupe Island fed mainly on jumbo squid at higher latitudes, whereas those from San Benito Archipelago fed mostly on opalescent squids at lower latitudes (Amador-Capitanachi et al. 2017). These distinct feeding patterns may be related to the different age classes that are predominant at each location (mature individuals at Guadalupe Island and juveniles at San Benito Archipelago) or also may be a result of strategies used to reduce intraspecific potential competition of resources by age classes (Amador-Capitanachi et al. 2017). Temporal shifts in prey availability are important to consider as well. Jumbo squid was not a dominant prey item for the Guadalupe fur seal in previous studies (e.g., Gallo-Reynoso 1994), but it was in 2013–2016 (Juárez-Ruiz et al. 2018, Amador-Capitanachi et al. 2020), likely due to changes in the range of this squid species in the California Current Ecosystem during the last decade (Field et al. 2007). Inter-annual variations in prey occurrence give insight into the opportunistic feeding behavior of the Guadalupe fur seal as these animals take advantage of what is available at a given time under certain conditions. Even within one season, there are individual foraging differences among adult female Guadalupe fur seals from Guadalupe Island, with at least three different groups that feed on different species and use different foraging grounds, which help reduce resource overlap. This segregation behavior is likely ecologically relevant in the current recovery process of the Guadalupe fur seal because it may reduce pressure among individuals (Juárez-Ruiz et al. 2018).

Threats

Given that the Guadalupe fur seal has only one reproductive colony (Guadalupe Island), understanding factors that threaten the population is particularly vital. Because there are no other breeding sites at different latitudes or in environments with different conditions, the species as a whole is highly vulnerable to certain factors.

Hurricanes

For instance, in 1992, a hurricane ("Darby") caused 33% pup mortality at Guadalupe Island, due partly to abnormally high waves that swept pups away to sea, caused injuries, and modified maternal attendance behavior (Gallo-Reynoso 1994).

Climate Change

Recent events have increased our understanding of threats to the Guadalupe fur seal. An unusual mortality event (UME) of young (mostly 9–15 month old) fur seals has been recorded along the coast of California from 2015 to 2019. To date, around 400 animals have stranded in this region, typically in poor body condition and sometimes exhibiting high parasite loads and opportunistic bacterial infections. There also have been several unusual stranding events in Washington and Oregon, most recently in 2019, with approximately 90 strandings in these two states. Current research is attempting to fully understand the cause(s) of this Guadalupe fur seal UME; and there is strong evidence that the event is related to warm oceanographic anomalies, such as the North Pacific marine heatwave termed "The Blob" and the 2015–2016 El Niño (Bond et al. 2015; NOAA Fisheries 2019). These warm water anomalies impact primary productivity, and hence prey availability across the trophic web, including for pinnipeds and other top predators (Trillmich and Ono 1991; Kovacs et al. 2012; Amador-Capitanachi et al. 2020).

Between 2015 and 2016, emaciated Guadalupe fur seals in the Bay of La Paz (southern Gulf of California) began to be observed (Elorriaga-Verplancken et al. 2016b). Interestingly, monthly surveys at the California sea lion rookery in La Paz Bay have been made since 2012, but Guadalupe fur seals were only recorded at this location during the UME along the US west coast, suggesting longer movements by these animals under anomalous oceanographic conditions, not only toward high latitudes (California, Washington, and Oregon) but also southward into the Gulf of California.

Anomalous warming of ocean temperatures can have a marked effect on individuals, as well as the colony, as was revealed by an approximately 50% reduction of the San Benito Archipelago colony during the 2015–2016 El Niño and an increase in the isotopic niche due to increased foraging effort of the few females that gave birth in that location (Elorriaga-Verplancken et al. 2016a). This effect also was observed in juvenile fur seals from San Benito Archipelago, which is the dominant class at this recolonized area. Stable isotopic ratios of nitrogen ($\delta^{15}N$) and carbon ($\delta^{13}C$) are proxies for trophic position and habitat use (e.g., inshore vs. offshore foraging habits, respectively; Newsome et al. 2007). Juveniles seemed to extend their foraging range with greater convex hull values in 2015, which is consistent with an increase in foraging effort (Landa-Garza 2020), compared to 2017 (Fig. 5.8).

More evidence regarding the impact of warming conditions on Guadalupe fur seals is provided by Gálvez et al. (2020), which recorded the lowest body masses in neonates in 2014–2015, as well as a high neonatal mortality (15%) in 2015 compared to 2013 and 2016, which was likely related to reduced maternal foraging success that led to greater numbers of starved neonates during the most anomalously warm years (Gálvez 2015).

Interactions with Fisheries

Other threats to Guadalupe fur seals include recreational or sport fishing. Sport fishing commonly occurs around Guadalupe Island and occasionally results in Guadalupe fur seals being harmed by fishing hooks or entangled in fishing gear (Fig. 5.9). The National Commission for Natural Protected Areas (CONANP) that

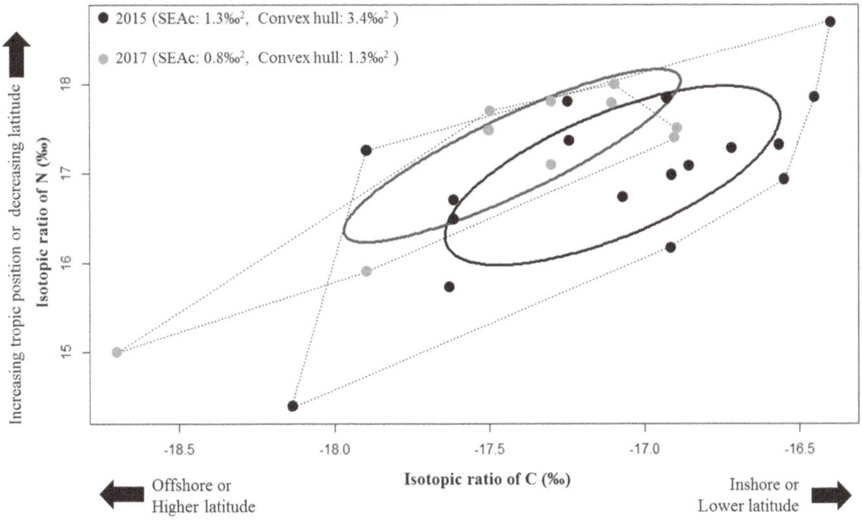

Fig. 5.8 Isotopic niche of Guadalupe fur seal juveniles from San Benito Archipelago, Mexico, in 2015 and 2017. The areas reflect foraging area in terms of trophic position ($\delta^{15}N$) and habitat use ($\delta^{13}C$). Convex hulls (polygons) include all values within each year, whereas the standard ellipses areas corrected for small sample sizes (SEAc) include the most representative data within each year (taken from Landa-Garza 2020)

Fig. 5.9 Guadalupe fur seal at Guadalupe Island, Mexico, with a sport fishing hook in the mouth as evidenced by the trailing line

manages the Guadalupe Island Biosphere Reserve has made important efforts to regulate human activities at the Guadalupe Biosphere Reserve and eventually nullify any anthropogenic effect. These efforts include the prohibition of sport fishing in certain areas. Commercial fisheries also may impact Guadalupe fur seals, with a greater proportion of Guadalupe fur seals that strand on California beaches entangled in netting (25%) than any other pinniped species (3.2%; Moore et al. 2009).

Diseases and Biotoxins

Pathogens and other health risks, including harmful algal blooms, cannot be ignored (Smith et al. 2009); but, there is extremely limited information in this species. A systematic examination of over 100 dead pups found at Guadalupe Island in 2013 and 2014 showed that the most common cause of death for this age class was trauma by crushing, caused by larger individuals, followed by starvation, drowning, and bacterial respiratory disease (Gálvez 2015). Furthermore, this species appears to be susceptible to infection by a variety of pathogens, some of which also infect other pinniped species (Gerber et al. 1993; Ziehl-Quirós et al. 2017; Seguel et al. 2018; Gálvez 2020). Given the recent anomalous climatic events that have been shown to impact the immune competence of neonatal California sea lion pups (Banuet-Martinez et al. 2017), it is pressing to investigate the prevalence and health implications of known and emerging pathogens and biotoxins in Guadalupe fur seals. This is particularly important considering that the species' breeding range is restricted to one well-established rookery and its wide-ranging at-sea distribution (Norris et al. 2017; Amador-Capitanachi 2018; García-Aguilar et al. 2018). Thus, virulent pathogens and toxigenic algal bloom, which may have expanded niches as a result of climate change, could increase the risk of extinction for Guadalupe fur seals (McCallum 2012; Gobler et al. 2017; Harvell et al. 2019).

Predation

This factor is not considered as a significant threat or concern to the Guadalupe fur seal population. However, it is important to point out that, given Guadalupe fur seals primarily are found on land at Guadalupe Island, they are subject to predation by different shark species [e.g., white shark (*Carcharodon carcharias*)] that occur in waters around the island. Guadalupe fur seals at Guadalupe Island also are commonly observed with superficial cookiecutter shark (*Isistius brasiliensis*) bites (Gallo-Reynoso and Figueroa-Carranza 1992).

Final Remarks

The Guadalupe fur seal has recovered from the brink of extinction with the implementation of important conservation measures by Mexico and the United States. Along with these recovery efforts, important applied and basic research (population trends, foraging ecology, distribution, health, etc.) is being conducted in a collaborative initiative between the two countries, in order to investigate the effects of climate change, which is the most significant threat to this species. Climate change likely already is impacting the population as evidenced by a recent unusual mortality event in the United States (2015–2019); a decrease in recolonization at San Benito Archipelago colony in recent years, especially since 2015; and a decrease in neonatal body mass at Guadalupe Island during recent warming anomalies (2014–2015). Because over 99% of Guadalupe fur seals breed at a single island (Guadalupe Island), this species is particularly vulnerable to the impacts of climate change, and even more attention has to be given in the coming years, in terms of both conservation and research efforts, to ensure the continued recovery of this population.

References

Amador-Capitanachi, MJ (2018) Variabilidad en el nicho isotópico (2013–2016) y análisis del desplazamiento mediante telemetría en el lobo fino de Guadalupe (*Arctocephalus philippii townsendi*). M.Sc. Thesis, CICIMAR-IPN, La Paz, Baja California Sur, México

Amador-Capitanachi MJ, Moreno-Sánchez XG, Ventura-Domínguez PD, Juárez-Ruiz A, González-Rodríguez E, Gálvez C, Norris T, Elorriaga-Verplancken FR (2020) Ecological implications of unprecedented warm water anomalies on interannual prey preferences and foraging areas of Guadalupe fur seals. Mar Mamm Sci 36(4):1254–1270. https://doi.org/10.1111/mms.12718

Amador-Capitanachi MJ, Moreno-Sánchez XG, Juárez-Ruiz A et al (2017) Trophic variation between the two existing Guadalupe fur seal colonies on Guadalupe Island and the San Benito Archipelago, Mexico. Aquat Mamm 43(1):14–25. https://doi.org/10.1578/AM.43.1.2017.14

Aurioles-Gamboa D (2015) *Arctocephalus townsendi* The IUCN red list of threatened species 2015. https://doi.org/10.2305/IUCN.UK.2015-2.RLTS.T2061A45224420.en. Accessed 30 Apr 2019

Aurioles-Gamboa D, Camacho-Ríos FJ (2007) Diet and feeding overlap of two otariids, *Zalophus californianus* and *Arctocephalus townsendi*: implications to survive environmental uncertainty. Aquat Mamm 33(3):315–326. https://doi.org/10.1578/AM.33.3.2007.315

Aurioles-Gamboa D, Hernández-Camacho CJ, Rodriguez-Krebs E (1999) Notes on the southernmost records of the Guadalupe fur seal, *Arctocephalus townsendi*, in Mexico. Mar Mamm Sci 15(2):581–583. https://doi.org/10.1111/j.1748-7692.1999.tb00825.x

Aurioles-Gamboa D, Elorriaga-Verplancken FR, Hernández-Camacho CJ (2010) The current population status of Guadalupe fur seal (*Arctocephalus townsendi*) on the San Benito Islands, México. Mar Mamm Sci 26(2):402–408. https://doi.org/10.1111/j.1748-7692.2009.00350.x

Aurioles-Gamboa D, Pablo-Rodríguez N, Rosas-Hernández P et al (2017) Guadalupe fur seal population expansion and its post-breeding male migration to the Gulf of Ulloa, Mexico: bioecology, threats and conservation. In: Alava JJ (ed) Tropical pinnipeds, 1st edn. CRC Press, Boca Raton, pp 91–119. https://doi.org/10.1201/9781315151588-7

Banuet-Martinez M, Espinosa-de Aquino W, Elorriaga-Verplanken FR et al (2017) Climatic anomaly affects the immune competence of California Sea lion pups. PLoS One 12(6):e0179359. https://doi.org/10.1371/journal.pone.0179359

Berdegué AJ (1956) La foca fina, el elefante marino y la ballena gris en Baja California y el problema de su conservación. Ediciones del Instituto Mexicano de Recursos Naturales Renovables 14:1–38

Bond NA, Cronin MF, Freeland H et al (2015) Causes and impacts of the 2014 warm anomaly in the NE Pacific. Geophys Res Lett 42:3414–3420. https://doi.org/10.1002/2015GL063306

Boness DJ, Bowen WD (1996) Evolution of maternal care in pinnipeds. Bioscience 46(9):645–654. https://doi.org/10.2307/1312894

Brunner S (2004) Furs seals and sea lions (Otariidae): identification of species and taxonomic review. Syst Biodivers 1:339–439

Elorriaga-Verplancken FR, Sierra-Rodríguez GE, Rosales-Nanduca H et al (2016a) Impact of the 2015 El Niño-southern oscillation on the abundance and foraging habits of Guadalupe fur seals and California sea lions from the San Benito archipelago, Mexico. PLoS One 11(5):e0155034. https://doi.org/10.1371/journal.pone.0155034

Elorriaga-Verplancken FR, Rosales-Nanduca H, Robles R (2016b) Unprecedented records of Guadalupe fur seals in La Paz Bay, Southern Gulf of California, Mexico, as a possible result of warm conditions in the northeastern Pacific. Aquat Mamm 42(3):261–267. https://doi.org/10.1578/AM.42.3.2016.261

Etnier MA (2002) Occurrences of Guadalupe fur seals (*Arctocephalus townsendi*) on the Washington coast over the past 500 years. Mar Mamm Sci 18(2):551–557. https://doi.org/10.1111/j.1748-7692.2002.tb01056.x

Field JC, Baltz KEN, Walker W (2007) Range expansion and trophic interactions of the jumbo squid, *Dosidicus gigas*, in the California current. Calif Coop Ocean Fish Investig Rep 48:131–146

Figueroa-Carranza AL (1994) Early lactation and attendance behavior of the Guadalupe fur seal females (*Arctocephalus townsendi*). Ph.D. Dissertation. University of California, Santa Cruz, California, United States

Fleisher LA (1978a) Guadalupe fur seal. In: Haley D (ed) Marine mammals of eastern North Pacific and Arctic waters. Pacific Search Press, Seattle, pp 160–165

Fleisher LA (1978b) The distribution, abundance, and population characteristics of the Guadalupe fur seal, *Arctocephalus townsendi* (Merriam, 1897). Ph.D. dissertation, University of Washington, Seattle, Washington, USA

Gallo-Reynoso JP (1994) Factors affecting the population status of the Guadalupe fur seal, *Arctocephalus townsendi* (Merriam, 1897), at Isla Guadalupe, Baja California, Mexico. Ph.D. dissertation, University of California, Santa Cruz, California, USA

Gallo-Reynoso JP, Figueroa-Carranza AL (1992) A cookie shark wound on a Guadalupe fur seal male. Mar Mammal Sci 8(4):428–430. https://doi.org/10.1111/j.1748-7692.1992.tb00060.x

Gallo-Reynoso JP, Figueroa-Carranza AL (1996) Size and weight of Guadalupe fur seals. Mar Mamm Sci 12(2):318–321. https://doi.org/10.1111/j.1748-7692.1996.tb00584.x

Gallo-Reynoso JP, Figueroa-Carranza AL (2010) Pup growth of the Guadalupe fur seal, *Arctocephalus townsendi*. Therya 1(1):75–90

Gallo-Reynoso JP, Figueroa-Carranza AL, Le Boeuf BJ (2008) Foraging behavior of lactating Guadalupe fur seal females. In: Lorenzo C, Espinoza E, Ortega J (eds) Avances en el estudio de los mamíferos de México, Publicaciones Especiales, vol II. Asociación Mexicana de Mastozoologia AC, Mexico City, pp 505–614

Gálvez C (2015) Causas de mortalidad en neonatos de lobo fino de Guadalupe *Arctocephalus townsendi* en Isla Guadalupe, B.C., México: temporada reproductiva 2013 y 2014. M.Sc. thesis, CICIMAR-IPN, La Paz, Baja California Sur, México

Gálvez C (2020) Salud del lobo fino de Guadalupe (*Arcrocephalus philippii townsendi*). Ph.D. dissertation, CICIMAR-IPN, La Paz, Baja California Sur, México

Gálvez C, Pardo MA, Elorriaga-Verplancken FR (2020) Impacts of extreme ocean warming on the early development of a marine top predator: the Guadalupe fur seal. Prog Oceanogr 180:102220. https://doi.org/10.1016/j.pocean.2019.102220

García-Aguilar MC, Gutiérrez-García D, de la Cueva H (2013) Terrestrial habitat segregation between the Guadalupe fur seal (*Arctocephalus townsendi*) and the California sea lion (*Zalophus californianus*) at islas San Benito, Mexico. Aquat Mamm 39(1):54–60. https://doi.org/10.1578/AM.39.1.2013.54

García-Aguilar MC, Elorriaga-Verplancken FR, Rosales-Nanduca H et al (2018) Population status of the Guadalupe fur seal (*Arctocephalus townsendi*). J Mammal 99(6):1522–1528. https://doi.org/10.1093/jmammal/gyy132

García-Capitanachi B, Schramm Y, Heckel G (2017) Population fluctuations of Guadalupe fur seals (*Arctocephalus philippii townsendi*) between the San Benito Islands and Guadalupe Island, Mexico, during 2009 and 2010. Aquat Mammals 43(5):492–500. https://doi.org/10.1578/AM.43.5.2017.492

Gerber JA, Roletto J, Morgan LE et al (1993) Findings in pinnipeds stranded along the central and northern California coast, 1984-1990. J Wildl Dis 29(3):423–433. https://doi.org/10.7589/0090-3558-29.3.423

Gobler CJ, Doherty OM, Hattenrath-Lehmann TK et al (2017) Ocean warming since 1982 has expanded the niche of toxic algal blooms in the North Atlantic and North Pacific oceans. Proc Natl Acad Sci 114(19):4975–4980. https://doi.org/10.1073/pnas.1619575114

Hamilton A (1951) Is the Guadalupe fur seal returning? Nat Hist 60:90–96

Hanni KD, Long DJ, Jones RE et al (1997) Sightings and strandings of Guadalupe fur seals in central and northern California, 1988–1995. J Mammal 78:684–690. https://doi.org/10.2307/1382920

Harvell CD, Montecino-Latorre D, Caldwell JM et al (2019) Disease epidemic and a marine heat wave are associated with the continental-scale collapse of a pivotal predator (*Pycnopodia helianthoides*). Sci Adv 5(1):eaau7042. https://doi.org/10.1126/sciadv.aau7042

Hubbs CL (1956) Back from oblivion, Guadalupe fur seal: still a living species. Pac Discov 9:14–21

Hubbs CL (1979) Guadalupe fur seal. Mammals in the Seas: report, 2, p.24.. Hudson PJ, Rizzoli AP, Grenfell BT, Heesterbeek JAP, Dobson AP, 2002. Ecology of wildlife diseases

Jefferson TA, Webber MA, Pitman RL et al (2015) Marine mammals of the world: a comprehensive guide to their identification. Academic Press, London

Juárez-Ruiz A, Elorriaga-Verplancken FR, Moreno-Sánchez XG et al (2018) Diversification of foraging habits among Guadalupe fur seals from their only well-established breeding colony, Guadalupe Island, Mexico. Mar Biol 165:86. https://doi.org/10.1007/s00227-018-3339-6

Kovacs KM, Aguilar A, Aurioles-Gamboa D et al (2012) Global threats to pinnipeds. Mar Mamm Sci 28(2):414–436. https://doi.org/10.1111/j.1748-7692.2011.00479.x

Landa-Garza MC (2020) Efectos de las anomalías oceanográficas sobre la recolonización del lobo fino de Guadalupe en el Archipiélago San Benito, Baja California, México. B.Sc. Thesis, Universidad Autónoma de Baja California Sur, La Paz, Baja California Sur, México

Lowry MS, Condit R, Hatfield B et al (2014) Abundance, distribution, and population growth of the northern elephant seal (*Mirounga angustirostris*) in the United States from 1991 to 2010. Aquat Mamm 40:20–31. https://doi.org/10.1578/AM.40.1.2014.20

McCallum H (2012) Disease and the dynamics of extinction. Philos Trans R Soc Lond Ser B Biol Sci 367(1604):2828–2839. https://doi.org/10.1098/rstb.2012.0224

Melin SR, DeLong RL (2006) Observations of a Guadalupe fur seal (*Arctocephalus townsendi*) female and pup at San Miguel Island, California. Mar Mamm Sci 15(3):885–888. https://doi.org/10.1111/j.1748-7692.1999.tb00854.x

Merriam CH (1897) A new fur-seal or sea-bear (*Arctocephalus townsendi*) from Guadalupe island, off Lower California. P Biol Soc Wash 11:175–178

Moore E, Lyday S, Roletto J et al (2009) Entanglements of marine mammals and seabirds in Central California and the north-west coast of the United States 2001–2005. Mar Pollut Bull 58(7):1045–1051. https://doi.org/10.1016/j.marpolbul.2009.02.006

Newsome SD, Del Rio CM, Bearhop S et al (2007) A niche for isotopic ecology. Front Ecol Environ 5:429–436. https://doi.org/10.1890/060150.1

NOAA Fisheries (2019) Guadalupe fur seal unusual mortality event in California. https://www.fisheries.noaa.gov/national/marine-life-distress/2015-2019-guadalupe-fur-seal-unusual-mortality-event-california. Accessed 25 Apr 2018

Norris T, Fauquier D, Field C et al (2017) At-sea distribution and habitat use of free-ranging and rehabilitated Guadalupe fur seals. The 6th International Bio-logging Science Symposium, Konstanz, Germany

Ortega-Ortiz CD, Vargas-Bravo MH, Olivos-Ortiz A et al (2019) Guadalupe fur seal encounters in the Mexican Central Pacific during 2010-2015: a dispersion pattern related to the species recovery? Aquat Mamm 45(2):244–252. https://doi.org/10.1578/AM.45.2.2019.246

Repenning CA, Peterson RS, Hubbs CL (1971) Contributions of the systematics of the southern fur seals, with particular reference to the Juan Fernandez and Guadalupe species. In: Burt WH (ed). Antarctic Pinnipedia, Antarct Res Ser 18:1–226

Seagars DJ (1984) The Guadalupe fur seal: a status review. National Marine Fisheries Service, Southwest Region, 29 p

Seguel M, Nadler S, Field C et al (2018) Vasculitis and thrombosis due to the sea lion lungworm, *Parafilaroides decorus*, in a Guadalupe fur seal (*Arctocephalus philippii townsendi*). J Wildl Dis 54:638–641. https://doi.org/10.7589/2017-12-291

SEMARNAT-NOM-059 (2010) Protección ambiental-Especies nativas de México de flora y fauna silvestre. Categorías de riesgo y especificaciones para su inclusión, exclusión o cambio. Lista de especies en riesgo. Norma oficial mexicana. Diario Oficial de la Federación

Smith KF, Acevedo-Whitehouse K, Pederson A (2009) The role of infectious diseases in biological conservation. Anim Conserv 12:1–12

Townsend CH (1931) The fur seal of the California islands with new descriptive and historical matter. Zoologica 9:442–457

Trillmich F, Ono KA (1991) Pinnipeds and El Niño: responses to environmental stress. Springer, Berlin

Walter PL, Craig S (1979) Archaeological evidence concerning the prehistoric occurrence of sea mammals at Poit Bennet, San Miguel Island, California. Fish and Game 65:50–54

Weber DS, Stewart BS, Lehman N (2004) Genetic consequences of a severe population bottleneck in the Guadalupe fur seal (*Arctocephalus townsendi*). J Hered 95(2):144–153. https://doi.org/10.1093/jhered/esh018

Ziehl-Quirós EC, García-Aguilar MC, Mellink E (2017) Colony-level assessment of *Brucella* and *Leptospira* in the Guadalupe fur seal, Isla Guadalupe, Mexico. Dis Aquat Org 122:185–193. https://doi.org/10.3354/dao03073

Chapter 6
South American Sea Lion (*Otaria flavescens*, Shaw 1800)

Enrique Alberto Crespo, Larissa Rosa de Oliveira, and Maritza Sepúlveda

Abstract The South American sea lion is distributed almost continuously from Zorritos, Peru, on the Pacific Ocean to Torres, Brazil, on the Atlantic Ocean. The total population was estimated at 445,000 individuals, and the species is considered as *Least Concern* in the IUCN Red List. Two evolutionarily significant units were recognized based on molecular data, a Pacific and an Atlantic population, with at least two management units in each ocean basin. The species is polygynous and sexually dimorphic, with males being larger and heavier than females. Reproduction occurs during the austral summer, between mid-December and early February. Longevity is close to 20 years for both sexes in free and captive life. The sea lion is an opportunistic species that predates mainly on pelagic, demersal, and benthic species of fish, squids, and crustaceans, some of them of commercial importance. Sea lions were used by aboriginal people as far as 8000 YBP and more recently by Europeans and local inhabitants until the first half of the twentieth century, mainly for food, leather, and oil. After the cease of hunting, the current major threat for sea lions has been the interaction with all types of fisheries and salmon farms. These include the huge extraction of fish biomass of which many target species are prey of sea lions. Fishermen claim across the entire distribution of the species that sea lions are their competitors, which cause them significant economic loss. Environmental education and sea lion watching tourism at haul-out sites are recommended in order to promote conservation of the species.

E. A. Crespo (✉)
Laboratorio de Mamíferos Marinos, CESIMAR-CCT CENPAT-CONICET, Puerto Madryn, Argentina
e-mail: kike@cenpat-conicet.gob.ar

L. R. de Oliveira
Mammal Ecology Lab, Universidade do Vale do Rio dos Sinos, São Leopoldo, Brazil
e-mail: larissaro@unisinos.br

M. Sepúlveda
Facultad de Ciencias, Universidad de Valparaíso, Valparaíso, Chile
e-mail: maritza.sepulveda@uv.cl

G. Heckel, Y. Schramm (eds.), *Ecology and Conservation of Pinnipeds in Latin America*, https://doi.org/10.1007/978-3-030-63177-2_6

Keywords Distribution · South American sea lion · *Otaria flavescens/byronia* · Population dynamics · Exploitation · Genetic structure · Foraging ecology · Threats

Introduction

The scientific name of the South American sea lion has a long history of controversy, and the epithet of the species still remains the subject of debate, because the holotypes were lost during the Second World War. The South American sea lion has been referred to as *Otaria flavescens* (Shaw 1800) or *Otaria byronia* (De Blainville 1820), but the former one is particularly used by South American scientists and will be followed by the authors of this chapter. The descriptions of both holotypes have problems. The type locality of *O. byronia* was the Tinian Islands (Marianas Archipelago), where the species does not occur. Shaw (1800), based on Pennant (1793), described the *O. flavescens* holotype as a small specimen of an otariid collected in the Strait of Magellan, probably a young individual, which lacks typical diagnostic traits of adult sea lions.

A huge and long-standing backstage discussion about the specific name of the South American sea lion, genus *Otaria*, took a new breath with the publications of Case 3058 and also by Opinion 1962 by the Bulletin of Zoological Nomenclature – Gardner and Robbins (1999) and ICZN (2000), respectively. In Case 3058, the International Commission of Zoological Nomenclature, which is the commission that moderates every nomenclature debate in zoological taxonomy, received the suggestion by Gardner and Robbins (1999) that *Otaria byronia* (De Blainville 1820) is a valid name. Later, on 1 March 2000, the members of the Commission were invited to vote on the Case 3058 proposal published in the Bulletin of Zoological Nomenclature (BZN 56: 139–140). At the close of the voting period on 1 June 2000, its result was published as Opinion 1962 in BZN 57:93–95, with 20 votes in favor of the designation *Otaria byronia*, one negative, one abstention, and two votes not received (ICZN 2000) (for an explanation of the procedure, see BZN 54:53–54). Under the plenary power, the Commission placed the name *byronia* on the Official List of Specific Names in Zoology as published in the binomen *Phoca byronia* (first available subjective synonym of *Phoca leonina* Molina, 1782, a junior primary homonym of *P. leonina* Linnaeus, 1758), the type species of *Otaria* Péron, 1816 (ICZN 2000).

Many researchers in South America did not agree with this decision of the International Commission of Zoological Nomenclature and still use the specific name *flavescens* (e.g., Cabrera 1940, Vaz-Ferreira 1982, Crespo 1988, Rodriguez and Bastida 1993), although the International Commission of Zoological Nomenclature rules that *byronia* is the valid name. During the last South American Meetings of Aquatic Mammals (*Reunião de Trabalho de Especialistas em Mamíferos Aquáticos da América do Sul*) that occurred after the Opinion 1962 publication (from 2002 to 2018), it is possible to see in the submitted abstracts that researchers are divided; some used the name *byronia*, while others used *flavescens*. Gardner and Robbins' (1999) proposal was based on a poor review, ignoring the most important works about the specific name *flavescens*, such as King (1978), Oliva (1988), Rodriguez and Bastida (1993), and also Rice (1998). Oliva (1988) analyzed the description of the *O. flavescens* holotype collected by Pennant (1793), and she supported the epithet *O. byronia*, since the author attributed this holotype to a neonatal specimen from the Arctocephalinae subfamily based on "…(1) the total length of the holotype; (2) the size of the external ear; and (3) the colour and length of the specimen's fur…." On the other hand, Rodriguez and Bastida (1993), analyzing the same traits of Shaw's holotype, concluded that they fit with the features of a South American sea lion pup.

The International Commission of Zoological Nomenclature, as we said before, is the ultimate international organization that rules, evaluates, and judges worldwide cases on zoological nomenclature and is followed by every biologist that practices zoological taxonomy. In addition, this committee had a democratic position opening the discussion, but no different official opinion was sent between the publishing of Case 3058 and Opinion 1962. In this context, as South American researchers and knowers of this problematic, we are responsible now to officially reopen this discussion, presenting valid information and arguments in favor of the name *flavescens* in order to change the International Commission of Zoological Nomenclature decision. Otherwise we shall resign and use correctly the name *byronia* according to the International Commission of Zoological Nomenclature.

Very recently, Lucero et al. (2019) proposed a neotype for *Otaria flavescens*, since *O. byronia* has a wrong locality, and theoretically the authors solved the taxonomic discussion on the name. However, they reopened the discussion of differences between Pacific and Atlantic populations, but without mentioning molecular or morphological studies to support it.

Geographic Range

The South American sea lion (*Otaria flavescens*) is widely distributed on both Atlantic and Pacific coasts of South America, occurring almost continuously from southern Brazil to Cape Horn in the Atlantic coast and from north up to the west coast of the continent to Zorritos, northern Peru, in the Pacific coast (Vaz-Ferreira 1982; Crespo 1988; Crespo et al. 2012a) (Fig. 6.1). On the Atlantic side, they also occur in the Falkland (Malvinas) Islands (Cárdenas-Alayza et al. 2016). Some specimens have been seen as far north as Río de Janeiro in southeast Brazil (Vaz-Ferreira 1982; Pinedo 1990; Rosas et al. 1993). However, breeding does not occur in Brazil; therefore a seasonal displacement of individuals, mainly males, moves from the breeding colonies in Uruguay after their breeding period (Pinedo 1990; Rosas et al. 1994; Oliveira 2013). The northernmost breeding rookery in the Atlantic Ocean is in Uruguay at Isla Verde and Isla La Coronilla (33°56′S 53°29′W), east of Cabo

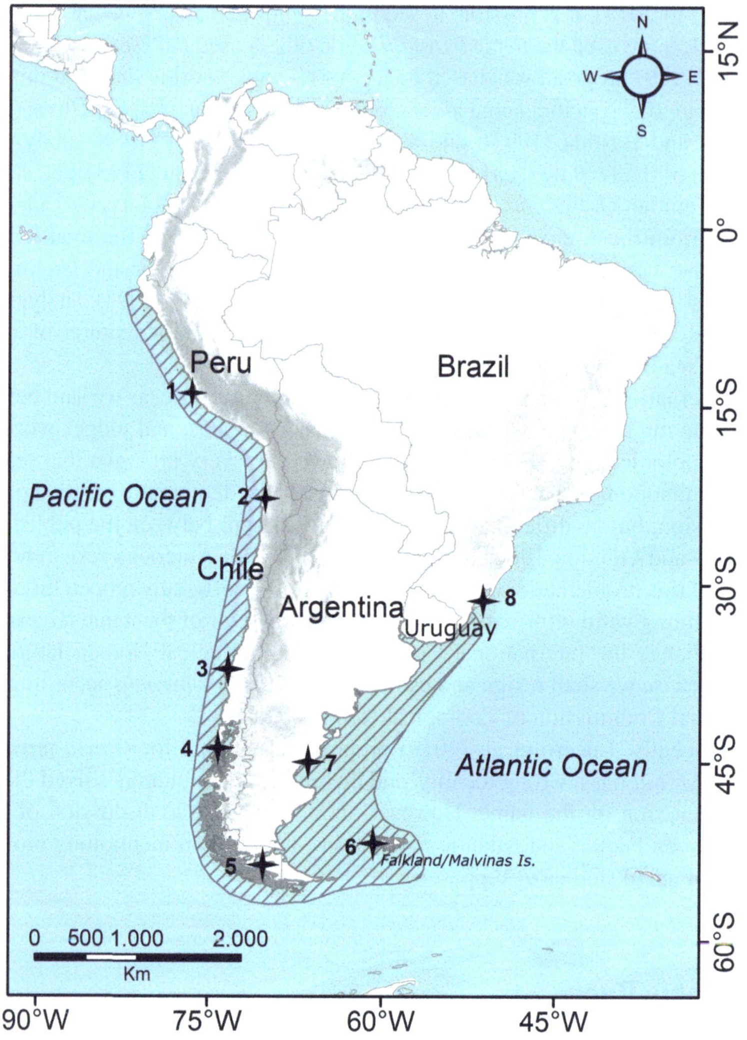

Fig. 6.1 Geographic distribution of South American sea lions. Numbers correspond to sampling locations for genetic structure analysis (see text)

Polonio (34°24′S) (Vaz-Ferreira 1975). On the Pacific coast, the northernmost breeding location is Zorritos, Peru (03°40′S; Crespo et al. 2012a), although some individuals have been found in Ecuador and Colombia, and even forming small rookeries (Félix et al. 1994; Capella et al. 2002; Cárdenas-Alayza et al. 2016).

On both the Atlantic and Pacific coasts, the colonies show an almost continuous distribution with the exception of the coast of Buenos Aires Province (Valette 1928). The lack of breeding colonies in Buenos Aires Province appears to be related to the large-scale pattern of human settlement that occurred at the end of the nineteenth century. Nevertheless, an aggregated number of colonies are located in the vicinity of Península Valdés, the northern coast of Golfo San Jorge, northern Santa Cruz province, and Tierra del Fuego and Isla de los Estados, where high primary and secondary productivity are found.

South American sea lions are found in waters over the continental shelf and slope. Males are more oceanic and follow fishing vessels as long as 100 nautical miles. They can travel more than 320 km from the coast (Campagna et al. 2001; Crespo et al. 2007; Hückstädt et al. 2014; Cárdenas-Alayza et al. 2016), as well as along the coast (Giardino et al. 2016; Hevia 2013; Sepúlveda et al. 2015), suggesting that they have a main role in the gene flow among colonies (Feijoo et al. 2011; Oliveira et al. 2017). This species also concentrates near continental waters and rivers (Schlatter 1976). Females instead move along and very close to the coast in small groups, with pups and juveniles (Crespo 1988), although some females traveled more than 100 km from the coast in an east-west direction (Hevia 2013).

History of Exploitation and Recovery

The South American sea lion is the most conspicuous marine mammal along the South American coast, where it was heavily exploited between the 1930s and 1960s. As a consequence of this exploitation, many of its populations were decimated during the early and mid-twentieth century. However, Patagonian rookeries showed a clear recovery in the last few decades. These rookeries represent approximately 72% of the species' abundance in the Atlantic Ocean. An understanding of the underlying processes and comprehensive history of population growth after a harvest-driven depletion was necessary when assessing the long-term effectiveness of management and conservation strategies. The interest was focused in how the population responded at low densities, how human-induced mortality interplays with natural mechanisms, and how density dependence may have regulated population growth. The observed population trajectory of South American sea lions in northern and central Atlantic Patagonian coast showed a nonlinear relationship with density, recovering with a maximum increase rate of 5.5%. However, 50 years after hunting cessation, the population still represents only 40% of its pre-exploitation abundance with the recovery process occurring at the same time as increasing development of human coastal activities (Romero et al. 2017). The huge amount of biomass extracted by the fishery since the 1990s today precludes the South American sea lions reach their former carrying capacity (Drago et al. 2010a; Romero et al. 2017).

The responses to this population decline have been different in diverse regions of the species' distribution range. Some populations are still decreasing (e.g., in Uruguay), some remain stable (e.g., central Chile), and some are slowly recovering (e.g., Atlantic Patagonia and southern Chile). In the southwestern Atlantic, with the exception of Uruguay, sea lion populations are now increasing. There was an increase in the number of sea lions in all colonies and a change in the social composition of many colonies, but no new breeding colonies were found so far. The population trajectory in southern Patagonia was similar to that observed in the rest of the South Atlantic stocks, but the recovery and recolonization processes are still in progress (Grandi et al. 2015).

Population recovery was not immediate after hunting ceased in 1962. The population of the species was stable until 1989 (Crespo and Pedraza 1991) and since then has grown at an annual rate of increase of 5.7% (Dans et al. 2004) (Fig. 6.2). Along with this growth, there was an increase of the juvenile fraction and changes in the social composition of colonies, which could be related to changes in some population parameters. Breeding rookeries are mainly composed of females and pups (around 80%), with the remaining specimens including adult and subadult males and juveniles of both sexes. During the 1980s, there was an increase of new colonies with a different social structure to that of traditional sites (Crespo 1988; Grandi et al. 2008). New groups, composed mainly of juveniles, split from traditional breeding colonies and settled a few kilometers away. These groups differ from traditional sites in the sense that they have higher rates of increase, different age, size, and social structure and a higher mortality rate of newborn pups. Juveniles grow and become sexually mature and reproduce within a different social structure. Newborn pups are exposed to frequent contact with subadult males, which carry out them with high risks and mortality. The rate of increase of new colonies is always over 10%, reaching almost 20% with immigration. By contrast, rates of increase in traditional colonies are never higher than 3–4%. Higher mortality in new groups may be explained by means of a higher potential for infanticide by solitary or subadult males (Campagna et al. 1988b) and less experience of younger females in nursing pups.

The global recovery of the population is probably mediated by an increase in the juvenile fraction and other changes in the social composition of colonies which

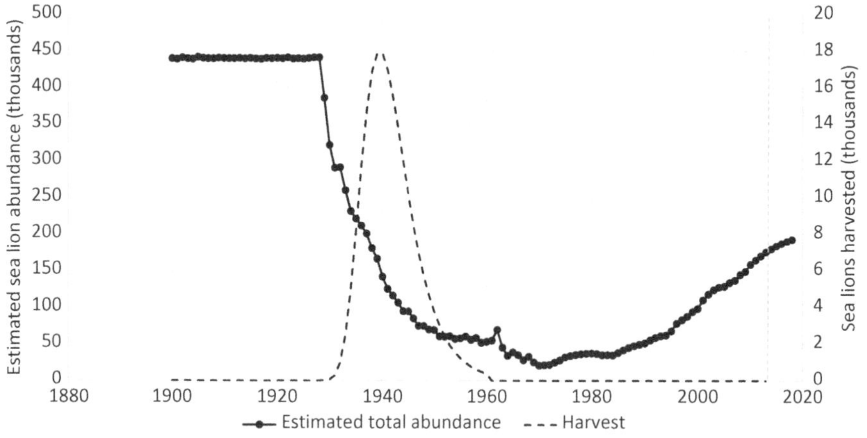

Fig. 6.2 Population trajectory and number of South American sea lions harvested in northern and central Patagonian Atlantic coast

likely relate to changes in population vital rates like survivorship. The comparisons of survivorship curves of males and females, obtained from the life tables of two periods with different population trends, 1981–1987 (stationary) and 2000–2008 (recovering), showed that there were differences in survivorship between sexes, where recent male and female age-specific survival was higher than that of males at any age. Both juveniles and adults, both male and female, from the recent period showed higher survival than those of the 1980s. This improvement in the survivorship could be one of the essential factors that drove population recovery in the last decades. The survivorship curves were more a Type II curve during the 1980s and transformed gradually to a Type I curve in the 2000s, more typical of a K-strategist population. The process may continue at present days. In fact, survivorship of mature males increased to 10% and for mature females 20% (Grandi et al. 2016). The reason of this fact may be that the harvest-selected individuals skewed in age and sex of the age structure of the population. The result was an age structure far from the original one of the species. The absence of harvest rates and time is recomposing the original proportions of age and sex classes.

On the Pacific coasts of Chile and Peru, there were similar events. The first antecedents found in Chile about the hunting of sea lions date back to the colonial period, where some chroniclers who toured the national coasts and islands were amazed by its abundance and the use that the aborigines made of their meat, oil, and skins. An example of these stories is the one made by Rosales (1877) (cited by Albert 1901) in General History of the Kingdom of Chile; Indonesian Flanders, who in the chapter referring to "Sea beasts, and other beasts that live in the sea and land called amphibian animals or doubtful," made comments on the behavior and feeding and kind number of sea lions throughout the coast of the country, and the trade started in Lima with their skins, which were used to make saddles, hats, trunks, and other gear and decorations of great value. During the eighteenth century and the beginning of the following century, the exploitation of sea lions in the seas and coasts of Chile was carried out by vessels from North America and Europe, many of which, according to Medina (1952), tried to obtain permits from the authorities of the country for fishing and hunting. Around 1860 their capture was taken over by Chilean catchers, active until 1907, when the Chilean government forbade it (Sielfeld 1999). Exploitation was restarted in 1976, mostly directed to pups, with variable regional quotas. The first specific protection measure for the South American sea lion was issued in 1953, but only partially, since its closure included a part of the country's coastline, remedying that situation in 1966 by including the entire coastline of the country (Torres et al. 1999).

Population Figures and Trends at a Global Scale

The South American sea lion is the most abundant marine mammal occurring along the Pacific and Atlantic Ocean basins (Crespo et al. 2012a). The population in Uruguay is concentrated at Isla de Lobos (35°01′S) and Cabo Polonio (34°24′S) and is estimated roughly at 12,000–13,000 individuals, with an average of 1500 pups born every year (Páez 2006; Franco-Trecu 2015). On the northern coast of Argentina, there are only four haul-out sites (about 2500 individuals), while the Patagonian region has both reproductive and non-reproductive colonies (about 120,700 individuals). In addition, 7500 animals are found in the Falkland (Malvinas) Islands (Crespo et al. 2012a). Baylis et al. (2015) reported a minimum estimate of 4443 pups born at the Falklands/Malvinas in 2014. A recent survey conducted in 2019 by Oliva and colleagues indicates that the total population of the South American sea lion is estimated at 128,000 individuals along the entire Chilean coast, only excluding the Magallanes Region. The 2019 census covered almost 3400 linear kilometers of coastline (Oliva et al. 2020). There are at least 105,000 individuals in Peru (IMARPE 2013; Cárdenas-Alayza et al. 2016), and no more than 200 on the Brazilian coast (Sanfelice et al. 1999; Pavanato et al. 2013). Therefore, the total global population is at least 445,000 individuals (Cárdenas-Alayza et al. 2016).

Population trends on the Atlantic coast are contrasting. While most of the populations increase in abundance, the Uruguayan has been decreasing. Negative trends for both sexes and all age classes of the breeding population were reported by Páez (2006), with −1.4% per year for adult males, −2.1% for adult females, and −4.5% for pups. Results from population modeling by Páez (2006) showed a 2% decline per year for total population size and a 3% decline in birth rates. This data coincides with recent findings by Franco-Trecu (2015), who estimated a −2% (CI −1.1% to −2.5%) decline in pup production using pup count data from 1956 to 2013. Although the reasons for the population decline in Uruguay are still unknown, it is suspected that it could be related with potential mortality related to local fishery activity (Crespo et al. 2012a; Riet-Sapriza et al. 2013). Other possible causes that have been mentioned are the long-term effects of harvest, the cumulative effects of population extractions, pup harvesting (~50,000), sales for zoos and aquaria (144 young males and 285 young females), and small population size. All these mechanisms could have disrupted the social structure of the Uruguayan population to the point where Allee effects could be limiting the post-harvesting population recovery at Isla de Lobos (Franco-Trecu et al. 2015).

On the other hand, numbers are increasing in most of the species' range. In northern Patagonia, Rio Negro (41°03′S), and northern Chubut (43°34′S) provinces, the rate of increase is around 5.7–6% per year (Dans et al. 2004). In central (43°57′S) and southern Chubut (45°23′S) province, the numbers of South American sea lions are also increasing at a rate of 6% (Reyes et al. 1999) and around 8% in Santa Cruz and Tierra del Fuego (Milano et al. 2020). At the Falkland (Malvinas) Islands, there was a 95% decline in the population from >380,000 to <30,000 animals (from 80,555 pups in the mid-1930s to 5506 pups in 1965; Hamilton 1939; Strange 1979). The number of pups estimated in 2014 for the Falkland (Malvinas) Islands was 6% that estimated in the 1930s (Baylis et al. 2015). Different hypotheses have been proposed to explain this decline, including commercial sealing and environmental changes (Strange 1979; Thompson et al. 2002; Baylis et al. 2015). However, the trend has been positive since 1990, with an 8.5% annual increase from 1990 to 1995 and a 3.8% annual increase between 1995 and 2003 (Crespo et al. 2012a).

On the Pacific coast, South American sea lion population trends along the Chilean coast are not homogenous. In northern and central Chile, the populations are stable, whereas the numbers are increasing in southern Chile (Oliva et al. 2020). Between the latitudes 15°56′S and 48°40′S, the population increased from 137,000 to 197,000 in only 7 years (Oliva et al. 2012; Contreras et al. 2014). For the Magallanes Region, the population trend is still unknown because only one survey has been conducted in that area (Venegas et al. 2002).

Due to the 1997–1998 El Niño-Southern Oscillation (ENSO), the Peruvian population of South American sea lions drastically declined from 144,087 in December 1997 to 27,991 animals in December 1998, a reduction of 81% (Arias-Schreiber and Rivas 1998). This was probably due to a combined effect of mortality and dispersal from historically surveyed breeding and haul-out sites. After this dramatic reduction, there was a recovery of 76.3%, with an estimated 118,220 individuals by 2006 (IMARPE 2006; Cárdenas-Alayza et al. 2016). The recovery of this population of sea lions on the coast of Peru was probably due to improved reproductive levels as a consequence of an increase in food availability as well as migration from northern Chile breeding colonies (Oliveira et al. 2012). However, stronger and more frequent ENSO events appear to be occurring along the Peruvian coast, and they may put the population in Peru at greater risk (Soto et al. 2004). The majority of subpopulations in the southwestern Atlantic Ocean are increasing, although the trends are not homogeneous. However, in contrast to what is observed on the Peruvian Pacific coast, the population sizes do not show large inter-annual fluctuations (Crespo et al. 2012a).

Genetic Structure of South American Sea Lion Populations

The analysis of population genetic structure and the evolutionary history of the species along its distribution were carried out by the analyses of mitochondrial DNA (mtDNA) and ten nuclear microsatellite loci including the entire species distribution, represented by its six main populations (Peru, northern Chile, southern Chile, Uruguay (including Brazil), Argentina, and Falkland (Malvinas) Islands) (Fig. 6.1). Sampled localities were grouped into the following studied areas: Peru – Punta San Juan (1); northern Chile – Punta Negra and Punta de Lobos (2); southern Chile – between Ritoque Beach and Isla Mocha (hatched lines 3), Guafo Island (4), and Punta Carrera (5); Falkland (Malvinas) Islands (6); Argentina (hatched lines 7) – Argentine central coast; and Uruguay (hatched lines 8) – southern Brazilian coast (Oliveira et al. 2017).

The mtDNA phylogeographic reconstruction of the sequences of populations of South American sea lions showed that haplotypes from the two ocean basins are two very divergent lineages, with around one million years of limited or low inter-oceanic female gene flow, due to high female philopatry. However, Bayesian analysis of biparental nuclear loci supported a less pronounced genetic structure than mtDNA, between Pacific and Atlantic populations, suggesting some level of inter-oceanic gene flow mediated by males. Higher male migration rates were found in the intra-oceanic population comparisons, and management units were proposed to the populations that share the same ocean basin, which implicates in future integrated conservation actions. Demographic analyses of molecular data showed that populations from both oceans went through a large population expansion ~10,000 years ago, suggesting a very similar influence of historical environmental factors, such as the last glacial cycle, in both regions. Results supported the proposition that the Pacific and Atlantic populations of the South American sea lion should be considered distinct evolutionarily significant units, with at least two management units in each ocean (Oliveira et al. 2017).

At the regional level, several major breeding areas have been defined along the Atlantic Ocean including the Uruguayan and Patagonian coasts. Together with a documented and severe reduction in population sizes caused by commercial hunting in the last century, these areas show opposite population trends. While Patagonian populations were recovering since hunting ceased, Uruguayan populations were declining (Feijoo et al. 2011). In this context, population genetic structure and genetic diversity were studied using both nuclear (microsatellites) and mitochondrial (control region) markers together. Alternative scenarios were found for both markers. While the mitochondrial marker showed geographically structured populations between Uruguayan and Patagonian populations, the nuclear loci showed a lack of geographical structure. These opposite patterns in genetic structure could be explained by female philopatry and high male dispersion. The reduction in population size caused by commercial hunting did not leave a detectable genetic bottleneck (Feijoo et al. 2011).

Effective Population Size Estimated for Northern Patagonia

Effective population size (Ne) is a parameter of central importance in evolutionary biology and conservation (Wright 1931; Sinclair et al. 2006). The effective population size (Ne) reflects the number of individuals responsible for the maintenance of the genetic diversity of the species as well as its evolutionary potential, so it can be used as a way to determine the conservation status of a species or population. It may be defined as the number of breeding individuals that an idealized population would show the same amount of dispersion of the allele frequencies under random genetic drift or the same amount of inbreeding as the population under consideration (Wright 1931; Sinclair et al. 2006). Factors such as unequal sex ratios of breeding individuals, periodic fluctuation in population size, and variance in reproductive success can affect the Ne in general. At present, South American sea lions from northern Patagonia, Argentina, belong to one of the several populations that are recovering from overhunting which occurred in the early twentieth century. Ne for this population takes into account the effects of their polygynous mating system and variation in population size through time. The resultant overall Nes were 4171 ± 2450 or 4745 ± 2681 breeding animals depending on the inclusion or not of peripheral adult males (Grandi et al. 2012a). The estimated Nes are not critical, because they are close to the mean minimum viable population for vertebrates (5000 breeding adults) (Reed et al. 2003, Traill et al. 2007). Even though the northern Patagonian population is increasing, its abundance is still far below its historical numbers (Romero et al. 2017). The Ne estimated should be considered the minimum range to be maintained, especially in light of bycatch related to fishery interactions along the Patagonian coast which has been estimated between 1 and 2% of the total population size (Crespo et al. 1997).

Feeding Habits

Food habits of the South American sea lion have been studied all along the distribution range. In early years the studies were based on stomach content analysis or scats; more recently they were based on stable isotopes and fatty acids.

The first study of diet in northern and central Patagonia was carried out by means of stomach content analysis. The samples came from individuals found dead on beaches and from animals recovered in incidental catches of the fishery. Forty-one prey species (including fish, cephalopods, crustaceans, gastropods, polychaetes, sponges, and tunicates) were identified; the most important were Argentine hake (*Merluccius hubbsi*), red octopus (*Enteroctopus megalocyathus*), Argentine shortfin squid (*Illex argentinus*), "raneya" (*Raneya brasiliensis*), Patagonian squid (*Loligo gahi*), and Argentine anchovy (*Engraulis anchoita*). All preys with the exception of "raneya" were commercial species. Differences in diet were found between sexes but not between sampled geographical areas or periods, nor source of samples. Females fed mostly on benthic species, whereas males fed mostly on demersal-pelagic species (Koen Alonso et al. 2000). The difference in diet between sexes was associated with different feeding grounds or different home ranges and could be produced by different constraints in the feeding behavior of each sex. These different constraints and restrictions could lead females to feed in more coastal and shallower waters than those waters where males feed. Some of the important prey were commercial species (Argentine hake, Argentine shortfin squid, Patagonian squid) consumed at both commercial and noncommercial sizes by sea lions.

In Uruguay and Buenos Aires province, the most abundant prey of sea lions included cephalopods like the whitemouth croaker (*Micropogonias furnieri*), large head hairtail (*Trichiurus lepturus*), Brazilian codling (*Urophysis brasiliensis*), Argentine croaker (*Umbrina canosai*), and striped weakfish (*Cynoscion guatucupa*) and squids of the family Ommastrephidae (Riet-Sapriza et al. 2013). On the southern Brazilian coast, sea lions predate mainly on fish from the Sciaenidae family such as *Macrodon ancylodon*, *T. lepturus*, *Paralonchurus brasiliensis*, *Pomatomus saltatrix*, and *C. guatucupa*, with almost no overlap with the target species by the local fishery (Oliveira et al. 2008). However, an increase in prey overlap between sea lions and fisheries was observed between 2004 and 2014, along with an increase in prey diversity, richness, and niche breadth of the sea lions' diet (Machado et al. 2018).

On the Pacific coast of Peru, sea lions prey mostly on anchoveta (*Engraulis ringens*), mote sculpin (*Normanichthys crockeri*), lumptail searobin (*Prionotus stephanophrys*), Peruvian hake (*Merluccius gayi*), red squat lobster (*Pleuroncodes monodon*), and cephalopods (family Loliginidae) (Paredes and Arias-Schreiber 1999). In Chile, temporal and spatial diet plasticity was found by Muñoz et al. (2011) and Guerrero et al. (2020). In northern Chile the main prey species for South American sea lions are anchoveta, Patagonia squid (*Loligo gahi*), Cabinza grunt (*Isacia conceptionis*), and corvina (*Cilus gilberti*) (Sielfeld et al. 2018; Guerrero et al. 2020; Sarmiento et al. 2020). In central Chile the main prey are South Pacific hake (*Merluccius gayi gayi*), snoek (*Thyrsites atun*), and Araucanian herring (*Strangomera bentincki*) (George-Nascimento et al. 1985; Muñoz et al. 2011), whereas in southern Chile, the main species are southern hake (*Merluccius australis*) and snoek. In southern Chile farm-raised salmonids are also important in the diet, suggesting that South American sea lions are capable of modifying their dietary habits in response to variation in abundance and/or accessibility of prey (Muñoz et al. 2013; Sepúlveda et al. 2015, 2017; Guerrero et al. 2020). Interestingly, a recent study using blubber fatty acids shows males displayed a more diverse fatty acid composition than females, suggesting a wider trophic niche and consequently a more diverse range of prey items (Guerrero et al. 2020).

In summary, diet analyses indicated that South American sea lions feed primarily on demersal and benthic species although in northern Chile they consume pelagic species. Therefore, in general terms, they use resources according to their environmental availability. Nevertheless, all these studies were done at a regional and wide scale. Diet on a local scale is relatively scarce. Search for evidence of geographical variations in the diet of sea lions was studied in Golfo San Matías (GSM) and Golfo Nuevo (GN), localities which are separated by 250 km. These bays are quite different regarding their physiography, oceanography, and anthropic pressures. The dietary composition differed between the two gulfs with respect to the most common species, the zoological groups, and the sizes of the primary prey. In GSM, sea lions fed on demersal-pelagic fish, while in GN the diet exhibited a high contribution of benthic species (Jarma et al. 2019). Differences in the items and sizes of prey consumed at each area enforce the idea of the use of differential prey availability along with the opportunistic feeding behavior of this predator. These differences, in addition to the species' main prey as targets for major fisheries, must be taken into account in order to understand the different population dynamics of this apex predator and thus ensure its conservation.

At this point it is important to consider it would be interesting to relate diet composition with foraging behavior. In Chile, for example, South American sea lions instrumented in northern Chile show a mean and maximum diving depth of 17 and 120 m, respectively, whereas individuals instrumented in Southern Chile show a mean and maximum diving depth of 75 and 325 m, respectively (Hückstädt et al. 2016). Variation in medium and maximum diving depth is mostly associated with the shallow dives of animals from northern Chile compared with animals from southern Chile. Likewise, the diving behavior of the sea lions varied between the sites. Individuals from northern Chile are largely epipelagic as they did not dive deeper than 200 m (boundary between epi- and mesopelagic), using the upper 10% of the water column. Deep-diving (> 200 m) sea lions from southern Chile displayed an intermediate level in their diving behavior (using 35.5% of the water column, respectively), indicating a mesopelagic and benthic behavior (Sepúlveda et al., unpublished data).

Stable Isotopes in the Study of Sea Lions

Traditional trophic ecology was based mainly on stomach content or scat analysis or alternative washing stomachs. However, all these alternatives consider only the last or few last meals. Instead, stable isotope analysis has become a valuable tool for the study of marine mammal ecology, allowing the reconstruction of dietary changes through longer periods of time depending of the tissue analyzed. Serum integrates information on the diet of the last week, blood cells for 1–2 months, as well as skin samples, while bone collagen for several years (Newsome et al. 2010). Stable isotope ratios of N and C are based on a predictable relation between the isotopic composition of a consumer and its diet (DeNiro and Epstein 1978, 1981). For aquatic organisms, $\delta^{15}N$ values can provide data on a trophic level, while $\delta^{13}C$ values can reveal information on feeding locations, including the relative use of benthic vs. pelagic and coastal vs. oceanic prey (Hobson et al. 1994).

During the last 15 years, the use of stable isotopes has been implemented in several studies about the ecology and feeding habits of South American sea lions. The first study used carbon and nitrogen isotopic analyses of skull bone to investigate how sealing and the development of industrial fishing have affected the diet of the sea lions in northern Patagonia (Drago et al. 2009a). The main conclusion of this study was that diet could vary enormously with the abundance of the main predator. In lower abundance condition, the animals feed close to the coast and on coastal prey like octopuses, while at higher abundances, the animals move far from the coast in search of demersal and benthic prey like common hake (Drago et al. 2009a). Stable isotopes of carbon and nitrogen in the skull bones were also used to determine whether their feeding habits changed during ontogeny. The analysis showed that $\delta^{13}C$ steadily increased in males and females with their developmental stage (young, first adult, adult, and senile), except in senile males whose $\delta^{13}C$ decreased to a value close to that of first adults (Drago et al. 2009b). The hypothesis that female sea lions shift from offshore, pelagic prey to coastal, benthic prey after parturition was also tested (Drago et al. 2010b). This happens in order to reduce the foraging trip duration and hence the time pups remain unattended on the beach during early lactation, the stage at which the pups suffer of infanticide by subadult males (Campagna et al. 1988b). Isotopic ratios revealed a generalized increase in the consumption of coastal-benthic prey after parturition (Drago et al. 2010b).

Stable isotopes in bone collagen were also used to reconstruct and compare the isotopic niches of adult sea lions and South American fur seals (*Arctocephalus australis*), two sympatric otariid species with marked morphological differences in the Rio de la Plata estuary and the adjacent Atlantic Ocean during the second half of the twentieth century and the beginning of the twenty-first century (Drago et al. 2017). Samples from the middle Holocene were also included in order to provide a reference point for characterizing resource partitioning before major anthropogenic modifications of the environment. It was found that fur seals and sea lions had distinct isotopic niches during the middle Holocene. Isotopic niche segregation was similar at the beginning of the second half of the twentieth century but has diminished over time. The progressive convergence of the isotopic niches of these two otariids during the second half of the twentieth century and the beginning of the twenty-first century is most likely due to the increased reliance of South American fur seals on demersal prey (Drago et al. 2017).

Change in the isotopic signal was also analyzed throughout the food web from the La Plata River to southern Patagonia in a time scale of 5000 YBP for three top predators: the South American sea lion, the South American fur seal, and the Magellanic penguin. It was found that ancient food webs were shorter, more redundant, and more overlapping than the current ones, both in northern-central Patagonia and southern Patagonia (Saporiti et al. 2014). After the sealing period, pinnipeds were and still are well below the ecosystem's carrying capacity, which resulted in a release of intraspecific competition and a shift toward larger and higher trophic-level prey (Saporiti et al. 2014).

It is interesting to see how plastic the feeding behavior is in the sense of inter-annual variations that have occurred in the composition of the sea lion diet. An example is the central zone of Chile, where sea lions changed from feeding on Chilean hake (George-Nascimento et al. 1985) to sardine (Muñoz et al. 2013) and then to jack mackerel (Hückstädt et al. 2014). Muñoz et al. (2013) found variations between tissues, as well as with what was recorded in previous studies, suggesting that this species is capable of adapting to variations, both intra- and inter-annual of the presence/absence of its prey. Not surprisingly, this temporal variation in diet composition is closely related to the exploitation levels of these resources by commercial fisheries in Chile.

It is also remarkable that stable isotopes have been key in estimating the importance of salmonids in the sea lion diet in southern Chile. This is because sea lions only feed on the soft parts, so these preys do not appear in studies of stomach or fecal contents (Sepúlveda et al. 2017). Stable isotopes have been also useful to show plasticity in the salmon consumption by sea lions, with a low consumption of this item in the last years, presumably related to an improvement in management procedures by the industry, including a much lower density of floating salmon pens (Niklitschek et al. 2013).

Predation

Predation on South American sea lions was documented including killer whales (*Orcinus orca*) (Hoelzel 1991; Grandi et al. 2012b; Hückstädt and Antezana 2004) and broadnose sevengill sharks (*Notorynchus cepedianus*) (Crespi Abril et al. 2004). At Punta Norte rookery on Península Valdés, killer whales are known to strand on the beach and take mainly newborn pups or juveniles swimming in the surf zone. Interactions between killer whales and sea lions and other marine mammals became more frequent in the last few decades and have been documented at sea by captains and fishing officers. These attacks increased at least five times during the last 40 years possibly related to the increase of the sea lion population (Laboratorio de Mamíferos Marinos, CENPAT-CONICET, unpublished information). The broadnose shark is known by its preference to prey on marine mammals, which includes in addition to sea lions, fur seals, elephant seals, dusky and Commerson's dolphins, and southern right whales (Crespi Abril et al. 2004).

Reproductive Biology, Development, and Behavior

Growth and age at sexual maturity of South American sea lions in the southwestern Atlantic were studied by means of the examination of 219 females and males collected between 1989 and 2008. Individuals were aged based on counting growth layer groups in tooth sections, standard body length measure, and macroscopically and histologically analyses of male and female reproductive organs to establish individual sexual maturity. Maximum recorded length for males and females was 264 cm and 200 cm, respectively, and maximum ages were 19 and 21 years (Fig. 6.3). Age at sexual maturity was defined as the age where 50% of females are mature, and it was estimated at 4.8 ± 0.5 years, corresponding to a mean standard length of 140.28 cm, about 80% of their asymptotic length. First observed ovulation occurred during the fourth year; first birth may occur between 4 and 5 years old. Males are physiologically mature between 4 and 6 years, but the size of the testes showed that all males became sexually mature by the age of 9 years when they reach a mean standard length of 212.19 cm, about 86% of their asymptotic body length (Grandi et al. 2012c), age at which they can hold females in harems (Crespo 1988).

Regarding reproductive behavior, males arrive at the breeding sites usually during the second half of December and set up territories, at least in Patagonian shores (Campagna 1985; Crespo 1988). After the male's arrival, females attempt to settle near established males in order to avoid peripheral males. Females give birth about 2–3 days after arriving on the rookery (Campagna and Le Boeuf 1988). Pups usually are born mostly on cephalic position. The placenta is expelled sometimes with the pup after parturition. Females do not eat the placenta (Fig. 6.4), which is disputed by gulls and other seabirds. Births occur from mid-December to early February (Campagna 1985), and the peak of newborn pups is usually during the

Fig. 6.3 Adult male, female, and pup of South American sea lion

Fig. 6.4 Parturition and placenta of female sea lion

second week of January. Some differences exist in the timing of the breeding season between Atlantic and Pacific populations. At Punta Negra rookery located in northern Chile, reproductive events extend from the third week of December until May, with a peak of newborn pups and mating late in January and mid-February (Acevedo et al. 2003). At Punta Lobería rookery in the central-south coast of Chile, the peak of births and mating occur between the last week of January and mid-February (Pavés et al. 2005).

Females daily nurse their pups until coming into estrus, usually 6 or 7 days after giving birth. Males use to sniff the females' vulva before mounting (Campagna and Le Boeuf 1988), and the peak of mating is 1 week after that of births. Once a male has mounted a female, he usually ends copulating with her. Most copulations take place on land in the breeding area, while the rest occur on the periphery. The mean duration of copulations is about 10 min (Campagna and Le Boeuf 1988). Around 2 days after mating, the females move out to sea to forage and leave their pups in the colony. Foraging trips last for 2 or 3 days and remain a similar time nursing their pups (Muñoz et al. 2011). Weaning takes place every year, but the male offspring are usually weaned earlier than the female offspring, which leads to a higher mortality (Crespo 1988; Grandi et al. 2016).

An alternative reproductive strategy is found in South American sea lions. Non-territorial males invade the breeding area, attempting to displace the resident males and establish in their territories and possessing their females or to abduct females from the territories (Vaz-Ferreira 1965, 1975; Campagna et al. 1988a). Raids are disruptive for the breeding animals, resulting in interference with copulations in progress, mother-pup separations, pup abductions, and sometimes with pup death (Vaz-Ferreira 1965; Campagna et al. 1988b). In addition, a lek-like breeding system was described for South American sea lions in Peru (Soto and Trites 2011). This one is a rare alternative to the male strategies in which they defend females or resources like ponds with water and is likely an evolutionary product of their highly skewed sex ratio, extended breeding season, and subtropical weather where they breed.

Threats, Conservation Status, and Management

The South American sea lion is the most conspicuous marine mammal in South American waters in the sense of abundance and interactions with human activities. It is mostly coastal and interacts with all kinds of fisheries and salmon aquaculture farming but also at a lesser degree with tourism. In the southwestern Atlantic and southeastern Pacific, the species interacts with trawling, purse-seine, gillnet, and longline vessels (for a review, see Crespo et al. 2012a). In the first two, it is incidentally killed as a consequence of entering in the nets for feeding; in the remaining two, it takes the catch or the bait but is rarely entangled (Fig. 6.5; Crespo et al. 1997, 2007).

Since the development of the trawling fishery for hake and shrimp in Patagonia and in central-southern Chile, during the 1970s, sea lions have drowned in nets.

Fig. 6.5 Subadult male caught alive in bottom trawler in northern Patagonia

Their opportunistic behavior of taking advantage of new food sources makes them vulnerable. In some cases the animals come out alive when retrieving the net, but it is more frequent they are accidentally killed by asphyxiation from remaining underwater for >45 min or by physical trauma of being crushed by the tons of catch entering the net during the trawl (Reyes et al. 2013). In Patagonian hake and shrimp trawls, mortality is roughly between 1 and 2% of the population size (Crespo et al. 1994, 1997), which is today well below a higher rate of increase. Sea lions are still increasing and have not yet reached the carrying capacity. Given the huge amount of biomass taken by the trawling and jigging fisheries, in which the main targets are also prey of sea lions, the current and future carrying capacity may be very much lower than the past one.

In Chile, incidental catch has been reported in purse-seine and trawling commercial fisheries. Hückstädt and Antezana (2003) reported a total of 20 sea lions that were caught or injured (0.64 sea lions per set) during purse-seine operations for jack mackerel in central Chile. For the trawling industrial fishery of Chilean hake in south-central Chile, Reyes et al. (2013) reported a much higher bycatch of sea lions, with a total of 82 individuals incidentally caught (1.2 sea lions per trawl). Interestingly, Hückstädt and Antezana (2003) mentioned that only juveniles were caught or injured, probably because of the inexperience of young animals in recognizing the risks. Besides industrial fisheries, bycatch and interaction between sea lions and artisanal fisheries have been reported in several areas of the country (e.g., Goetz et al. 2008, Bartheld et al. 2008, de la Torriente et al. 2010, Sepúlveda et al. 2018). In general, all of these studies coincide in interactions being more frequent at night, during the austral fall and winter months, caused by a small group of animals and that most of the animals that interacted with fishing gears were males. Also, most of these studies coincide that the interaction is below to the one perceived by fishers and that the catch per unit effort does not show a significant difference with or without the presence and number of sea lions interacting with the gear. Thus, interaction with South American sea lions is not a determining factor in explaining reduced catch by artisanal fisheries (Sepúlveda et al. 2018).

In southern Chile, in addition to industrial and artisanal fisheries, the South American sea lion interacts heavily with salmon farming. Sea lions typically bite salmon through the nets (cage and anti-predator nets), killing a variable number of fish. An animal attacks by pushing the anti-predator net (a net hanging from each side of the cage and beneath it) until it attaches to the cage net, allowing it to reach its prey. Attacks are usually from the bottom or sides of the nets (Sepúlveda and Oliva 2005). In addition to killing fish, sea lions sometimes rip the nets, liberating some or all of the salmon in the raft cage, with consequent ecological, economic, and social problems (Sepúlveda et al. 2013). Attacks are mainly concentrated during winter and spring months (Sepúlveda and Oliva 2005; Crespo et al. 2012b). This was confirmed by Vilata et al. (2010), who reported that the number of sea lions at night was double than during the day, behavior which may be related to the movement of sea lions among colonies and with the intensification of feeding during the non-reproductive period of the species. Chile attempted a number of measures that have been adopted to prevent sea lion attacks, like acoustic systems, sound generators above and in the water, sounds and models of killer whales, and anti-predator nets with different results. The only protection system currently in use is the anti-predator net (Sepúlveda and Oliva 2005), although new attempts are being made with limited results.

In Brazil, the first official record of incidental catch of South American sea lions occurred in 2008, a juvenile during a commercial pair trawl fishing trip along the southern coast (32°57′S, 52°31′W), in a depth ranging from 15 to 20 m (Machado et al. 2015). However, interactions with South American sea lions are considered usual and frequent by local fishermen on the southern Brazilian coast, mainly during winter and spring months, when the species arrives from Uruguayan rookeries. Interactions between sea lions and local fisheries were observed onboard, where only in 24% of the fishing operations sea lions interacted, and mainly with boats fishing with bottom gillnets on the northern coast of Rio Grande do Sul State (Machado et al. 2016). In Uruguay there was a systematic assessment of the mortality rates of marine mammals caused by the industrial coastal bottom trawl fisheries by means of an onboard data collection program. Three species (franciscana dolphin, South American sea lion, and South American fur seal) were found facing conservation problems either at the regional or local scale. Mortalities estimated for the South American sea lion were in the order of 0.8% of the local population (Machado et al. 2015).

Taking into account this discrepancy of evidences about the conflict with sea lions, fishermen's perceptions on the South American sea lions and their interactions with the local fishery were studied close to the Wildlife Refuge of Ilha dos Lobos, a marine protected area in southern Brazil. Sea lions prey upon the same resources targeted by the fishermen. They repeatedly hunt on the nets and consequently damage them. In response, fishermen chase sea lions (Machado et al. 2015). However, in conflicts with high-profile animals, the perceived damage often exceeds the actual evidence. Results from 100 interviews revealed that fishermen's perception of damage and their attitudes were affected by age; hierarchical position in the crew, if fishing was the only source of income; and level of formal education. Greater perception of damage and more negative attitude were found among older, less-educated sailor fishermen who had no other sources of income besides fishing. The average fisherman had a relatively good knowledge about sea lions but also a negative attitude toward them (Pont et al. 2015). These negative attitudes may be diminished addressing actions through environmental education, with emphasis on adjusting exaggerated perceptions of impact and the potential of the species for wildlife tourism, as a vital step toward the conciliation of sustainable fisheries and sea lion conservation in the southern Brazilian coast or elsewhere.

There is a new but relevant threat that has been registered in a reproductive colony in central Chile, which is the increase in the stranding of newborn calves as a result of the increase in the intensity and frequency of storm surges due to climate change. This fact in recent years has corresponded to more than 10% of the total number of pups born in that colony (Sepúlveda et al. 2020).

In general terms the South American sea lion has been related to the human population since the very beginning of the colonization of South America by aboriginal people. There is a huge piece of evidence all along the distribution range. At that time the use was mainly for food, leather, and oil. When Europeans started conquering the continent, the species was increasingly used with the same purposes but at the industrial level taking the population to dramatically reduced numbers. When conservation measures were taken in more recent times during the twentieth century, recovery was recorded in many areas, but direct takes changed toward incidental catches and overlapping in the use of target prey with the development of fisheries. The near future will require management measures that allow an agreed balance between fishing interests and population levels.

Acknowledgments The authors want to thank CONICET, ANPCyT, FAPESP, CAPES, CNPq, Amnéville Zoo, UNP, and UNISINOS for supporting us and our students for almost 37 years of research and conservation. To our colleagues from CENPAT and Grupo de Estudos de Mamíferos Aquáticos do Rio Grande do Sul (GEMARS) for sharing their field work and papers. We thank Doris Oliva and René Durán for sharing their unpublished information on the present status of sea lion and fur seal populations. To Cesar Drehmer for his collaboration in the analysis of the scientific name of the South American sea lion and to Néstor A. García, Fernando Lopes, and Alejandra Romero for their help in drawing figures. Two anonymous reviewers increased the quality of the original manuscript.

References

Acevedo RJ, Aguayo-Lobo A, Sielfeld W (2003) Eventos reproductivos del león marino común, *Otaria flavescens* (Shaw 1800), en el norte de Chile (Pacífico suroriental). Rev Biol Mar Oceanogr 38(2):69–75

Albert F (1901) Apreciaciones sobre la caza y pesca de los lobos marinos en los territorios australes de Chile, 1901. Soph Aust 22 P Arenas dic 2018. https://doi.org/10.4067/S0719-56052018000200071

Arias-Schreiber M, Rivas C (1998) Distribución, tamaño poblacional y estructura de las poblaciones de lobos marinos (*Arctocephalus australis* y *Otaria byronia*) en el litoral peruano en noviembre de 1996 y marzo de 1997. Inf Prog 73:16–32

Bartheld J, Pavés H, Contreras F et al (2008) Cuantificación poblacional de lobos marinos en el litoral de la I a IV región. Inf Fin Proy FIP 2006-50, 124 pp

Baylis MM, Orben RA, Arnould JPY et al (2015) Decline of sea lions. Bull Ecol Soc Am 96(4):635–638

Cabrera A (1940) Notas sobre carnívoros sudamericanos. Not Mus la Plata (Zool) 5(29):1–22

Campagna C (1985) The breeding cycle of the Southern sea lion, *Otaria byronia*. Mar Mamm Sci 3:210–218

Campagna C, Le Boeuf BJ (1988) The breeding biology of the Southern sea lion. Behaviour 104:233–261

Campagna C, Le Boeuf BJ, Cappozzo HL (1988a) Group raids: a mating strategy of male Southern sea lions. Behaviour 05(3–4):224–249

Campagna C, Cappozzo HL, Le Boeuf BJ (1988b) Pup abductions, and infanticide in Southern sea lions. Behaviour 107:44–60

Campagna C, Werner R, Karesh W et al (2001) Movements and location at sea of South American sea lions (*Otaria flavescens*). J Zool (Lond) 257:205–220

Capella JJ, Flórez-González L, Falk-Fernández et al (2002) Regular appearance of otariid pinnipeds along the Colombian Pacific coast. Aq Mamm 28(1):67–72

Cárdenas-Alayza S, Crespo EA, Oliveira LR (2016) *Otaria byronia*. The IUCN red list of threatened species e.T41665A61948292. https://doi.org/10.2305/IUCN.UK.2016

Contreras F, Bartheld J, Montecinos M et al (2014) Cuantificación poblacional de lobo marino común (*Otaria flavescens*) en el litoral de la XV, I y II Regiones. Inf Fin Proy 2012-6-FAP-1 86 pp + Anex

Crespi Abril A, García NA, Crespo EA et al (2004) Marine mammals' consumption by broadnose sevengill shark *Notorynchus cepedianus* in the northern and central Patagonian shelf. Lat Am J Aq Mamm 2(2):101–107

Crespo EA (1988) Dinámica poblacional del lobo marino del sur *Otaria flavescens* (Shaw, 1800) en el norte del litoral Patagónico. Doc Th, FCEN UBA, Buenos Aires, Argentina

Crespo EA, Pedraza SN (1991) Estado actual y tendencia de la población de lobos marinos de un pelo (*Otaria flavescens*) en el litoral norpatagónico. Ecol Aust 2(1):87–95

Crespo EA, Corcuera J, López-Cazorla A (1994) Interactions between marine mammals and fisheries in some fishing areas of the coast of Argentina. Gillnets and cetaceans. IWC Sp Iss 15:283–290

Crespo EA, Pedraza SN, Dans SL et al (1997) Direct and indirect effects of the high seas fisheries on the marine mammal populations in the northern and central Patagonian coast. J NW Atl Fish Sci 22:189–208

Crespo EA, Dans SL, Koen Alonso M et al (2007) Interacciones entre mamíferos marinos y pesquerías en la costa argentina. El Mar Argentino y sus recursos pesqueros, pp 151–169. Tomo 5, El ecosistema marino

Crespo EA, Oliva D, Dans SL et al (2012a) Estado de situación del lobo marino común en su área de distribución. Editorial Universidad de Valparaíso, Valparaíso

Crespo EA, Sepúlveda M, Szteren D (2012b) Capítulo 5. Interacciones entre el lobo marino común y las actividades pesqueras y de acuicultura. In: Crespo EA, Oliva D, Dans SL, Sepúlveda M

(eds) Estado de situación del lobo marino común en su área de distribución. Current situation of the South American Sea Lion along the distribution range. Universidad de Valparaíso, Valparaíso, pp 63–107
Dans SL, Crespo EA, Pedraza SN et al (2004) Recovery of the South American sea lion population in northern Patagonia. Can J Fish Aq Sci 61:1681–1690
De Blainville, HMD (1820) Sur quelques Crânes de Phoques. J De Phys. Paris, XCI
De la Torriente A, Quiñones RA, Miranda-Urbina DA et al (2010) South American sea lion and spiny dogfish predation on artisanal catches of southern hake in fjords of Chilean Patagonia. ICES J Mar Sci 67:294–303
DeNiro MJ, Epstein S (1978) Influence of diet on the distribution of carbon isotopes in animals. Geoch Cosmoch Act 42:495–506. https://doi.org/10.1016/0016-7037(78)90199-0
DeNiro MJ, Epstein S (1981) Influence of diet on the distribution of nitrogen isotopes in animals. Geoch Cosmoch Act 45(3):341–351. https://doi.org/10.1016/0016-7037(81)90244-1
Drago M, Crespo EA, Aguilar A et al (2009a) Historic change of the diet of the South American sea lion in Patagonia as revealed by isotopic analysis. Mar Ecol Prog Ser 384:273–286
Drago M, Cardona L, Crespo EA et al (2009b) Ontogenic dietary changes in South American sea lions (*Otaria flavescens*). J Zool 279:251–261
Drago M, Cardona L, Crespo EA et al (2010a) Reduction of skull size in South American sea lions reveals density dependent growth during population recovery. Mar Ecol Prog Ser 420:253–261
Drago M, Cardona L, Crespo EA et al (2010b) Change in the foraging strategy of female South American sea lions (Carnivora: Pinnipedia) after parturition. Sci Mar 74(3):589–598
Drago M, Cardona L, Franco-Trecu V et al (2017) Isotopic niche partitioning between two apex predators over time. J Anim Ecol. https://doi.org/10.1111/1365-2656.12666
Feijoo M, Lessa EP, Loizaga de Castro R et al (2011) Mitochondrial and microsatellite assessment of population structure of South American sea lion (*Otaria flavescens*) in the Southwestern Atlantic Ocean. Mar Biol 158:1857–1867
Félix F, Hasse BI, Samaniego J et al (1994) New evidence of the presence of South American sea lion *Otaria flavescens* (Carnivora, Pinnipedia) in ecuadorean waters. Estud Oceanol 13:85–88
Franco-Trecu V (2015) Tácticas comportamentales de forrajeo y apareamiento y dinámica poblacional de dos especies de otáridos simpátricas con tendencias poblacionales contrastantes. PhD Thesis UdelaR Montev, Urug
Franco-Trecu V, Drago M, García Olazabal MD et al (2015) Post-harvesting population dynamics of the South American sea lion (*Otaria byronia*) in the Southwestern Atlantic. Mar Mam Sci 31:963–978
Gardner AL, Robbins CB (1999) Case 3058. *Arctocephalus* F. Cuvier, 1826 and *Callorhinus* Gray, 1859 (Mammalia, Pinnipedia): proposed conservation by the designation of *Phoca pusilla* Schreber, [1775] as the type species of *Arctocephalus*; and *Otaria* Péron, 1816 and *Eumetopias* Gill, 1866: proposed conservation by the designation of *Phoca leonina* Molina, 1782 as the type species of *Otaria*. Bull Zool Nom 56(2):136–141
George-Nascimento M, Bustamante R, Oyarzún C (1985) Feeding ecology of the South American sea lion *Otaria flavescens*: food contents and food selectivity. Mar Ecol Prog Ser 21:135–143
Giardino GV, Mandiola MA, Bastida J et al (2016) Travel for sex: long-range breeding dispersal and winter haulout fidelity in southern sea lion males. Mamm Biol 81:89–95
Goetz S, Wolff M, Stotz W et al (2008) Interactions between the South American sea lion (*Otaria flavescens*) and the artisanal fishery off Coquimbo, northern Chile. ICES J Mar Sci 65:1739–1746
Grandi MF, Dans SL, Crespo EA (2008) Changes in social composition and spatial distribution of colonies in an expanding population of South American sea lions, *Otaria flavescens*. J Mam 89(5):1218–1228
Grandi MF, Oliveira LR, Dans SL et al (2012a) Effective population size for South American sea lions along the Northern Patagonian coast: a hunted population in recovery. An Biol 62:433–450
Grandi MF, Loizaga de Castro R, Crespo EA (2012b) Killer whale attacking on South American sea lion associated to a fishing vessel: predator and prey tactics. Lat Am J Aquat Res 40(4):1072–1076

Grandi, MF, Dans SL, Crespo EA et al (2012c) Sexual maturity and corporal growth of the South American sea lion, *Otaria flavescens*, in Patagonia. Estado de situación del lobo marino común en su área de distribución. 2012, Ed Univ Valp, Valp Chile, 144p

Grandi MF, Dans SL, Crespo EA (2015) The recovery process of a population is not always the same: the case of *Otaria flavescens*. Mar Biol Res 11(3):225–235

Grandi MF, Dans SL, Crespo EA (2016) Improvement in survivorship: the key for population recovery? Zool Stud 55:9. https://doi.org/10.6620/ZS.2016.55-09

Guerrero A, Pavez G, Santos-Carvallo M et al (2020) Foraging behaviour of the South American sea lion (*Otaria byronia*) in two disparate ecosystems assessed through blubber fatty acid analysis. Sci Rep 10:5725

Hamilton JE (1939) A second report on the southern sea lion (*Otaria byronia*) (de Blainville). Discov Rep 19 Univ Press

Hevia KA (2013) Áreas de alimentación del lobo marino común *Otaria flavescens* en la zona norte de Chile. Trab titul Ing Amb. Univ Valp

Hobson KA, Piatt JF, Pitocchelli J (1994) Using stable isotopes to determine seabird trophic relationships. J Anim Ecol 63:786–798. https://doi.org/10.2307/5256

Hoelzel AR (1991) Killer whale predation on marine mammals at Punta Norte, Argentina: food sharing, provisioning and foraging strategy. Behav Ecol Sociobiol 29:197–204

Hückstädt LA, Antezana T (2003) Behaviour of the southern sea lion (*Otaria flavescens*) and consumption of the catch during purse-seining for jack mackerel (*Trachurus simmetricus*) off Central Chile. ICES J Mar Sci 60:1–9

Hückstädt LA, Antezana T (2004) Behaviour of southern sea lions in presence of killer whales during fishing operations in Central Chile. Sci Mar 68(2):295–298

Hückstädt LA, Quiñones RA, Sepúlveda M et al (2014) Movement and diving patterns of juvenile male South American sea lions off the coast of Central Chile. Mar Mam Sci 30:1175–1183

Hückstädt LA, Tift MS, Riet-Sapriza F et al (2016) Regional variability in diving physiology and behavior of a widely distributed air-breathing marine predator, the South American sea lion *Otaria byronia*. J Exp Biol 219:2320–2330

ICZN (International Comission of Zoological Nomenclature) (2000) Opinion 1962. *Arctocephalus* F. Cuvier, 1826 and *Callorhinus* Gray, 1859 (Mammalia, Pinnipedia): proposed conservation by the designation of *Phoca pusilla* Schreber, [1775] as the type species of *Arctocephalus*; and *Otaria* Péron, 1816 and *Eumetopias* Gill, 1866: proposed conservation by the designation of *Phoca leonina* Molina, 1782 as the type species of *Otaria*. Bull Zool Nom 57(3):193–195

IMARPE (2006) Censo Nacional de Lobo Chusco (*Otaria flavescens*) (2006) Unidad de Investigación de Depredadores Superiores. Inf Int

IMARPE (2013) Anuario Científico y Tecnológico. IMARPE 13:2013

Jarma DA, Romero MA, García NA et al (2019) Small-scale variation in the diet of South American Sea lion (*Otaria flavescens*) in northern Patagonia (Argentina). Reg Stud Mar Sci 28:1–11

King JE (1978) On the specific name of the southern sea lion (Pinnipedia, Otariidae). J Mam 59(4):861–863

Koen Alonso M, Crespo EA, Pedraza SN et al (2000) Food habits of the South American sea lion, *Otaria flavescens*, off Patagonia, Argentina. Fish Bull 98(2):250–263

Linnaeus C (1758) Phoca ursina. Sys Nat Ed 10(1):37

Lucero S, Rodríguez SM, Teta P et al (2019) Solving a long-standing nomenclatorial controversy: designation of a neotype for the Southern Sea Lion *Otaria flavescens* (Shaw, 1800). Zootaxa 4555(2):296–300

Machado R, Oliveira LR, Montealegre-Quijano S (2015) Incidental catch of South American sea lion in a pair trawl off southern Brazil. Neot Biol Cons 10(1):43–47

Machado R, Ott PH, Moreno IB et al (2016) Operational interactions between South American sea lions and gillnet fishing in southern Brazil. Aqu Cons Mar Freshw Ecosyst 26:108–120

Machado M, Oliveira LR, Ott PH et al (2018) Changes in the feeding ecology of South American sea lions on the southern Brazilian coast over the last two decades of excessive fishing exploration. Hydrobiologia 819:17–37. https://doi.org/10.1007/s10750-018-3618-8

Medina JT (1952) Cosas de la colonia. Apuntes para la crónica del siglo XVIII en Chile. Imprent Univ, Santiago
Milano VN, Grandi MF, Schiavini ACM et al (2020) Sea lions (*Otaria flavescens*) from the end of the world: insights of a recovery. Polar Biol. https://doi.org/10.1007/s00300-020-02672-9
Muñoz L, Pavéz G, Inostroza P et al (2011) Foraging trips of female South American sea lions (*Otaria flavescens*) from Isla Chañaral, Chile. Lat Am J Aq Mamm 9(2):140144
Muñoz L, Pavez G, Quiñones RA et al (2013) Diet plasticity of the South American sea lion in Chile: stable isotope evidence. Rev Biol Mar Oceanogr 48(3):613–622
Newsome SD, Clementz MT, Koch PL (2010) Using stable isotope biogeochemistry to study marine mammal ecology. Mar Mam Sci 26(3):509–572. https://doi.org/10.1111/j.1748-7692.2009.00354.x
Niklitschek EJ, Soto D, Lafon A et al (2013) Southward expansion of the Chilean salmon industry in the Patagonian Fjords: main environmental challenges. Rev Aquacult 4:1–24
Oliva D (1988) *Otaria byronia* (de Blainville, 1820), the valid scientific name for the southern sea lion. J Nat Hist 22:767–772
Oliva D, Sepúlveda M, Durán LR et al (2012) Cuantificación poblacional de lobos marinos en las Regiones X –XI y propuesta de escenarios de manejo. Inf Fin Proy FAP ID 4728-46-LP11
Oliva D, Durán LR, Sepúlveda M et al (2020) Estimación poblacional de lobos marinos e impacto de la captura incidental. Inf Fin Proy FIP 2018-54, 184 pp + Anex
Oliveira LR (2013) Carnívoros Marinhos. In: Weber MM, Roman C, Cáceres NC (eds) Mamíferos do Rio Grande do Sul. Editora UFSM, Santa Maria, pp 405–427
Oliveira LR, Ott PH, Malabarba LR (2008) Ecologia alimentar dos pinípedes do sul do Brasil e uma avaliação de suas interações com atividades pesqueiras. In: Reis NR, Peracchi AL, Santos GASD (eds) Ecologia de mamíferos. Technical Books Ed, Londrina, pp 93–109
Oliveira LR, Fraga LD, Majluf P (2012) Effective population size for south American sea lions along the Peruvian coast: the survivors of the strongest El Niño event in history. J Mar Biol Assoc UK 92(8):1835
Oliveira LR, Gehara MCM, Fraga LD et al (2017) Ancient female philopatry, asymmetric male gene flow, and synchronous population expansion support the influence of climatic oscillations on the evolution of South American sea lion (*Otaria flavescens*). PLoS One 12(6):e0179442. https://doi.org/10.1371/journal.pone.0179442
Páez E (2006) Situación de la administración del recurso lobos y leones marinos en Uruguay. In: Conde D, Scarabino F, Menafra R, Rodriguez-Gallego L (eds) Bases para la conservación y el manejo de la costa Uruguaya. Vida Silvestre, Soc Urug Cons Nat, Montevideo, pp 577, 667 p–583
Paredes R, Arias-Schreiber M (1999) Dieta del lobo fino (*Arctocephalus australis*) y lobo chusco (*Otaria byronia*) en la costa peruana durante mayo y junio de 1999. Rep Subdir Invest Mam Mar. Inst Mar Perú, Callao
Pavanato HA, Silva KG, Estima SC et al (2013) Occupancy dynamics of South American Sea-Lions in Brazilian haul-outs. J Biol 74(4):855–862
Pavés HJ, Schlatter RP, Espinoza CI (2005) Patrones reproductivos del lobo marino común, *Otaria flavescens* (Shaw 1800), en el centro-sur de Chile. Rev Chil Hist Nat 687(78):687–700
Pennant J (1793) History of quadrupeds, 3rd edn. B & J White, London
Péron (1816) Histoire de l'éléphant marin, ou phoque à trompe [*Phoca proboscide*, N.]: pêches des Anglois aus Terres Australes in: Voyage de découvertes aux Terres Australes, exécuté sur les Corvettes le Géographe, le Naturaliste, et la Goëtte le Casuarina, pendant les années 1800, 1801, 1802, 1803 et 1804. (2):37 footnote
Pinedo MC (1990) Ocorrência de Pinípedes na costa brasileira. Garcia de Orla, Sér Zool 15(2):3748
Pont AC, Marchini S, Engel MT et al (2015) The human dimension of the conflicts between fishermen and sea lions in southern Brazil. Hydrobiology 770:89–104. https://doi.org/10.1007/s10750-015-2576-7

Reyes L, Crespo EA, Szapkievich V (1999) Distribution and population size of the Southern Sea lion (*Otaria flavescens*) in central and Southern Chubut, Patagonia, Argentina. Mar Mam Sci 15:478–493

Reyes P, Hucke-Gaete R, Torres-Florez JP (2013) First observations of operational interactions between bottom-trawling fisheries and South American sea lion, *Otaria flavescens* in south-Central Chile. J Mar Biol Ass UK 93:489–494

Reed DH, Grady JJO, Brook BW et al (2003) Estimates of minimum viable population sizes for vertebrates and factors influencing those estimates. Biol Conserv 113:23–24

Rice DW (1998) Marine mammals of the world – systematics and distribution. Soc Mar Mam Sp Pub

Riet-Sapriza FG, Costa DP, Franco-Trecu V et al (2013) Foraging behavior of lactating South American sea lions (*Otaria flavescens*) and spatial–temporal resource overlap with the Uruguayan fisheries. Deep Sea Res Part II: Top Stud Ocean 88–89:106–119

Rodriguez DH, Bastida RO (1993) The southern sea lion, *Otaria byronia* or *Otaria flavescens*? Mar Mam Sci 9(4):327–381

Romero MA, Grandi MF, Koen-Alonso M et al (2017) Analysing the natural population growth of a large marine mammal after a depletive harvest. Sci Rep 7:5271. https://doi.org/10.1038/s41598-017-05577-6

Rosas FCW, Haimovici M, Pinedo MC (1993) Age and growth of the south American sea lion, *Otaria flavescens* (Shaw, 1800), in southern Brazil. J Mamm 74:141–147

Rosas FCW, Pinedo MC, Marmontel M et al (1994) Seasonal movements of the South American sea lion (*Otaria flavescens*, Shaw) off the Rio Grande do Sul coast, Brazil. Mamm 58(1):51–59

Rosales D (1877) Historia General del Reino de Chile, Flandes Indiano. Imp Merc, Valparaíso

Sanfelice D, Vasques VC, Crespo EA (1999) Ocupação sazonal por Otariidae (MammaliaPinnipedia) da Reserva Ecológica da Ilha dos Lobos, Torres, Rio Grande do Sul, Brasil. Iher Ser Zool 87:101–110

Saporiti F, Bearhop S, Silva L et al (2014) Longer and less overlapped food webs in anthropogenically disturbed marine ecosystems: confirmations from the past. PLoS One 9(7):e103132

Sarmiento R, Sepúlveda M, Pavez G et al (2020). Diet composition of an opportunistic predator from an upwelling area in the Southerneast Pacific. Aust Ecol. https://doi.org/10.1111/aec.12944

Schlatter R (1976) Penetración del lobo marino común, *Otaria flavescens* Shaw, en el río Valdivia y afluentes. Medio Ambiente 2:86–90

Sepúlveda M, Oliva D (2005) Interactions between South American sea lions *Otaria flavescens* (Shaw) and salmon farms in southern Chile. Aquac Res 36(11):1062–1068

Sepúlveda M, Arismendi I, Soto D et al (2013) Escaped farmed salmon and trout in Chile: incidence, impacts, and the need for an ecosystem view. Aquacult Env Interact 4:273–283

Sepúlveda M, Newsome SD, Pavez G et al (2015) Using satellite tracking and isotopic information to characterize the impact of South American sea lions on salmonid aquaculture in southern Chile. PLoS One 10(8):e0134926. https://doi.org/10.1371/journal.pone.0134926

Sepúlveda M, Pavez G, Santos-Carvallo M et al (2017) Spatial, temporal, age, and sex related variation in the diet of South American sea lions in southern Chile. Mar Mam Sci 33(2):480–495. https://doi.org/10.1111/mms.12379

Sepúlveda M, Martínez T, Oliva D et al (2018) Spatial and temporal variation in the interaction of South American sea lions and the artisan gillnet fishery in Chile. Fish Res 208:147–152

Sepúlveda M, Quiñones R, Esparza C et al (2020) A novel approach to assess the vulnerability of a top marine predator to coastal storms in Central Chile. Sci Rep 10:12807

Sielfeld W (1999) Estado del conocimiento sobre conservación y preservación de *Otaria flavescens* (Shaw, 1800) y *Arctocephalus australis* (Zimmerman, 1783) en las costas de Chile. Estud Oceanol 18:81–96

Shaw G (1800) Seals. 28: 1800–1826. In: General zoology, or systematic natural history. G. Kearsley, London

Sielfeld W, Barraza J, Amado N (2018) Patrones locales de alimentación del león marino sudamericano *Otaria byronia*: el caso de Punta Patache, Norte de Chile. Rev Biol Mar Oceanogr 53:307–319

Sinclair ARE, Fryxell JM, Caughley G (2006) Wildlife ecology, conservation and management. Wiley, Oxford

Soto K, Trites A (2011) South American sea lions in Peru have a lek-like mating system. Mar Mam Sci 27(2):306–333

Soto K, Trites A, Arias-Schreiber M (2004) The effects of prey availability on pup mortality and the timing of birth of South American sea lions (*Otaria flavescens*) in Perú. J Zool (Lond) 264:419–428

Strange I (1979) Sea lion survey in the Falklands. Oryx 15:175–184

Thompson D, Strange I, Riddy M et al (2002) The size and status of the population of Southern sea lions *Otaria flavescens* in the Falkland Islands. Biol Conserv 121:357–367

Torres D, Aguayo-Lobo A, Acevedo J (1999) Los Mamíferos Marinos de Chile: II. Carnivora. Ser. Cient. INACH N° 50: 25–103

Traill LW, Bradshaw CJA, Brook BW (2007) Minimum viable population size: a meta-analysis of 30 years of published estimates. Biol Conserv 139:159–166

Valette L (1928) El lobo marino de un pelo. Consultas Dirección General de Agricultura, Ganadería e Industrias (Prov Buenos Aires), Circ Nro 97:6 pp.

Vilata J, Oliva D, Sepúlveda M (2010) Predation of farmed salmon by south American sea lions (*Otaria flavescens*) in southern Chile. ICES J Mar Sci 67:475–482

Vaz-Ferreira R (1965) Comportamiento antisocial en machos sub-adultos de *Otaria byronia* (de Blainville), (lobo marino de un pelo). Rev Fac Hum Cienc Montevideo, Uruguay 22:203–207

Vaz-Ferreira R (1975) Behavior of the Southern Sea Lion *Otaria flavescens* (Shaw) in the Uruguayan Islands. V Reun Cons Int Explor Mar 169:219–227

Vaz-Ferreira R (1982) *Otaria flavescens* (Shaw, 1800), South American sea lion. Mammals in the seas, vol IV: small cetaceans, seals, sirenians and otters, pp 477–495. FAO Fish Ser 5

Venegas C, Gibbons J, Aguayo A et al (2002) Distribución y abundancia de lobos marinos (Pinnipedia: Otariidae) en la Región de Magallanes, Chile. An Inst Patag Ser Cs Nat 30:67–82

Wright S (1931) Evolution in Mendelian populations. Genetics 16:97–159

Chapter 7
California Sea Lion (*Zalophus californianus*, Lesson 1828)

Claudia J. Hernández-Camacho, Lili Pelayo-González, and Martha P. Rosas-Hernández

Abstract The California sea lion is distributed from British Columbia, Canada, to the Islas Marías and Gulf of California, Mexico. The population is divided into five subpopulations: one each on the Pacific coasts of the United States and Baja California, Mexico, and three within the Gulf of California, Mexico. The California sea lion has a polygynous lek reproductive system; the breeding season occurs during the summer (May–August). Females give birth to one pup per season, and the lactation period can last from 11 months up to a year or more, with females alternating feeding trips at sea with short stays on land to nurse their young. Adult females have higher survival rates than their male counterparts. The current total population is 327,157–334,205 individuals, with 80% in the United States, 14% on the Pacific coast of Baja California, and 6% in the Gulf of California, Mexico. The US subpopulation has increased rapidly; in contrast, the four subpopulations in Mexico are at risk of disappearing. California sea lions are regularly impacted by warming events like El Niño that affect prey availability around their rookeries, as bycatch in fishing nets, and through illegal extraction presumably for their genitals.

Keywords Population dynamics · Life history · Foraging ecology · Conservation · Gulf of California

---

C. J. Hernández-Camacho (✉) · L. Pelayo-González
Laboratorio de Ecología de Pinnípedos "Burney J. Le Boeuf", Centro Interdisciplinario de Ciencias Marinas, Instituto Politécnico Nacional, La Paz, Baja California Sur, Mexico
e-mail: jcamacho@ipn.mx

M. P. Rosas-Hernández
Marine Macroecology Department, Centro de Investigación Científica y de Educación Superior de Ensenada, Unidad La Paz, La Paz, Baja California Sur, Mexico
e-mail: mrosas@cicese.mx

G. Heckel, Y. Schramm (eds.), *Ecology and Conservation of Pinnipeds in Latin America*, https://doi.org/10.1007/978-3-030-63177-2_7

Morphology and Systematics

The California sea lion is an otariid characterized by pronounced sexual dimorphism. Although the morphological differences between sexes are more evident by 4–5 years of age (time of sexual maturity), males are slightly larger than females since before birth (Lluch-Belda 1969; Ono and Boness 1996). The length ratio of male and female adults is 4:1 (Ono and Boness 1996). The newborn pups of both sexes have dark gray or dark brown fur and on average measure between approximately 75–78 cm for males and 71–73 cm for females (Luque and Aurioles-Gamboa 2001; Laake et al. 2016) (Fig. 7.1). Adult males are dark brown, and some have lighter-colored fur on the back, giving them a silverish appearance (Heath 1989); they measure 2–2.5 m in length and weigh 200–300 kg. Males are characterized by a thickened neck and sagittal crest; the latter is not found in any other otariid (Peterson and Bartholomew 1967; Lluch-Belda 1970). Subadult males have fur of similar color but are smaller (1.5–2 m long), and neither their neck nor crest is fully developed (Orr et al. 1970; Le Boeuf et al. 1983). Adult females are tan to light brown in color and measure 1.5–2 m in length and weigh 50–100 kg; adult females do not exhibit a thick neck or pronounced sagittal crest (Fig. 7.2).

The California sea lion was considered a subspecies of *Zalophus californianus* (*Z. c. californianus*) along with the Galapagos sea lion (*Z. c. wollebaeki*) and the now-extinct Japanese sea lion (*Z. c. japonicus*). Rice (1998) considered them to be three different species based on comparison of cranial morphology between the Japanese sea lion and the California sea lion by Itoo (1985) and between the California sea lion and the Galápagos sea lion by Sivertsen (1953). Subsequently and based on morphological evidence, Brunner (2004) recognized the Japanese sea

Fig. 7.1 California sea lion newborn pup cooling in a tidal pool. (Photographer: Armando Martínez Castro)

Fig. 7.2 California sea lion male and females during the breeding season at Los Islotes rookery, Gulf of California. (Photographer: Armando Martínez Castro)

lion and the California sea lion as distinct species, and Wolf et al. (2007) classified the Galápagos sea lion and the California sea lion as true species based on molecular evidence, resulting in the current classification of three distinct species in the genus.

Distribution

The geographical range of the distribution of California sea lions extends from British Columbia, Canada, to the Islas Marías and the Gulf of California, Mexico (Peterson and Bartholomew 1967; Lowry and Maravilla-Chavez 2005). The most extreme records of this species in latitudinal terms are the Aleutian Islands in Alaska and Punta San Pedrillo, Costa Rica (Maniscalco et al. 2004; Aurioles-Gamboa and Hernández-Camacho 2015).

California sea lion colonies are located in highly productive coastal areas of the eastern Pacific Ocean. The main breeding colonies are found on the Channel Islands in California, USA: San Miguel, San Nicolas, Santa Barbara, and San Clemente (Bartholomew and Boolootian 1960; Lowry and Forney 2005); two small colonies were established in the 1990s in central California at Año Nuevo Island and the Farallon Islands (Lowry et al. 2017). In Mexico, reproduction occurs on islands and islets on both coasts of the Baja California peninsula. The breeding colonies on the Pacific coast of Baja California are Santa Margarita, Asunción, Natividad, Cedros, the San Benito Archipelago, Guadalupe, San Jerónimo, San Martín, Todos Santos, and Los Coronados; meanwhile, there are 13 reproductive colonies distributed throughout the Gulf of California (Fig. 7.3) (Peterson and Bartholomew 1967; Le Boeuf et al. 1983; Lowry and Maravilla-Chavez 2005).

The reproductive colonies are mainly populated by adult females, as they are philopatric, and some also exhibit fine-scale site fidelity continuously returning to nurse their young at the same place within the colony where their pups were born (Hernández-Camacho 2001; Rayas-Estrada and Hernández-Camacho 2019). This behavior means that nearby colonies share certain characteristics due to genetic flow while differing from the rest of the colonies farther away (González-Suárez et al. 2009; Schramm et al. 2009). Based on the structure and genetic variability of the California sea lion, five subpopulations have been identified: (1) Temperate Pacific (the United States), (2) Tropical Pacific (Pacific coast of the Baja California peninsula), and three subpopulations within the Gulf of California: (3) North, (4)

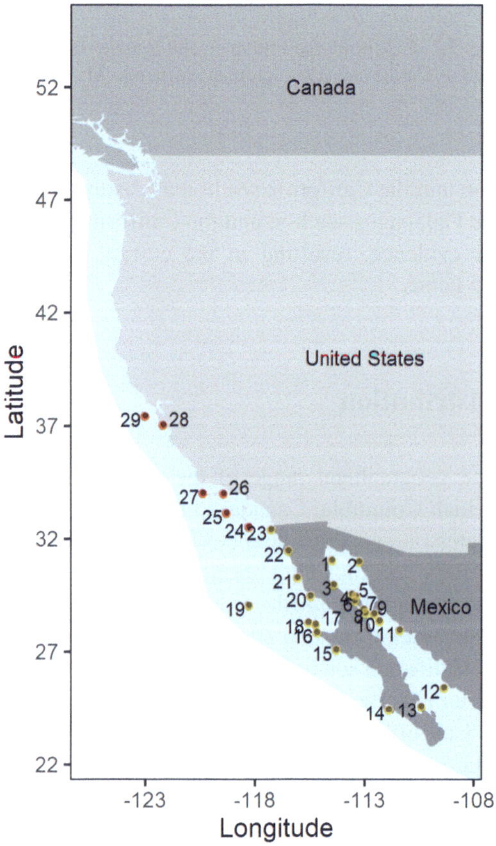

Fig. 7.3 Geographical location of the California sea lion breeding colonies throughout their distribution (blue polygon): (1) Roca Consag, (2) San Jorge, (3) Lobos, (4) Granito, (5) Cantiles, (6) Los Machos, (7) El Partido, (8) Rasito, (9) San Esteban, (10) San Pedro Mártir, (11) San Pedro Nolasco, (12) Farallón de San Ignacio, (13) Los Islotes, (14) Santa Margarita, (15) Asunción, (16) Natividad, (17) Cedros, (18) San Benito Archipelago, (19) Guadalupe, (20) San Jerónimo, (21) San Martín, (22) Todos Santos, (23) Los Coronados, (24), San Clemente, (25) San Nicolas, (26), Santa Barbara, (27) San Miguel, (28) Año Nuevo, and (29) Farallon Islands

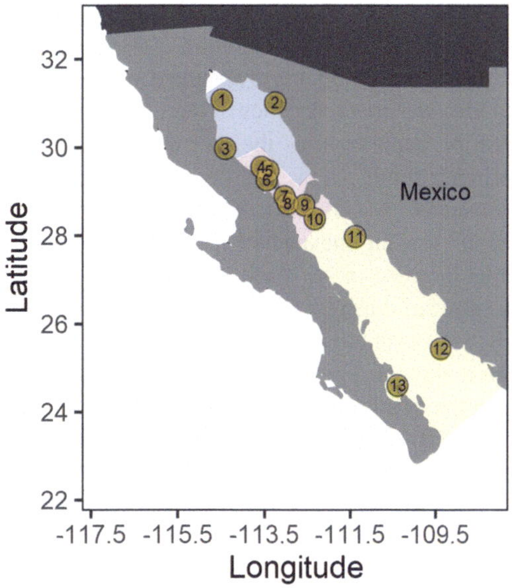

Fig. 7.4 Northern (blue), central (pink), and southern (yellow) subregions of the Gulf of California based on ecological and environmental variables as well as the population parameters of the California sea lion colonies

Central, and (5) South (González-Suárez et al. 2009; Schramm et al. 2009; Aurioles-Gamboa et al. 2011). The division of the Gulf of California into three subpopulations is supported with information that indicates distinct population dynamics based on population parameters (abundance and number of births); ecological variables like diet diversity, stable isotopes, and diseases (presence of leptospirosis, osteoarthritis, and trace metals); and environmental variables (Fig. 7.4) (Szteren et al. 2006; González-Suárez et al. 2009; Schramm et al. 2009; Ward et al. 2010; Aurioles-Gamboa et al. 2011; Pelayo-González 2018).

Environment

The largest California sea lion rookeries are found along the eastern coast of the Pacific Ocean where their biology is influenced by the California Current, which persists year-round. During the spring and summer, winds that blow southward along the shore intensify the upwelling and transport of nutrients to the surface, triggering an increase in primary productivity and the presence of pelagic species like sardines, anchovies, and squid, all important prey for the California sea lion (Lowry and Carretta 1999; Weise and Harvey 2008; Checkley Jr and Barth 2009). Prey availability on the Pacific coast of the Baja California peninsula is also modulated by tides and upwellings, which are more intense in spring and summer due to the effect of local winds. In addition to local dynamics and topography, both the California Current and the California Subcurrent influence the oceanographic conditions of this area (Zaytsev et al. 2003; Durazo 2015).

California sea lion reproductive colonies are regularly affected by warming events like El Niño (King et al. 2011; McClatchie 2016). El Niño conditions begin at the equator and are propagated north through currents and teleconnections that modify the characteristics of the California Current. Among the main consequences are the depression of the thermocline and the weakening of upwelling, lowering primary productivity, and decreasing the availability of the prey species primarily consumed by California sea lions (Jacox et al. 2016; Robinson et al. 2018). "The Blob" was an unexpected warming phenomenon that was identified as an anomalous increase in sea surface temperature in the Gulf of Alaska in 2013, spreading southward during subsequent years. This event coincided with one of the strongest El Niño events (2015–2016), which resulted in positive sea surface temperature anomalies of 2–5 °C. This warming triggered massive strandings of pups (6 to 9 months of age) and California sea lion yearlings off the coast of California during 2013–2017; the individuals showed signs of dehydration and very low weight due to changes in prey availability (https://www.fisheries.noaa.gov/national/marine-life-distress/2013-2017-california-sea-lion-unusual-mortality-event-california). Moreover, at the San Benito Archipelago rookeries in Mexico, pups displayed poor body condition, low blood glucose levels, impaired immune response capacity, alterations in erythrocyte morphology, and marked iron deficiency (Banuet-Martínez et al. 2017; Flores-Morán et al. 2017).

Changes in prey availability associated with ocean warming may lead to California sea lions investing more energy during their feeding trips as they must travel longer distances and perform deeper dives to find sufficient food (Weise et al. 2006; Melin et al. 2008). This nutritional stress has resulted in high mortality rates of young sea lions and decreases in both the number of births and the weight of dependent pups (Greig et al. 2005; Melin et al. 2012b; Laake et al. 2018). Although the California sea lion population grew in the United States in the last decades, those colonies were seriously affected when phenomena like El Niño occur. The same is observed at the California sea lion colonies on the western coast of Baja California, where there have been declines in the number of births when strong El Niño events (e.g., 1982–1983 and 2015–2016) occur (Aurioles-Gamboa and Le Boeuf 1991; Elorriaga-Verplancken et al. 2016).

Unlike the populations of the western coast of the United States and the Baja California peninsula, the California sea lion population in the Gulf of California apparently has not been affected during El Niño events (Aurioles-Gamboa and Le Boeuf 1991; Pelayo-González 2018). Dynamic oceanographic processes associated with the Gulf of California could explain the reduced impact of El Niño events on the sea surface temperature and chlorophyll-a concentrations (Escalante et al. 2013). The Gulf of California is characterized by tidal changes and wind-induced upwelling in the northern and central regions that make them highly productive areas compared to the south (Álvarez-Borrego and Lara-Lara 1991; Lavín and Marinone 2003; Lluch-Cota et al. 2007). Upwellings take place on the east side of the gulf during winter and are caused by winds from the northwest and on the peninsula side in summer with winds from the southeast. In the south, the winds have no effect due to the marked stratification of the water column (Santamaría-del-Ángel et al. 1999;

Álvarez-Borrego 2010); however, prey diversity is higher, and prey species consumed by California sea lions converge in the area (Brusca et al. 2005). Thus, the various characteristics of the Gulf of California create a productive environment that favors the availability of prey consumed by California sea lions (Sánchez-Velasco et al. 2000; Rubio-Rodríguez et al. 2018).

Population Dynamics

Due to the extensive geographical distribution and number of California sea lion rookeries (Fig. 7.3), there are no systematic counts for all colonies; thus, it is difficult to estimate the size of the global population and its growth trend. Most population counts have been made during the breeding season and at the breeding colonies of the Channel Islands in California (Barlow et al. 1995; Forney et al. 2000; Carretta et al. 2015; Lowry et al. 2017) and the Gulf of California in Mexico (Lluch-Belda 1969; Orr et al. 1970; Le Boeuf et al. 1983; Aurioles-Gamboa and Zavala-González 1994; Szteren et al. 2006; Adame-Fernández 2018; Gallo Reynoso 2019). The counts for colonies on the Pacific coast of the Baja California peninsula include mainly the colonies on the San Benito Archipelago and Cedros Island and are more recent (Bartholomew and Hubbs 1952; Rice et al. 1965; Brownell et al. 1974; Mate 1977; Le Boeuf et al. 1983; Lowry and Maravilla-Chavez 2005; Milanés Salinas 2012; Angell 2014; Elorriaga-Verplancken et al. 2015). For all counts, correction factors are used to account for animals onshore that researchers failed to see due to substrate or other factors and for animals at sea during a survey. Some studies have used the number of newborn pups as an index of population size because pups remain at the colony during their first few weeks of life. The relationship between the number of pups and the total population depends on the population age structure and vital rates and has limited utility if pup births are highly variable in relation to the older-age classes in the population (Berkson and DeMaster 1985; Lowry and Maravilla-Chavez 2005; Milanés Salinas 2012).

Globally, the California sea lion population has increased in recent decades (Aurioles-Gamboa and Hernández-Camacho 2015); however, growth trends vary considerably between colonies and regions (Szteren et al. 2006; Ward et al. 2010; Pelayo-González 2018). Le Boeuf et al. (1983) estimated that the California sea lion world population during the 1971–1981 reproductive seasons was 145,000 individuals, distributed 46% and 16% on the Pacific coast of Baja California and in the Gulf of California, respectively, 35% in the United States, and 3% in Canada. Subsequently, Lowry and Maravilla-Chavez (2005) used pup counts at colonies in California and on the western coast of Baja California, as well as counts from colonies in the Gulf of California (1980–1990) reported by Aurioles-Gamboa and Zavala-González (1994) to estimate that the California sea lion world population had increased to 344,000–359,000 individuals, but now 67–69% of the population was in the United States, 22–24% on the Pacific coast of Baja California, and 9% in the Gulf of California. As yet, there is no information for Canada.

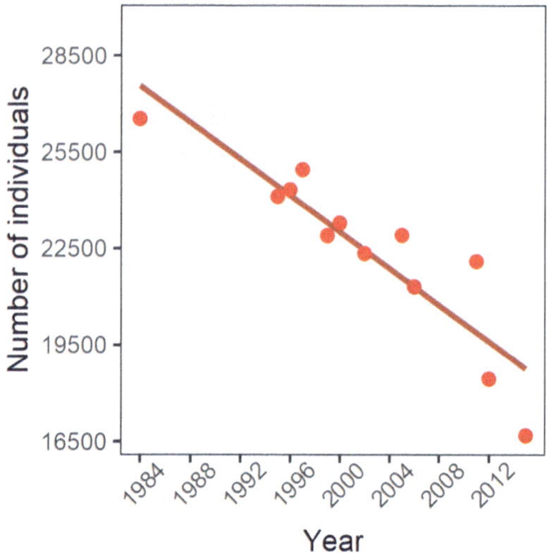

Fig. 7.5 Growth trend of the California sea lion population in the Gulf of California. The red line represents the trend; the shaded area is the 95% confidence interval

The California sea lion population in the United States was estimated to be 257,606 individuals in 2014 and apparently has stopped growing (Laake et al. 2018). In contrast, colonies in Mexico declined. Milanés Salinas (2012) estimated that the population of the west coast of Baja California was 52,846–54,482 during the 2010 breeding season. For the same period, the population in the Gulf of California was estimated to be 16,705–22,117 individuals (Fig. 7.5) (Hernández-Camacho, unpublished data). Thus, the current world population is around 327,157–334,205 individuals, of which 80% are found in the United States, 14% on the Pacific coast of Baja California, and only 6% in the Gulf of California.

The growth rate of the California sea lion population in the United States from 1975 to 2014 was 7% (Laake et al. 2018). The population of the Pacific coast of Mexico grew at a rate of 0.6% between 1979 and 2010 (Milanés Salinas 2012). Although the population on the Pacific coast decreased significantly since the abundance estimate by Lowry and Maravilla-Chavez (2005), the growth rate was positive because the estimate covered periods of growth as well as decline and probably also due to the fact that the annual abundance was estimated based on the California sea lion life table of a growing colony (Szteren et al. 2006; Hernández-Camacho et al. 2015). Between 1984 and 2015, the population of the Gulf of California decreased at a rate of 2% per year ($\lambda = 0.98$ CI 0.96–1.005).

Foraging Behavior and Diet

Research on California sea lion feeding habits, known as trophic studies, have been carried out by defining the diet based on the identification of hard structures in California sea lion scats, including fish otoliths and cephalopod beaks. The

California sea lion diet in the United States consists of 13 cephalopod species and 45 fish species (Melin et al. 2012b; Robinson et al. 2018). At their rookeries in Mexico, the California sea lion diet composition is characterized by smaller pelagic fish, demersal fish, squid, and octopus (Table 7.1). The diet on the western coast of the Baja California peninsula primarily includes demersal fish and squid (Aurioles-Gamboa and Camacho-Ríos 2007; Espinosa de los Reyes 2007). In the Gulf of California, the California sea lion diet varies latitudinally; in the north, prey are from pelagic and benthic habitats (Romero-Saavedra 2000; Porras Peters 2004; Porras-Peters et al. 2008); in the central region, the diet consists of smaller pelagic fish as well as some demersal and myctophid species (Orta-Dávila 1988; Sánchez-Arias 1992; Bautista-Vega 2000; Porras-Peters et al. 2008); and in the south, the diet includes more than 80 demersal and benthic species, but only 20 of them are

Table 7.1 Main prey species at California sea lion breeding colonies

California	Mexican Pacific	Gulf of California
Cololabis saira	*Argentina sialis*	**North**
Doryteuthis opalescens	*Citharichthys sordidus*	*Cetengraulis mysticetus*
Merluccius productus	*Engraulis mordax*	*Engraulis mordax*
Octopus rubescens	*Kathetostoma averruncus*	*Lolliguncula panamensis*
Sardinops sagax	*Loligo opalescens*	*Lutjanus aratus*
Trachurus symmetricus	*Merluccius angustimanus*	*Mugil cephalus*
Engraulis mordax	*Merluccius productus*	*Pomadasys panamensis*
Scomber japonicus	*Paralabrax clathratus*	*Porichthys analis*
Sebastes spp.	*Porichthys notatus*	*Porichthys notatus*
	Sardinops sagax	*Trichiurus lepturus*
	Serranus aequidens	**Center**
	Sthenoteuthis oualaniensis	*Engraulis mordax*
	Trachurus symmetricus	*Loliolopsis diomedeae*
		Merluccius angustimanus
		Merluccius productus
		Myctophids
		Sardinops sagax
		Scomber japonicus
		Trachurus symmetricus
		Trichiurus lepturus
		South
		Abraliopsis affinis
		Aulopus bajacali
		Engraulis mordax
		Myctophids
		Porichthys notatus
		Pronotogrammus eos
		Serranus aequidens
		Trachurus symmetricus

considered main prey (García Rodríguez 1999; Cárdenas Palomo 2003; Brassea Pérez 2014; Zavaleta-Romero 2015; Pelayo-González 2018).

The lack of historical information on diet has made it difficult to assess its effect on the population decline currently occurring in the Gulf of California. As a result, other indicators of prey availability have been used, including environmental variables and fishery data as some of the main prey items also are of commercial interest (e.g., *S. sagax*, *E. mordax*, *S. japonicus*, *M. productus*). Research has suggested that, at least in the California sea lion population in the central region, anomalous increases in sea surface temperature affect prey availability (Pelayo-González 2018).

Differences in age- and sex-class as well as reproductive status must be considered when describing the feeding behavior of California sea lions. During the reproductive season (May–August), territorial adult males do not feed because they must defend their reproductive territories (Peterson and Bartholomew 1967). After the reproductive season ends, adult males from California and Baja California migrate northward to locations in California, Oregon, Washington, and British Columbia (Bartholomew and Boolootian 1960; Hancock 1970; Bigg 1973; Mate 1975; Wright et al. 2010; Gearin et al. 2017). Adult males perform short-duration dives (average 1.86 ± 1.55 min) while feeding in epipelagic waters (average 32.2 ± 44.3 m). Biologging technology made it possible to document that under abnormally warm conditions, adult males perform deeper dives (>500 m) of longer duration (>20 min) than any other species of otariid yet registered (Costa et al. 2007). During these anomalous conditions, there is a decrease in prey abundance in traditional foraging areas; thus, these individuals venture more than 450 km from the coast to find food (Weise et al. 2006; Costa et al. 2007). Subadult males inhabiting colonies in the northern Gulf of California and on the Pacific coast of Baja California migrate to the Los Islotes colony in the southern Gulf of California during winter (Aurioles-Gamboa et al. 1983; Elorriaga-Verplancken et al. 2018).

On the other hand, juveniles display a different feeding strategy in order to meet their distinct energy requirements. Generally, they remain at or relatively close to their rookeries (Odell 1981) due to their life history, lack of experience, and their physiological and morphological limitations (Jeglinski et al. 2013; Leung et al. 2014; McHuron et al. 2018). Estimates of food consumption suggest that juveniles ingest about 4 kg of food daily; this represents 10% of their body mass and 1.4 times more than adult consumption rates (Innes et al. 1987; Aurioles-Gamboa et al. 2003). Juveniles feed by performing epipelagic dives, similar to those performed by adults, along the continental shelf (McHuron et al. 2018). However, because their body size is smaller and their blood and muscle oxygen reserves do not develop until 1.5–2.5 years and 4–6 years, respectively (Weise and Costa 2007), their dives are both shorter and shallower than those of adults (McHuron et al. 2018).

The greatest amount of data is available on the diet and feeding behavior of adult females as they tend to remain near their breeding colonies, facilitating investigation. Studies carried out with satellite telemetry have provided a glimpse at the variation in the distance traveled by females from the Channel Islands relative to those inhabiting the colonies of both coasts of the Baja California peninsula, Mexico. When the environmental conditions are favorable around San Miguel and

San Nicolas in the north, females undertake 1–3-day feeding trips in which they travel 100 km on average (Antonelis et al. 1990; Melin and DeLong 2000; Melin et al. 2008); meanwhile, on Santa Margarita Island in the south, feeding trips only last 20 h and cover distances of just 46 km. However, when conditions are abnormally warm, feeding trips tend to increase in both distance and duration (>150 km and up to 5 days of travel) (Villegas-Amtmann et al. 2011; Kuhn and Costa 2014; Rosas-Hernández et al. 2018). Females also employ a variety of diving strategies based on seasonal changes in prey availability and abundance. For example, during the cold season (February to May), female California sea lions in the Gulf of California typically perform coastal dives; during the warm season (July to August), they focus on benthic dives (Villegas-Amtmann et al. 2011).

Stable isotope analysis is a complementary tool employed in trophic studies that provides information on a broader time frame; once the isotopic signatures of the different trophic levels have been determined. This technique permits the identification of temporal patterns at different scales that vary based on the type of tissue analyzed (Porras Peters 2004; Dalerum and Angerbjörn 2005). Stable isotopes of N and C ($\delta^{15}N$ and $\delta^{13}C$) also allow the association of isotopic variations with changes in diet and food sources (Elorriaga-Verplancken 2009; Rosas-Hernández 2013; Rosas-Hernández et al. 2018).

Stable isotope analysis of C and N has revealed that females display fidelity to their feeding areas and that, within the same area, they select between different feeding strategies; this may serve to diversify their diet and reduce competition between females (McHuron et al. 2016; Rosas-Hernández et al. 2019). For example, using the isotopic values for different prey species consumed by California sea lions for reference, vibrissae isotope values suggest that at the group level adult females from Santa Margarita Island make use of a large pool of resources, both in terms of feeding areas and prey items; however, when evaluated individually, each female displays specialized feeding habits. The areas adjacent to Santa Margarita Island include five different marine ecosystems ranging from lagoons to coastal, pelagic, and transitional areas characterized by varying physical conditions like sea surface temperature and depth in which females exhibit fidelity to feeding area or prey items of a specific trophic level (Rosas-Hernández et al. 2019).

Reproductive Biology and Development

Zalophus californianus is a highly polygynous species displaying strong seasonal cyclicality in June and July (Peterson and Bartholomew 1967). During the breeding season, males establish territories as part of a lek-type mating system (Heath 1989) in which the male defends his territory from other males and females choose from among the males (Le Boeuf and Peterson 1969; Schusterman and Gentry 1971; Heath 1989). The use of physical aggression in defending territories intensifies in July and August when females are about to enter estrus (Fig. 7.6) (Peterson and Bartholomew 1967; Odell 1975). The territories provide adequate sites for nursing

Fig. 7.6 Fight between two California sea lion adult males during the breeding season. Photo taken at Los Islotes rookery during the 2019 breeding season. (Photographer: Armando Martínez Castro)

and resting, which influences mate choice by females (Heath 1989). Males can maintain these territories for up to six or more reproductive seasons (Ono 1991; Flatz et al. 2012; Rayas-Estrada and Hernández-Camacho 2019).

Although males invest a great deal of energy in the defense of their territories, this apparently does not guarantee reproductive success. In a study conducted at colonies in the Gulf of California, less than 30% of the males observed at the colony (territorial and non-territorial) had sired the pups born in the following breeding season (Flatz et al. 2012). Marginalized males or those with no territory may intercept and copulate with females when they return from their feeding trips or move through the colony (Flatz et al. 2012; Rayas-Estrada and Hernández-Camacho 2019). The defense of territories favors the reproductive success of males by increasing the survival of pups from related females with males present in those territories (Rayas-Estrada and Hernández-Camacho 2019).

Females bear the burden of pregnancy and the care of young. During the breeding season, they generally give birth to a pup between the end of May and the beginning of July (Peterson and Bartholomew 1967; García-Aguilar and Aurioles-Gamboa 2003). Subsequently, they enter estrus and begin mating regularly the fourth week after giving birth (Odell 1975; Heath 1989); a period considerably longer than that seen in most pinnipeds (Francis 1987; Riedman 1990). Thus, mating typically occurs between the end of June and early August (García-Aguilar and Aurioles-Gamboa 2003) in water or in shallow tidal pools (Odell 1975; Boness 1991). Following a nine-month gestation period that commences after a three-month delayed implantation period, the first births occur in May. A few days after giving birth, females take feeding trips at sea that last 1–3 days (Antonelis et al. 1990;

Fig. 7.7 California sea lion female nursing her pup. (Photographer: Armando Martínez Castro)

Melin and DeLong 2000; Villegas-Amtmann et al. 2011; Kuhn and Costa 2014; Rosas-Hernández et al. 2018). They then return to the colony to nurse and care for their pups for about 2 days before embarking on the next feeding trip as part of a feeding strategy known as the feeding cycle (Fig. 7.7) (Boness and Bowen 1996; Rosas-Hernández et al. 2018).

Neonate pups have very limited mobility and depend on their mother's milk during the first months of life in order to survive (Ono et al. 1987; Riedman 1990). Weaning occurs at approximately 11 months after birth at colonies in California (Melin et al. 2000; Orr et al. 2011) and after 1 year at colonies in Mexico (Elorriaga-Verplancken 2009; López Alvirde 2014), although some continue nursing as juveniles (Francis and Heath 1991). Estimates of nursing time indicate that 37% of juveniles on Santa Barbara and San Nicolas Islands nurse during their second year; in addition, females are more likely to exhibit this behavior (Francis and Heath 1991). On the San Benito Archipelago, Santa Margarita, Los Islotes, and San Esteban Islands, juveniles also have been observed nursing (Peterson and Bartholomew 1967). Analysis of the stable isotopes of N in vibrissae indicates that juveniles $\leq$14 months old consume a mixed diet of fish and milk (López Alvirde 2014). Researchers have speculated about the factors that might promote nursing by juveniles. From the perspective of the juveniles, as they grow, they become increasingly more capable of feeding themselves, satisfying their nutritional needs by consuming prey, and using milk to meet any remaining nutritional requirements. If environmental conditions make feeding difficult or if nursing opportunities are available, total or partial dependence on maternal milk may be beneficial to maximize growth. For example, at the colonies on the Channel Islands, weaning usually

Fig. 7.8 Adult female California sea lion nursing a pup and a juvenile (~1 year) concurrently. Photo taken at Los Islotes rookery during the 2019 breeding season. (Photographer: Armando Martínez Castro)

occurs at 11 months of age; however, if ocean conditions are abnormally warm, weaning can occur before 8 months (Melin et al. 2000; Melin et al. 2008) which often results in higher mortality. From the maternal perspective, the absence of a pup (females do not have a pup or lose their pup early in the pupping period) during the current breeding season may encourage an extended nursing period for the pup born during the previous breeding season. Moreover, there have been reports from some colonies on the Baja California peninsula of females nursing a 1-year-old juvenile and a pup simultaneously (Fig. 7.8). Rearing of young into the second year may provide a benefit in terms of maternal reproductive success if survival of the young is improved such that it exceeds that obtained through the investment in future pups (Trivers 1972).

Demography

Age- and sex-specific annual survival rates and age-specific birth rates in females were estimated for the California sea lion based on mark-recapture data from several cohorts of animals marked as pups in both the United States and Mexico (Hernández-Camacho et al. 2008a, b; Melin et al. 2012a; DeLong et al. 2017). The survival pattern for California sea lions is consistent with the age- and sex-specific patterns described for other large mammals; the typical mammalian survival pattern is characterized by low survival rates in young and old-age classes and more variation in survival of young than adults. Females have higher survival rates than males. The average annual survival is 0.60–0.99 for female pups, 0.55–0.99 for male pups, 0.7–0.9 for adult females, and 0.3–0.8 for adult males (Hernández-Camacho et al. 2008b; DeLong et al. 2017).

California sea lion juveniles are more susceptible to dying under conditions of limited food resources (Melin et al. 2010; Melin et al. 2012b). The higher mortality rate in males is due to (1) their higher growth rate and metabolic activity, which lead to increased susceptibility to nutritional stress and diseases; (2) intrasexual competition for resources, which increases risk of death by infections and diseases provoked by physical debilitation; and (3) the tendency to migrate, which exposes them to predation, diseases, and nutritional stress (Ralls et al. 1980; Clinton and Le Boeuf 1993).

The maximum lifespan observed is 19 years for males and 25 years for females (Hernández-Camacho et al. 2008b; DeLong et al. 2017). Birth rates are lower among young (0.59) and old females (0.11–0.35) and higher and constant (0.79) in middle-aged individuals (Melin et al. 2012a; Hernández-Camacho et al. 2008a).

Health

Environmental conditions indirectly influence the health of organisms. For example, sea temperature modulates California sea lion prey availability. During abnormal warming events, the distribution and abundance of different prey species vary. As a result, California sea lions experience nutritional stress, which is reflected in a low body condition, which in turn significantly affects their immune response capacity (Banuet-Martínez et al. 2017).

The health status of the California sea lion has been determined through clinical studies with samples taken mainly from pups and stranded animals. Pathogens like *Leptospira* have been identified in all California sea lion colonies in California. Several epizootic leptospirosis (*Leptospira pomona*) events were recorded in the California sea lion colonies in the United States, which mainly affected juvenile males with a large proportion of them developing renal disease (Gulland et al. 1996). The most frequent serotypes in California sea lion colonies in Mexico are *L. interrogans serovar Pomona* on the west coast of Baja California and *L. interrogans serovar hardjo* in the Gulf of California (Godínez et al. 1999; Avalos-Téllez et al. 2016). The prevalence of leptospirosis in the California sea lion colonies in the Gulf of California varies; at several colonies, pups were observed to have already developed immunity to certain serovars as a result of being in contact with the main vectors of these bacteria like bats and rodents (Acevedo-Whitehouse 2001; Acevedo-Whitehouse et al. 2003; Pedernera 2004).

Other types of pathogens have been detected at some California sea lion colonies in the Gulf of California. For example, juvenile and adult female California sea lions from San Esteban Island have been exposed to strains of *Brucella* (Ávalos-Téllez et al. 2014). *Morbillivirus* has been detected in California sea lion pups at the Granito reproductive colony, likely due to contact with invasive species (Suárez-Ramírez 2014).

Other diseases identified in the California sea lion population in the Gulf of California and US populations include liver and urogenital cancer (Acevedo-Whitehouse et al. 1999; Buckles et al. 2006; Browning et al. 2014). One of the possible causes for the development of carcinomas and neoplasms is the presence in tissues of high concentration of contaminants present in the marine environment, many of which are organochlorines (Kannan et al. 2004; Ylitalo et al. 2005). In addition, the intake of trace metals (As, Cd, Pb, Hg) in high quantities is related to premature births and miscarriages due to their toxic effect on the organism. These pollutants are acquired from the environment through their accumulation in the tissues of prey items consumed by California sea lions (Gilmartin et al. 1976; Szteren et al. 2006; Szteren and Aurioles-Gamboa 2013).

Conservation Status and Management

California sea lions are not listed as *Endangered* or *Threatened* under the Endangered Species Act (since 1986) or as *Depleted* under the Marine Mammal Protection Act because the population is within the range that US government agencies have defined as an optimum sustainable population (Carretta et al. 2015). In contrast, the species has been granted *Special Protection* under Mexican Official Standard NOM-059-SEMARNAT-2010 (Norma Oficial Mexicana 2010). Globally, California sea lions are considered to be of *Least Concern* according to the Red List of the International Union for Conservation of Nature (IUCN) (Aurioles-Gamboa and Hernández-Camacho 2015).

In light of the recent notable population decline at the colonies on the Pacific coast of the Baja California peninsula and in the Gulf of California, the species' vulnerability and risk assessment must be reevaluated in order to determine whether the current listings continue to be appropriate based on the criteria established by the IUCN. We conducted a preliminary population viability analysis based on counts (Dennis et al. 1991) using the abundance values available for the colonies of the central and northern regions of the Gulf of California (1980–2018, unpublished data); results indicate that most of the colonies are declining except Roca Consag, Rasito, and San Esteban, which are stable. According to the classification criteria (*Endangered*, *Threatened*, and *Vulnerable*) proposed by the IUCN (2012), most of these colonies currently should be classified as *Endangered* or *Vulnerable*. Small populations, such as those in these regions of the Gulf of California, are more vulnerable to environmental stochasticity events, demography, genetics, and catastrophic phenomena (Mills 2012). Thus, the colonies in this region are of great interest for the conservation of the species; in addition to experiencing declining populations, their individuals are genetically isolated from the rest of the colonies in the Gulf of California and on the Pacific coast of Baja California, making them especially vulnerable (González-Suárez et al. 2009; Schramm et al. 2009).

The populations of the Pacific coast of Mexico and the United States are particularly vulnerable to the reduction in prey availability as a result of warming phenomena such as the Pacific Decadal Oscillation (PDO) and the El Niño Southern Oscillation (ENSO) (Melin et al. 2012b; Elorriaga-Verplancken et al. 2016). The decrease in prey availability results in lower growth rates, a decrease in the number of births, and an increase in individual mortality (Melin et al. 2010; Laake et al. 2018). In the case of the Gulf of California colonies, mortality associated with entanglement, illegal catches, and diseases has the greatest impact on population dynamics (Acevedo-Whitehouse et al. 2003; Kovacs et al. 2012; Avalos-Téllez et al. 2016; Pelayo-González 2018; Gallo Reynoso 2019).

Acknowledgments Some of the research on California sea lions reported here was funded by the National Polytechnic Institute (IPN; Instituto Politécnico Nacional: SIP 20100911, SIP20110299, SIP 20120069, SIP 20130402, SIP 20140111, SIP 20150347, SIP 20160055, SIP 20170489, SIP 20180709, SIP20195945), the National Committee on Science and Technology (CONACYT; Consejo Nacional de Ciencia y Tecnología: #132415), and the National Commission on Protected Natural Areas (CONANP; Comisión Nacional de Áreas Naturales Protegidas: PROMOBI 2011, 2012, 2014, 2015; PROMANP 2016; PROCER 2018). We are grateful to Dr. Sharon Melin, Dr. Elizabeth Jaime, and one anonymous reviewer whose comments and suggestions improved the quality and clarity of our manuscript. We also thank Dr. Kristin Sullivan for translating the Spanish text.

References

Acevedo-Whitehouse KA (2001) Incidencia de leptospirosis en crías de *Zalophus californianus californianus* en siete colonias reproductivas del Golfo de California durante la temporada reproductiva del 2000. M.Sc. thesis, Centro de Investigación Científica y de Educación Superior de Ensenada Baja California

Acevedo-Whitehouse KA, Constantino-Casas F, Aurioles-Gamboa D et al (1999) Hepatic carcinoma with spleen metastasis in a California sea lion from the Gulf of California. J Wildl Dis 35(3):565–568

Acevedo-Whitehouse KA, De la Cueva H, Gulland FMD et al (2003) Evidence of *Leptospira interrogans* infection in California sea lion pups from the Gulf of California. J Wildl Dis 39(1):145–151

Adame-Fernández K (2018) Abundancia y hábitat terrestre del lobo marino de California en el golfo de California. M.Sc. thesis, Centro Interdisciplinario de Ciencias Marinas, Instituto Politécnico Nacional

Álvarez-Borrego S (2010) Physical, chemical, and biological oceanography of the Gulf of California. In: Brusca RC (ed) The Gulf of California: biodiversity and conservation. University of Arizona Press, Tucson, pp 24–48

Álvarez-Borrego S, Lara-Lara JR (1991) The physical environment and primary productivity of the Gulf of California. In: Dauphin PJ, Simoneit BRT (eds) Physical oceanography, primary productivity, sedimentology. American Association of Petroleum Geologist, Tulsa, pp 555–567

Angell OC (2014) Abundancia y estructura de las colonias de pinnípedos que habitan el Archipiélago San Benito, B. C., México, en el verano e invierno (2012–2013). B.Sc. thesis, Universidad Autónoma de Baja California Sur

Antonelis GA, Stewart BS, Perryman WF (1990) Foraging characteristics of female northern fur seals (*Callorhinus ursinus*) and California sea lions (*Zalophus californianus*). Can J Zool 68(1):150–158

Aurioles-Gamboa D, Camacho-Ríos FJ (2007) Diet and feeding overlap of two otariids, *Zalophus californianus* and *Arctocephalus townsendi*: implications to survive environmental uncertainty. Aquat Mamm 33(3):315–326

Aurioles-Gamboa D, García-Rodríguez F, Ramírez-Rodríguez M et al (2003) Interaction between the California sea lion and the artisanal fishery in La Paz Bay, Gulf of California, Mexico. Cienc Mar 29(3):357–370

Aurioles-Gamboa D, Godínez-Reyes C, Hernández-Camacho CJ et al (2011) Taller de análisis del estado de la población de lobo marino de California *Zalophus californianus* en México. Comisión Nacional de Áreas Naturales Protegidas, La Paz

Aurioles-Gamboa D, Hernández-Camacho CJ (2015) Zalophus californianus. The IUCN Red List of Threatened Species. https://doi.org/10.2305/IUCN.UK.2015-4.RLTS.T41666A45230310.en

Aurioles-Gamboa D, Le Boeuf B (1991) Effects of the El Niño 1982–83 on California sea lions in Mexico. In: Trillmich F, Ono A (eds) Pinnipeds and El Nino, 1st edn. Springer, Berlin, p 112–118

Aurioles-Gamboa D, Sinsel F, Fox C et al (1983) Winter migration of subadult male California sea lions (*Zalophus californianus*) in the southern part of Baja California. J Mammal 64(3):513–518

Aurioles-Gamboa D, Zavala-González A (1994) Ecological factors that determine distribution and abundance of the California sea lion *Zalophus californianus* in the Gulf of California. Cienc Mar 20(4):535–553

Avalos-Téllez R, Carrillo-Casas EM, Atilano-López D et al (2016) Pathogenic *Leptospira serovars* in free-living sea lions in the Gulf of California and along the Baja California coast of Mexico. J Wildl Dis 52(2):199–208

Ávalos-Téllez R, Ramírez-Pfeiffer C, Hernández-Castro R et al (2014) Infection of California sea lions (*Zalophus californianus*) with terrestrial *Brucella* spp. Vet J 202(1):198–200

Banuet-Martínez M, Espinosa-de Aquino W, Elorriaga-Verplancken FR et al (2017) Climatic anomaly affects the immune competence of California sea lions. PLoS One 12(6). https://doi.org/10.1371/journal.pone.0179359

Barlow J, Brownell Jr R, DeMaster D et al (1995) US Pacific marine mammal stock assessments. U. S. Department of Commerce. National Marine Fisheries Service, Southwest Fisheries Science Center, National Oceanic and Atmospheric Administration Technical Memorandum NOAA-TM-NMFSSWFSC-219, La Jolla, California

Bartholomew GA, Boolootian RA (1960) Numbers and population structure of the pinnipeds on the California Channel Islands. J Mammal 41(3):366–375

Bartholomew GA, Hubbs CL (1952) Winter population of pinnipeds about Guadalupe, San Benito, and Cedros Islands, Baja California. J Mammal 33(2):160–171

Bautista-Vega A (2000) Variación estacional en la dieta del lobo marino común, *Zalophus californianus*, en las Islas Ángel de la Guarda y Granito, Golfo de California, México. B.Sc. thesis, Universidad Nacional Autónoma de México

Berkson JM, DeMaster DP (1985) Use of pup counts in indexing population changes in pinnipeds. Can J Fish Aquat Sci 42(5):873–879

Bigg MA (1973) Census of California sea lions on southern Vancouver Island, British Columbia. J Mammal 54(1):285–287

Boness DJ (1991) Determinants of mating systems in the Otariidae (Pinnipedia). In: Renouf D (ed) The behaviour of pinnipeds, 1st edn. Springer Nature, Cham, pp 1–44

Boness DJ, Bowen W (1996) The evolution of maternal care in pinnipeds: new findings raise questions about the evolution of maternal feeding strategies. Bioscience 46(9):645–654

Brassea Pérez E (2014) Hábitos alimentarios de las hembras de lobo marino de California (*Zalophus californianus* Lesson, 1828) durante la lactancia en la Isla San Pedro Nolasco, Sonora. B.Sc. thesis, Universidad de Sonora

Brownell RL, DeLong RL, Schreiber RW (1974) Pinniped populations at Islas de Guadalupe, San Benito, Cedros, and Natividad, Baja California, in 1968. J Mammal 55(2):469–472

Browning HM, Acevedo-Whitehouse K, Gulland FM et al (2014) Evidence for a genetic basis of urogenital carcinoma in the wild California sea lion. Proc R Soc B-Biol Sci 281(1796):20140240

Brunner S (2004) Fur seals and sea lions (Otariidae): identification of species and taxonomic review. Syst Biodivers 1(3):339–439

Brusca RC, Findley LT, Hastings PA et al (2005) Macrofaunal diversity in the Gulf of California. In: Cartron J-LE, Ceballos G, Stephen RF (eds) Biodiversity, ecosystems, and conservation in northern Mexico. Oxford University Press, New York, pp 179–203

Buckles E, Lowenstine LJ, Funke C et al (2006) Otarine herpesvirus-1, not papillomavirus, is associated with endemic tumours in California sea lions (*Zalophus californianus*). J Comp Pathol 135(4):183–189

Cárdenas Palomo N (2003) Hábitos alimenticios y amplitud trófica de machos y hembras adultos del lobo marino de California (*Zalophus californianus californianus*) en Los Islotes, BCS, México. B.Sc. Thesis, Universidad Autónoma de Yucatán

Carretta JV, Oleson E, Weller DW et al (2015) US Pacific marine mammal draft stock assessments: 2014. U. S. Department of Commerce. National Marine Fisheries Service, Southwest Fisheries Science Center, National Oceanic and Atmospheric Administration Technical Memorandum NOAA-TMNMFS-SWFSC-549, La Jolla, California

Checkley DM Jr, Barth JA (2009) Patterns and processes in the California current system. Prog Oceanogr 83(1–4):49–64

Clinton WL, Le Boeuf BJ (1993) Sexual selection's effects on male life history and the pattern of male mortality. Ecology 74(6):1884–1892

Costa DP, Kuhn C, Weise M (2007) Foraging ecology of the California sea lion: diet, diving behavior, foraging locations, and predation impacts on fisheries resources. Sea grant final report-29 may 2007, Santa Cruz, California

Dalerum F, Angerbjörn A (2005) Resolving temporal variation in vertebrate diets using naturally occurring stable isotopes. Oecologia 144(4):647–658

DeLong RL, Melin SR, Laake JL et al (2017) Age-and sex-specific survival of California sea lions (*Zalophus californianus*) at San Miguel Island, California. Mar Mamm Sci 33(4):1097–1125

Dennis B, Munholland PL, Scott JM (1991) Estimation of growth and extinction parameters for endangered species. Ecol Monogr 61(2):115–143

Durazo R (2015) Seasonality of the transitional region of the California Current System off Baja California. J Geophys Res Oceans 120(2):1173–1196

Elorriaga-Verplancken FR (2009) Variación de $\delta^{15}N$ y $\delta^{13}C$ en colágeno dental de lobos marinos del género *Zalophus:* patrones ontogénicos y geográficos. Ph.D. dissertation, Centro Interdisciplinario de Ciencias Marinas, Instituto Politécnico Nacional

Elorriaga-Verplancken FR, Ferretto G, Angell OC (2015) Current status of the California sea lion (*Zalophus californianus*) and the northern elephant seal (*Mirounga angustirostris*) at the San Benito Archipelago, Mexico. Cienc Mar 41(4):269–281

Elorriaga-Verplancken FR, Sandoval-Sierra J, Paniagua-Mendoza A et al (2018) Seasonality and potential foraging grounds of migratory California sea lions from La Paz Bay, Southern Gulf of California, Mexico. Aquat Mamm 44(1):56–61

Elorriaga-Verplancken FR, Sierra-Rodríguez GE, Rosales-Nanduca H et al (2016) Impact of the 2015 El Niño-Southern Oscillation on the abundance and foraging habits of Guadalupe fur seals and California sea lions from the San Benito Archipelago, Mexico. PLoS One 11(5). https://doi.org/10.1371/journal.pone.0155034

Escalante F, Valdez-Holguín JE, Álvarez-Borrego S et al (2013) Temporal and spatial variation of sea surface temperature, chlorophyll a, and primary productivity in the Gulf of California. Cienc Mar 39(2):203–215

Espinosa de los Reyes AMG (2007) Variabilidad espacial de la dieta del lobo marino de California (*Zalophus californianus californianus*, Lesson 1828). M.Sc. thesis, Centro de Investigación Científica y de Educación Superior de Ensenada Baja California

Flatz R, González-Suárez M, Young JK et al (2012) Weak polygyny in California sea lions and the potential for alternative mating tactics. PLoS One 7(4). https://doi.org/10.1371/journal.pone.0033654

Flores-Morán A, Banuet-Martínez M, Elorriaga-Verplancken FR et al (2017) Atypical red blood cells are prevalent in California sea lion pups born during anomalous sea surface temperature events. Physiol Biochem Zool 90(5):564–574

Forney KA, Barlow J, Muto MM et al (2000) US Pacific marine mammal stock assessments, 2000. U. S. Department of Commerce. National Marine Fisheries Service, Southwest Fisheries Science Center, National Oceanic and Atmospheric Administration Technical Memorandum NOAA-TM-NMFS-SWFSC-300, La Jolla, California

Francis J (1987) Interfemale aggression and spacing in the northern fur seal *Callorhinus ursinus* and the California sea lion *Zalophus californianus*. Ph.D. dissertation, University of California

Francis J, Heath C (1991) The effects of El Niño on the frequency and sex ratio of suckling yearlings in the California sea lion. In: Trillmich F, Ono A (eds) Pinnipeds and El Nino, 1st edn. Springer, Berlin, pp 193–201

Gallo Reynoso JP (2019) Hacia el concepto ecológico de especies de interés mutuo. In: Azamar AA, Matus PJ (eds) Tendiendo puentes para una sustentabilidad integral, 1st edn. Universidad Autónoma Metropolitana, Mexico City, pp 128–239

García-Aguilar MC, Aurioles-Gamboa D (2003) Breeding season of the California sea lion (*Zalophus californianus*) in the Gulf of California, Mexico. Aquat Mamm 29(1):67–76

García Rodríguez FJ (1999) Cambios espaciales y estacionales en la estructura trófica y consumo del lobo marino de California, *Zalophus californianus*, en la región de las grandes islas, Golfo de California. M.Sc. thesis, Instituto Politécnico Nacional, Centro Interdisciplinario de Ciencias Marinas

Gearin PJ, Melin SR, DeLong RL et al (2017) Migration patterns of adult male California sea lions (*Zalophus californianus*). U. S. Department of Commerce. National Marine Fisheries Service, Alaska Fisheries Science Center, National Oceanic and Atmospheric Administration Technical Memorandum NMFS-AFSC-346, Seattle, Washington

Gilmartin WG, DeLong RL, Smith AW et al (1976) Premature parturition in the California sea lion. J Wildl Dis 12(1):104–115

Godínez CR, De Romillo BZ, Aurioles-Gamboa D et al (1999) Antibodies against *Leptospira interrogans* in California sea lion pups from Gulf of California. J Wildl Dis 35(1):108–111

González-Suárez M, Flatz R, Aurioles-Gamboa D et al (2009) Isolation by distance among California sea lion populations in Mexico: redefining management stocks. Mol Ecol 18(6):1088–1099

Greig DJ, Gulland FM, Kreuder C (2005) A decade of live California sea lion (*Zalophus californianus*) strandings along the Central California coast: causes and trends, 1991-2000. Aquat Mamm 31(1):11–22

Gulland F, Koski M, Lowenstine LJ et al (1996) Leptospirosis in California sea lions (*Zalophus californianus*) stranded along the Central California coast, 1981–1994. J Wildl Dis 32(4):572–580

Hancock D (1970) California sea lion as a regular winter visitant off the British Columbia coast. J Mammal 51(3):614–614

Heath C (1989) The behavioral ecology of the California sea lion, *Zalophus californianus*. Ph.D. dissertation, University of Santa Cruz, California

Hernández-Camacho CJ (2001) Tabla de Vida del lobo marino de California *Zalophus californianus californianus* en la lobera Los Islotes, B.C.S., México. M.Sc. Thesis, Centro interdisciplinario de Ciencias Marinas, Instituto Politécnico Nacional

Hernández-Camacho CJ, Aurioles-Gamboa D, Gerber LR (2008a) Age-specific birth rates of California sea lions (*Zalophus californianus*) in the Gulf of California, México. Mar Mamm Sci 24(3):664–676

Hernández-Camacho CJ, Aurioles-Gamboa D, Laake J et al (2008b) Survival rates of the California sea lion, *Zalophus californianus*, in Mexico. J Mammal 89(4):1059–1066
Hernández-Camacho CJ, Bakker VJ, Aurioles-Gamboa D et al (2015) The use of surrogate data in demographic population viability analysis: a case study of California sea lions. PLoS One 10(9). https://doi.org/10.1371/journal.pone.0139158
Innes S, Lavigne D, Earle W et al (1987) Feeding rates of seals and whales. J Anim Ecol 56(1):115–130
Itoo T (1985) New cranial materials of the Japanese sea lion, *Zalophus californianus japonicus* (Peters, 1866). J Mammal Soc Jpn 10:135–148
IUCN (2012) IUCN Red List categories and criteria: version 3.1. 2nd edn https://eee.iucnredlist.org/resources/categories-and-criteria. Accessed 21 Nov 2019
Jacox MG, Hazen EL, Zaba KD et al (2016) Impacts of the 2015–2016 El Niño on the California current system: early assessment and comparison to past events. Geophys Res Lett 43(13):7072–7080
Jeglinski JW, Goetz KT, Werner C et al (2013) Same size–same niche? Foraging niche separation between sympatric juvenile Galapagos sea lions and adult Galapagos fur seals. J Anim Ecol 82(3):694–706
Kannan K, Kajiwara N, Le Boeuf B et al (2004) Organochlorine pesticides and polychlorinated biphenyls in California sea lions. Environ Pollut 131(3):425–434
King JR, Agostini VN, Harvey CJ et al (2011) Climate forcing and the California current ecosystem. J Mar Sci 68(6):1199–1216
Kovacs KM, Aguilar A, Aurioles-Gamboa D et al (2012) Global threats to pinnipeds. Mar Mamm Sci 28(2):414–436
Kuhn CE, Costa DP (2014) Interannual variation in the at-sea behavior of California sea lions (*Zalophus californianus*). Mar Mamm Sci 30(4):1297–1319
Laake JL, Lowry MS, DeLong RL et al (2018) Population growth and status of California sea lions. J Wildl Manag 82(3):583–595
Laake JL, Melin SR, Orr AJ et al (2016) California sea lion sex-and age-specific morphometry. U. S. Department of Commerce. National Marine Fisheries Service, Alaska Fisheries Science Center, National Oceanic and Atmospheric Administration Technical Memorandum NMFS-AFSC-312, Seattle, Washington
Lavín M, Marinone S (2003) An overview of the physical oceanography of the Gulf of California. In: Velasco FOU, Sheinbau J, Ochoa J (eds) Nonlinear processes in geophysical fluid dynamics, 1st edn. Springer Nature, Cham, pp 173–204
Le Boeuf BJ, Aurioles D, Condit R et al (1983) Size and distribution of the California sea lion population in Mexico. Proc Calif Acad Sci 43(7):77–85
Le Boeuf BJ, Peterson RS (1969) Social status and mating activity in elephant seals. Science 163(3862):91–93
Leung ES, Chilvers BL, Nakagawa S et al (2014) Size and experience matter: diving behaviour of juvenile New Zealand sea lions (*Phocarctos hookeri*). Polar Biol 37(1):15–26
Lesson RP (1828) Phone. In: Bory de Saint-Vincent JBGM (ed) Dictionnarie classique d'historie naturelle, vol 13. Rey et Gravier, Paris, pp 400–426
Lluch-Belda D (1969) El lobo marino de California *Zalophus californianus californianus* (Lesson 1828) Allen, 1880. Observaciones sobre su ecología y explotación. Instituto Mexicano de Recursos Naturales Renovables, AC México, DF:1–71
Lluch-Belda D (1970) Crecimiento y mortalidad del lobo marino de California. An Esc Nac Cienc Biol 18:167–189
Lluch-Cota SE, Aragon-Noriega EA, Arreguín-Sánchez F et al (2007) The Gulf of California: review of ecosystem status and sustainability challenges. Prog Oceanogr 73:1):1–1)26
López Alvirde S (2014) Variación de δ^{13}C y δ^{15}N en vibrisas de lobo marino de California (*Zalophus californianus*) y su relación con el periodo de lactancia. M.Sc thesis, Centro Interdisciplinario de Ciencias Marinas, Instituto Politécnico Nacional

Lowry MS, Carretta JV (1999) Market squid (*Loligo opalescens*) in the diet of California sea lions (*Zalophus californianus*) in southern California (1981–1995). CalCOFI Rep 40:196–207

Lowry MS, Forney KA (2005) Abundance and distribution of California sea lions (*Zalophus californianus*) in central and northern California during 1998 and summer 1999. Fish Bull 103(2):331–343

Lowry MS, Maravilla-Chavez O (2005) Recent abundance of California sea lions in western Baja California, Mexico and the United States. Proceedings of the Sixth California Islands symposium, pp 485–497

Lowry MS, Melin SR, Laake JL (2017) Breeding season distribution and population growth of California sea lions, *Zalophus californianus*, in the United States during 1964–2014. U. S. Department of Commerce. National Marine Fisheries Service, Southwest Fisheries Science Center, National Oceanic and Atmospheric Administration Technical Memorandum NOAA-TM-NMFS-SWFSC 574, La Jolla, California

Luque SP, Aurioles-Gamboa D (2001) Sex differences in body size and body condition of California sea lion (*Zalophus californianus*) pups from the Gulf of California. Mar Mamm Sci 17(1):147–160

Maniscalco JM, Wynne K, Pitcher KW et al (2004) The occurrence of California sea lions (*Zalophus californianus*) in Alaska. Aquat Mamm 30(3):427–433

Mate BR (1975) Annual migrations of the sea lions *Eumetopias jubatus* and *Zalophus californianus* along the Oregon coast. Rapports et Proces-Verbaux des Reunions (Denmark)

Mate BR (1977) Aerial censusing of pinnipeds in the eastern Pacific for assessment of population numbers, migratory distributions, rookery stability, breeding effort, and recruitment. Marine Mammal Commission #MM5AC001.NTIS Cat. No. PB-265859, Springfield, Virginia

McClatchie S (2016) State of the California current 2015–16: comparisons with the 1997–98 El Niño. CalCOFI Rep 57:6–61

McHuron EA, Block BA, Costa DP (2018) Movements and dive behavior of juvenile California sea lions from Año Nuevo Island. Mar Mamm Sci 34(1):238–249

McHuron EA, Robinson P, Simmons S (2016) Foraging strategies of a generalist marine predator inhabiting a dynamic environment. Oecologia 182(4):995–1005

Melin SR, DeLong RL (2000) At-sea distribution and diving behavior of California sea lion females from San Miguel Island, California. In: Proceedings of the fifth California Islands symposium. Department of the Interior, Minerals Management Service, United States

Melin SR, DeLong RL, Siniff DB (2008) The effects of El Niño on the foraging behavior of lactating California sea lions (*Zalophus californianus californianus*) during the nonbreeding season. Can J Zool 86(3):192–206

Melin SR, Delong RL, Thomason JR et al (2000) Attendance patterns of California sea lion (*Zalophus californianus*) females and pups during the non-breeding season at San Miguel Island. Mar Mamm Sci 16(1):169–185

Melin SR, Laake JL, DeLong RL et al (2012a) Age-specific recruitment and natality of California sea lions at San Miguel Island, California. Mar Mamm Sci 28(4):751–776

Melin SR, Orr AJ, Harris JD et al (2012b) California sea lions: an indicator for integrated ecosystem assessment of the California current system. CalCOFI Rep 53:140–152

Melin SR, Orr AJ, Harris JD et al (2010) Unprecedented mortality of California sea lion pups associated with anomalous oceanographic conditions along the Central California coast in 2009. CalCOFI Rep 51:182–194

Milanés Salinas MA (2012) Abundancia y distribución de sitios de crianza del lobo marino de California, *Zalophus californianus*, en el Pacífico mexicano. M.Sc. thesis, Universidad Autónoma de Baja California

Mills LS (2012) Conservation of wildlife populations: demography, genetics, and management. Wiley, Montana

Norma Oficial Mexicana (2010) NOM-059-SEMARNAT-2010. Diario Oficial de la Federación. 30 de diciembre de 2010

Odell DK (1975) Breeding biology of the California sea lion, *Zalophus californianus*. Rapports et Proces-verbaux des Réunions Conseil International pour l. Éxploration de la Mer 169:374–378

Odell DK (1981) California sea lion *Zalophus californianus* (Lesson, 1828). In: Ridgway SH, Harrison RJ (eds) Handbook of marine mammals. Academic, New York, pp 67–97

Ono KA (1991) Introductory remarks and the natural history of the California sea lion. In: Trillmich F, Ono A (eds) Pinnipeds and El Nino, 1st edn. Springer, Berlin, pp 109–111

Ono KA, Boness DJ (1996) Sexual dimorphism in sea lion pups: differential maternal investment, or sex-specific differences in energy allocation? Behav Ecol Sociobiol 38(1):31–41

Ono KA, Boness DJ, Oftedal OT (1987) The effect of a natural environmental disturbance on maternal investment and pup behavior in the California sea lion. Behav Ecol Sociobiol 21(2):109–118

Orr AJ, Newsome SD, Laake JL et al (2011) Ontogenetic dietary information of the California sea lion (*Zalophus californianus*) assessed using stable isotope analysis. Mar Mamm Sci 28(4):714–732

Orr RT, Schonewald J, Kenyon KW (1970) The California sea lion: skull growth and a comparison of two populations. Proc Calif Acad Sci 37(11):381–394

Orta-Dávila F (1988) Hábitos alimentarios y censos globales del lobo marino (*Zalophus californianus*) en el Islote El Rasito, Bahía de las Ánimas, Baja California. México durante octubre 1986–1987. B.Sc. Thesis, Universidad Autónoma de Baja California

Pedernera C (2004) Regionalización de la presencia de anticuerpos contra Leptospira spp., niveles de cortisol y valores hemáticos en once colonias de lobos marinos *Zalophus californianus californianus* en el Golfo de California. M. Sc. Thesis, Universidad Autónoma de Baja California

Pelayo-González L (2018) Efecto de variables ambientales en el número de nacimientos de lobo marino de California (*Zalophus californianus*) en el golfo de California. M.Sc. Thesis, Centro Interdisciplinario de Ciencias Marinas, Instituto Politécnico Nacional

Peterson RS, Bartholomew GA (1967) The natural history and behavior of the California sea lion. The American Society of Mammalogists Special publication No 1, pp 1–79

Porras Peters HJ (2004) Nivel, amplitud y superposición trófica de las colonias de lobo marino *Zalophus californianus* del Golfo de California, México. M.Sc. Thesis, Instituto Politécnico Nacional, Centro Interdisciplinario de Ciencias Marinas

Porras-Peters HJ, Aurioles-Gamboa D, Cruz-Escalona VH et al (2008) Trophic level and overlap of sea lions (*Zalophus californianus*) in the Gulf of California, Mexico. Mar Mamm Sci 24(3):554–576

Ralls K, Brownell Jr RL, Ballou J (1980) Differential mortality by sex and age in mammals, with specific reference to the sperm whale. Report of the International Whaling Commission, Special (2):233–243

Rayas-Estrada A, Hernández-Camacho CJ (2019) Fine-scale site fidelity in California sea lions, *Zalophus californianus,* in the colony at Los Islotes, La Paz Bay, Mexico. Cienc Mar 45(4)

Rice DW (1998) Marine mammals of the world: systematics and distribution. The Society of Marine Mammalogy. Allen Press, Kansas

Rice DW, Kenyon K, Lluch-Belda D (1965) Pinniped populations at Islas Guadalupe, San Benito, and Cedros, Baja California, in 1965. Trans San Diego Soc Nat Hist 14:75–84

Riedman M (1990) The pinnipeds: seals, sea lions, and walruses. University of California Press, Berkeley

Robinson H, Thayer J, Sydeman WJ et al (2018) Changes in California sea lion diet during a period of substantial climate variability. Mar Biol 165(10):169

Romero-Saavedra A (2000) Dieta del lobo marino (*Zalophus californianus californianus*) en 1998 y 1999 en la Isla de San Jorge, Sonora. Ph.D. Dissertation, Centro de Investigación Científica y de Educación Superior de Ensenada Baja California

Rosas-Hernández MP (2013) Señal isotópica de $\delta^{13}C$ y $\delta^{15}N$ en vibrisas de hembras de lobo marino de California (*Zalophus californianus*) y su relación con la variabilidad ambiental en la isla San Esteban, Sonora, México. M.Sc. Thesis, Centro Interdisciplinario de Ciencias Marinas, Instituto Politécnico Nacional

Rosas-Hernández MP, Aurioles-Gamboa D, Hernández-Camacho CJ (2018) Differences in the isotopic niche and trophic position of female California sea lions (*Zalophus californianus*) in distinct oceanographic conditions. Aquat Mamm 44(4):374–388

Rosas-Hernández MP, Aurioles-Gamboa D, Hernández-Camacho CJ (2019) Specialized foraging habits of adult female California sea lions *Zalophus californianus*. Mar Mamm Sci 35(4):1463–1488

Rubio-Rodríguez U, Villalobos H, Nevárez-Martínez MO (2018) Acoustic observations of the vertical distribution and latitudinal range of small pelagic fish schools in the Midriff Islands Region, Gulf of California, Mexico. Lat Am J Aquat Res 46(5):989–1000

Sánchez-Arias M (1992) Contribución al conocimiento de los hábitos alimentarios del lobo marino de California *Zalophus californianus* en las Islas Ángel de la Guarda y Granito, Golfo de California, México. B.Sc. Thesis, Universidad Nacional Autónoma de México

Sánchez-Velasco L, Shirasago B, Cisneros-Mata MA et al (2000) Spatial distribution of small pelagic fish larvae in the Gulf of California and its relation to the El Niño 1997–1998. J Plankton Res 22(8):1611–1618

Santamaría-del-Ángel E, Álvarez-Borrego S, Millán-Núñez R (1999) Sobre el efecto débil de las surgencias de verano en la biomasa fitoplanctónica del Golfo de California. Rev Soc Mex Hist Nat 49:207–212

Schramm Y, Mesnick S, De la Rosa J et al (2009) Phylogeography of California and Galápagos sea lions and population structure within the California sea lion. Mar Biol 156(7):1375–1387

Schusterman RJ, Gentry RL (1971) Development of a fatted male phenomenon in California sea lions. Dev Psychobiol 4(4):333–338

Sivertsen E (1953) A new species of sea lion, *Zalophus wollebaeki*, from the Galapagos Islands. K Norske Vidensk Selsk Forh 26:1–3

Suárez-Ramírez L (2014) Identificación molecular y serológica de morbilivirus en mamíferos marinos y carnívoros terrestres de las costas de la península de Baja California, México. M.Sc. Thesis, Universidad Nacional Autónoma de México

Szteren D, Aurioles-Gamboa D (2013) Elementos traza en hueso de *Zalophus californianus* en el golfo de California: Una evaluación comparativa de áreas potencialmente contaminadas. Cienc Mar 39(3):306–315

Szteren D, Aurioles D, Gerber LR (2006) Population status and trends of the California sea lion (*Zalophus californianus californianus*) in the Gulf of California, Mexico. In: Trites AW, Atkinson AK, DeMaster DP et al (eds) Sea lions of the world. Alaska Sea Grant College Program, Fairbanks, pp 369–384

Trivers R (1972) Parental investment and sexual selection. In: Campbell BG (ed) Sexual selection and the descent of man, 1st edn. Transaction Publishers, Chicago, pp 52–95

Villegas-Amtmann S, Simmons SE, Kuhn CE et al (2011) Latitudinal range influences the seasonal variation in the foraging behavior of marine top predators. PLoS One 6(8). https://doi.org/10.1371/journal.pone.0023166

Ward EJ, Chirakkal H, González-Suárez M, Aurioles-Gamboa D et al (2010) Inferring spatial structure from time-series data: using multivariate state-space models to detect metapopulation structure of California sea lions in the Gulf of California, Mexico. J Appl Ecol 47(1):47–56

Weise MJ, Costa DP (2007) Total body oxygen stores and physiological diving capacity of California sea lions as a function of sex and age. J Exp Biol 210(2):278–289

Weise MJ, Costa DP, Kudela RM (2006) Movement and diving behavior of male California sea lion (*Zalophus californianus*) during anomalous oceanographic conditions of 2005 compared to those of 2004. Geophys Res Lett 33(22):1–6

Weise MJ, Harvey JT (2008) Temporal variability in ocean climate and California sea lion diet and biomass consumption: implications for fisheries management. Mar Ecol Prog Ser 373:157–172

Wolf JB, Tautz D, Trillmich F (2007) Galápagos and California sea lions are separate species: genetic analysis of the genus *Zalophus* and its implications for conservation management. Front Zool 4(1):20

Wright BE, Tennis MJ, Brown RF (2010) Movements of male California sea lions captured in the Columbia River. Northwest Sci 84(1):60–72

Ylitalo GM, Stein JE, Hom T et al (2005) The role of organochlorines in cancer-associated mortality in California sea lions (*Zalophus californianus*). Mar Pollut Bull 50(1):30–39

Zavaleta-Romero G (2015) Análisis de la dieta del lobo marino de California (*Zalophus californianus*) 1981–2013 en Los Islotes, B. C. S., México. B.Sc. thesis, Instituto Tecnológico de Boca del Río

Zaytsev O, Cervantes-Duarte R, Montante O et al (2003) Coastal upwelling activity on the Pacific shelf of the Baja California Peninsula. J Oceanogr 59(4):489–502

Chapter 8
Galápagos Sea Lion (*Zalophus wollebaeki*, Sivertsen 1953)

Oliver Krüger, Stephanie Kalberer, Kristine Meise, Jonas Schwarz, Eugene DeRango, Paolo Piedrahita, Diego Páez-Rosas, and Fritz Trillmich

Abstract The Galápagos sea lion (GSL, *Zalophus wollebaeki*) is the smallest sea lion species in the world, endemic to the Galápagos archipelago and formerly considered a subspecies of the California sea lion (*Zalophus californianus*). Breeding rookeries occur throughout the archipelago. Females are highly site faithful; hence genetic exchange between rookeries occurs predominantly through male dispersal. Females start their reproductive career rather late at a mean age of 6 years, give birth to a single pup, and nurse it unusually long for between 2 and 4 years. If a female bears another pup during lactation, the newborn's survival is seriously reduced. The extended period of maternal care lowers the reproductive rate to around 0.4 pups per adult female per year. Moreover, in years of strong El Niño events, pup mortality increases, and even older animals commonly die of starvation. Pups show distinct personalities such as shy and bold within their early social environment, thus shaping how they explore new situations and habitats. Once independent, diving strategies can also consistently differ between individuals: some forage via shallow dives, whereas others dive deep and/or mostly at night. Over the last 15 years, survival rates across all age classes have decreased in one important rookery located in the center of the species' distribution. Recent demographic models in the Caamaño rookery predict substantial population decrease in the coming years, likely to be further exacerbated by increasing sea surface temperatures due to global warming and more frequent El Niño events. Low abundance within a small distribution area makes the GSL vulnerable to a combination of anthropogenic and stochastic envi-

O. Krüger (✉) · S. Kalberer · K. Meise · J. Schwarz · E. DeRango · F. Trillmich
Department of Animal Behavior, Bielefeld University, Bielefeld, Germany
e-mail: oliver.krueger@uni-bielefeld.de; kristine.meise@uni-bielefeld.de

P. Piedrahita
Facultad de Ciencias de La Vida, Escuela Superior Politécnica Del Litoral, Guayaquil, Ecuador
e-mail: ppiedra@espol.edu.ec

D. Páez-Rosas
Galapagos Science Center, Universidad San Francisco de Quito, San Cristóbal, Islas Galápagos, Ecuador
e-mail: dpaez@usfq.edu.ec

G. Heckel, Y. Schramm (eds.), *Ecology and Conservation of Pinnipeds in Latin America*, https://doi.org/10.1007/978-3-030-63177-2_8

ronmental disturbances. Fishery interactions, pressures from increased tourism, and habitat intrusion along with the potential for introduction of diseases are presently great dangers which may pose a serious threat to this endangered species.

Keywords El Niño · Foraging strategy · Life history · Population dynamics · Vital rates

Morphology and Systematic Position

The Galápagos sea lion (GSL, *Zalophus wollebaeki*, Fig. 8.1) is the smallest sea lion species in the world. Adult males have been accurately weighed up to 158 kg and had body lengths up to 2.05 m. Maximum male mass and body length are estimated to be in the range of 200 kg and 2.1 m (Trillmich et al. 2014). Males have dark brown fur and develop a prominent sagittal crest when fully mature. Adult females are smaller and attain body lengths of 1.56–1.76 m and an average body mass of 60 kg with maximum body mass of around 95 kg (Trillmich et al. 2014). Their fur is generally lighter and more tan compared with adult males. Pups are born with a body weight of around 6 kg, with males already being heavier (6.7 kg) than females (5.8 kg, Kraus et al. 2013). The lanugo of pups is dark brown or black, and the molt starts by about 3 months of age and lasts to about 5 months. The resulting sexual dimorphism ratio in adult body weight of 2.6 is considerably smaller than in the sister species, the California sea lion (3.3, Krüger et al. 2014).

Taxonomic and phylogenetic analyses agree that the species is most closely related to the California sea lion (*Zalophus californianus*; Higdon et al. 2007; Wolf et al. 2007b), and indeed it had been considered a subspecies of the California sea lion until only recently (Wolf et al. 2007b). The separation date of the two species has been estimated from less than 1 million to around 2.3 million years ago (Wolf et al. 2007b; Schramm et al. 2009).

Distribution

The GSL is endemic to the Galápagos Islands and is distributed throughout the archipelago, settling its largest rookeries in the islands of the southeastern region (Trillmich 2015; Riofrío-Lazo et al. 2017) (Fig. 8.2), where waters are shallower and conditions are less cold than in the western parts. This species is highly gregarious, settling on islands, islets, and sites protected from predators. Thus, it is a non-migratory otariid, with vagrants recorded along the coasts of Mexico, Colombia, and Ecuador (Capella et al. 2002; Ceballos et al. 2010).

Females show very high natal philopatry, i.e., they almost always recruit into those rookeries where they were born (Wolf and Trillmich 2007, 2008; Wolf et al.

Fig. 8.1 Adult male Galápagos sea lion amidst females (top panel), adult female, and her pup establishing olfactory recognition shortly after birth (middle panel) and pups of about 1 month of age (bottom panel)

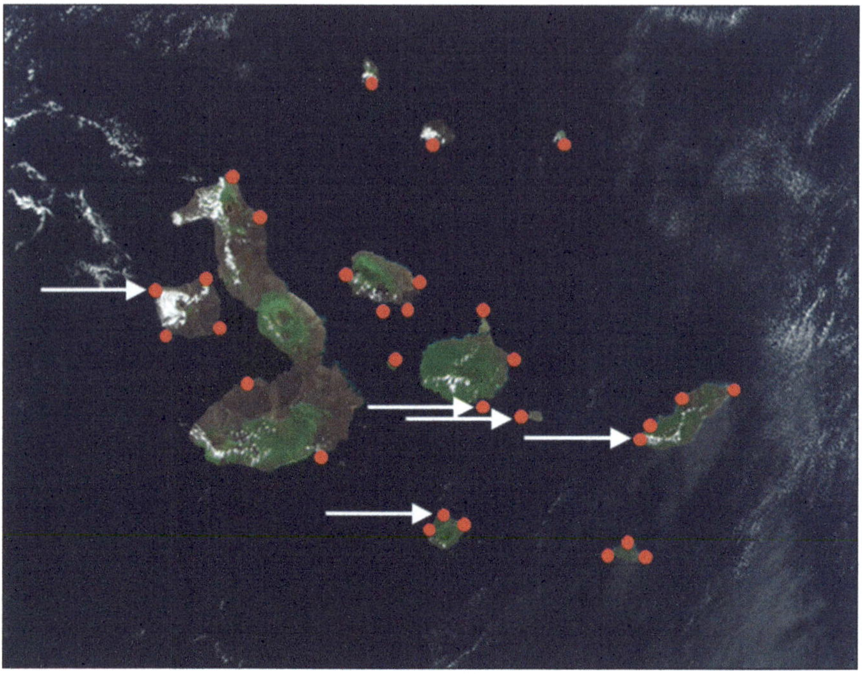

Fig. 8.2 Distribution of the Galápagos sea lion across the archipelago. Red circles show the approximate locations of the breeding colonies, and white arrows indicate which colonies have been scientifically investigated in detail. (Modified after Paez-Rosas, 2018)

2007a). This should lead to strong population structuring of matrilineal genetic markers, whereby most colonies, even if separated by a few kilometers, should be genetically distinguishable from another. However, detectable genetic differentiation of mitochondrial DNA was found mainly between central and western rookeries (Wolf et al. 2008). The largest breeding rookeries are found on San Cristóbal, Española, and Floreana islands (Fig. 8.2), which together harbor about half of the total population. Unusually for pinnipeds, the El Malecón rookery, at San Cristobal Island, is the largest in the archipelago and is located right in the town center of Puerto Baquerizo Moreno (Fig. 8.2). In this rookery, an increase of 2% per year has been observed from 2005 to 2015 (Riofrío-Lazo et al. 2017). However, despite this small increase, this rookery may still be affected by anthropogenic disturbances (Páez-Rosas and Guevara 2017).

GSL prefer to haul out on sandy beaches without large boulders but still require the opportunity to rest protected from the intense solar radiation of an equatorial habitat (Montero-Serra et al. 2014). Only a narrow fringe of coastline is useful as resting habitat, since most animals need access to water or at least shade at the hottest time of the day for thermoregulation (Wolf et al. 2005).

The Environment

The habitat of GSL consists of an area of the eastern tropical Pacific, the Galápagos archipelago. It has been described in detail in Chap. 3 about the Galápagos fur seal. In summary, the Galápagos archipelago is an unusually cold, productive marine environment right on the equator (Fernández-Álamo and Färber-Lorda 2006; Pennington et al. 2006). This results in rich marine life with species originating both from the tropics and from much more temperate regions, as is the case for both pinniped species of the archipelago. In addition, there is a profound seasonality with reduced cold upwelling resulting in relatively warm sea surface temperatures (SST) between December and April and a colder season with lower SST between May and November. This seasonality can be massively perturbed during an El Niño event, when warming continues over several months and SST increase significantly. As a result, the thermocline is depressed often several tens of meters, making prey much less accessible to sea lions. Consequences can be devastating for marine life (Arntz et al. 1991; Robinson and del Pino 1985; Trillmich and Limberger 1985; Trillmich and Dellinger 1991). If the intensity and duration of El Niño events will increase due to climate change (as seems likely, Cai et al. 2014, but see Collins et al. 2010), this could endanger the survival of both pinniped species which have already declined, most likely due to the recurrent El Niño events over the last three decades (Alava and Salazar 2006; Trillmich 2015; Trillmich et al. 2016, Riofrio-Lazo et al. 2017). On the positive side, El Niño events are often followed by La Niña (Timmermann et al. 1999) or unusually cold conditions when the thermocline gets close to the surface and marine productivity is increased, allowing marine life to recover.

Population Dynamics

In 1978, the GSL population was estimated to a number of about 40,000 individuals (Trillmich 1979). This estimate was based on a nearly complete survey of the coastlines of all Galápagos Islands conducted by observers onshore and in boats nearshore. This resulted in an actual count of 9093 individuals. The number counted was subsequently adjusted to account for missed animals or those not onshore at the time of the counts. In November 2001, 7942 sea lions were counted at known haulouts and colonies (Alava and Salazar 2006). By correcting for the probability that animals could be seen at a distance, total abundance was estimated to be 14–16,000. The corresponding number of mature individuals was likely about 26,400 in 1978 and 9200–10,600 in 2001, which indicates a substantial reduction of about 60–65% in population size. Census data from Caamaño rookery in the center of the archipelago, covering the years from 2003 to 2015 (Fig. 8.3), showed no significant trend in the total number of sea lions counted (mean ± SD, 244 ± 63) nor in the number of pups born (mean 103 ± 34.7), but substantial fluctuations around the mean

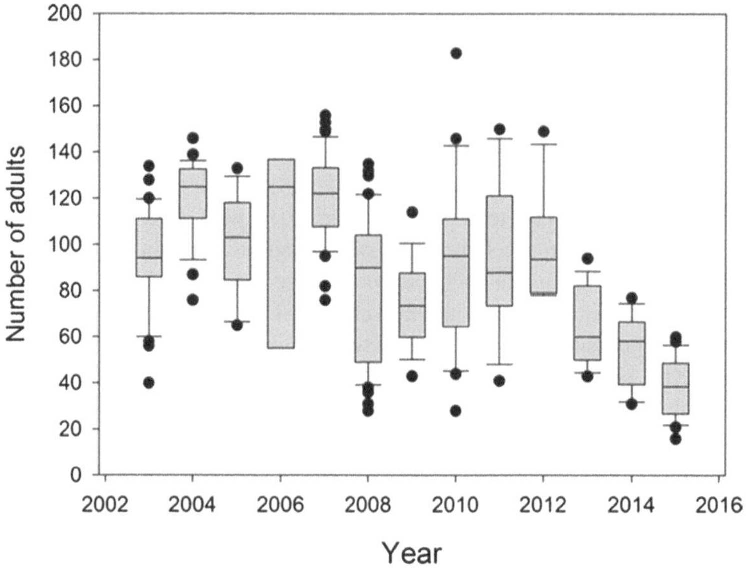

Fig. 8.3 Year-to-year cold-reproductive season variability in a total number of adults counted ashore. Boxes give medians and quartile ranges, barbs 10–90% ranges, dots indicate outliers. (Modified after Trillmich et al. 2016)

(Trillmich et al. 2016). This variability suggests that single counts as represented by the 1978 and 2001 census may be misleading about trends in total sea lion numbers.

The available data suggest that the GSL population declined greatly between 1978 and 2001 and further suggest that it has not recovered since then. If the number of adults in 2020 is still about 9,200–10,600, that would be a reduction of 60–65% in 40 years (four generations). Assuming an exponential decline pattern between 1978 and 2001 (with an annual decline of 3.9%), and a relatively stable population from 2001–2020, this would still represent a reduction of >50% (49–54%) over the last four generations (1984–2020). Recent results from population modeling of the Caamaño breeding rookery confirm that this population at least is most likely undergoing a severe decline (see *demography*). The causes of the reduction are partly understood (reduced survival rate), have not ceased, and may not be easily reversible. Recent global census data suggest the population at present comprises about 17,000 animals (Páez-Rosas et al. unpublished data). It may be worth to consider the historical long-term development of the population as well. Microsatellite data do not show strong evidence of a recent bottleneck, and no particularly low allelic richness or other measures of genetic diversity (Stoffel et al. 2018).

Foraging Behavior and Diet

The transition toward independent foraging starts late and is exceptionally slow compared to other pinnipeds, likely a consequence of the slow growth and development of physiological diving abilities found in this species (Mueller et al. 2011; Trillmich et al. 2008). Galápagos sea lion pups are terrestrial at birth and start swimming in shallow open water at around 2–3 months. Pups exhibit substantial individual variation in behavior that is both consistent and repeatable; hence they exhibit so-called personalities (Fig. 8.4, DeRango et al. 2019). Early differences in boldness and exploration may translate to pups utilizing their terrestrial habitat differently based on their personality type. Substantial diving activity starts at approximately 12 months of age, when juveniles dive on average to depths of 30 m and for 2.5 min (Jeglinski et al. 2012). Successful independent foraging supplementing continuing milk consumption seems to take place even later, on average at around 18 months. Diving performance increases with age and body mass, but 2-year-old juveniles still dive to only approximately 75% of the average diving depths of adult females (Jeglinski et al. 2012). Juvenile Galápagos sea lions therefore depend on average for the first 2 years of life on maternal input while slowly shifting from complete reliance on maternal milk to independent foraging, with profound differences between the sexes; females start independent foraging sooner than males (Fig. 8.5; Piedrahita et al. 2014). Development to independence varies considerably. Annual variation in marine productivity seems to cause differences between cohorts, but such cohort effects have been difficult to detect (Kalberer et al. 2018). During mild El Niño conditions, when food availability decreases, the onset of independent foraging was shifted beyond the age of 18 months (Jeglinski et al. 2012). Further, the onset of diving and independent foraging varies between

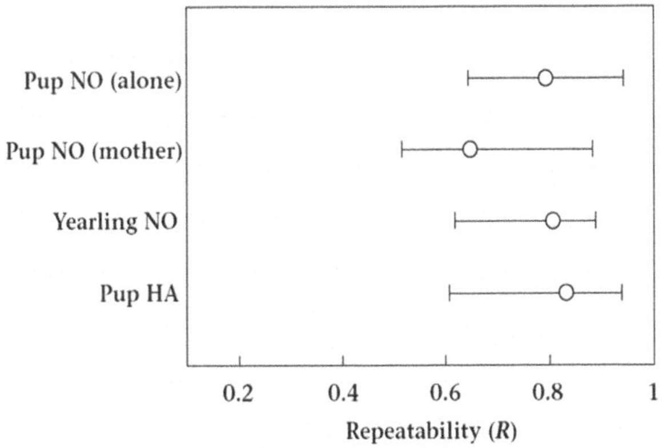

Fig. 8.4 Repeatability (R) of responses of GSL pups during seven trials of human approach (HA) tests and two trials of novel object (NO) tests during different scenarios: when alone, in the presence of their mother and again as a yearling. Error bars represent 95% confidence intervals calculated with parametric bootstrapping. After DeRango et al. (2019)

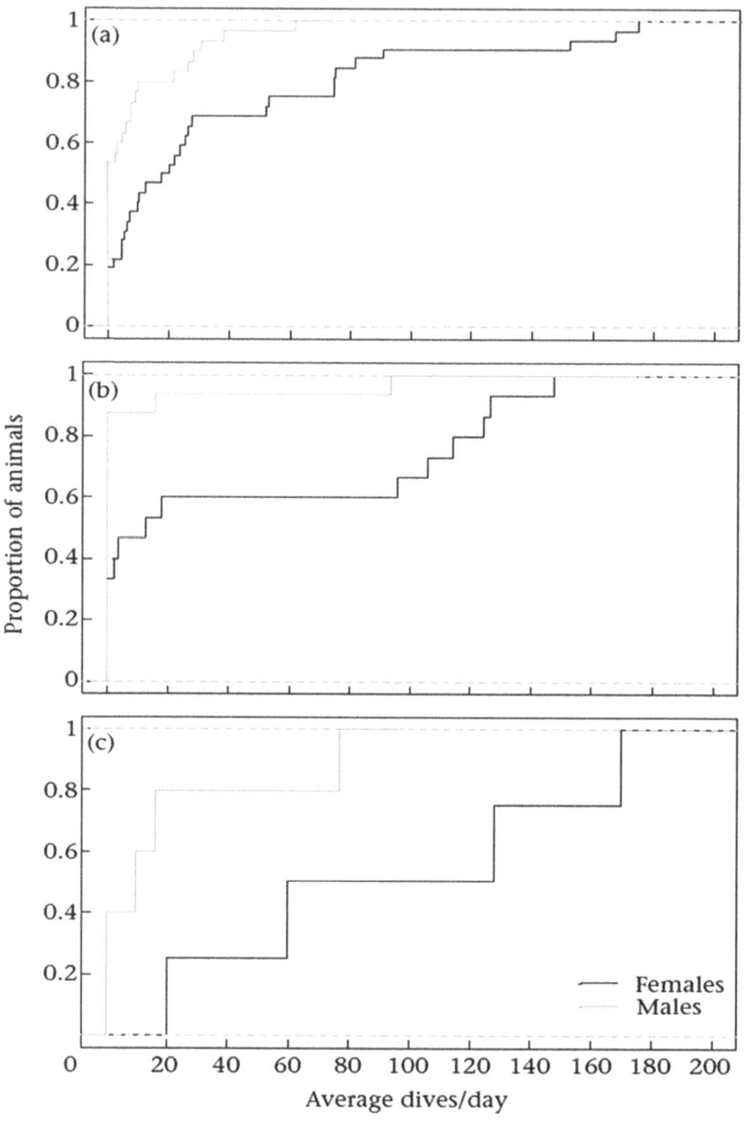

Fig. 8.5 Cumulative distribution of proportion of (**a**) yearlings (N = 62), (**b**) 1.5-year-olds (N = 31), and (**c**) 2-year-olds (N = 9) of each sex performing a given number of dives/day (only dives to ≥5 m). After Piedrahita et al. (2014)

colonies, possibly as a consequence of local differences in marine productivity (Jeglinski et al. 2012, 2013; Jeglinski 2013; Piedrahita and Trillmich, pers. obs.).

Information on spatial movements and diving behavior is available for adult females with dependent offspring and juveniles up to 2 years. Despite their small body size, compared to other sea lion species, Galápagos sea lions are exceptionally deep divers: 12-month-old juveniles already dive to a maximum depth of 367 m, and the deepest dive recorded for an adult female was 584 m (Jeglinski et al. 2012, 2013). However, adult females regularly dive to depths between 92 and 178 m for 3.3–4.7 min (Villegas-Amtmann et al. 2008; Villegas-Amtmann and Costa 2010; Jeglinski et al. 2012). Sea lions mix benthic and pelagic diving and forage both at day and at night, but there is considerable variation in foraging patterns within the species.

Juvenile sea lions dive predominantly at night and to shallow depths, a likely consequence of their small body size (Jeglinski et al. 2012, 2013). The diving behavior of adult females, especially diving depth and duration, as well as activity period, differs between individuals (Villegas-Amtmann et al. 2008; Villegas-Amtmann and Costa 2010). Recently, it has been found that individuals from the Caamaño rookery exhibit different diving and foraging strategies: some individuals forage closer to the coast in shallow water, some predominantly forage at night and utilize a broad spectrum of different depths, and others show consistently deeper dives in habitats further away from the colony (Fig. 8.6, unpublished data). Similar results have previously been observed in adult females of the El Malecón rookery on San Cristóbal island (Páez-Rosas et al. 2017). It is not yet clear if these

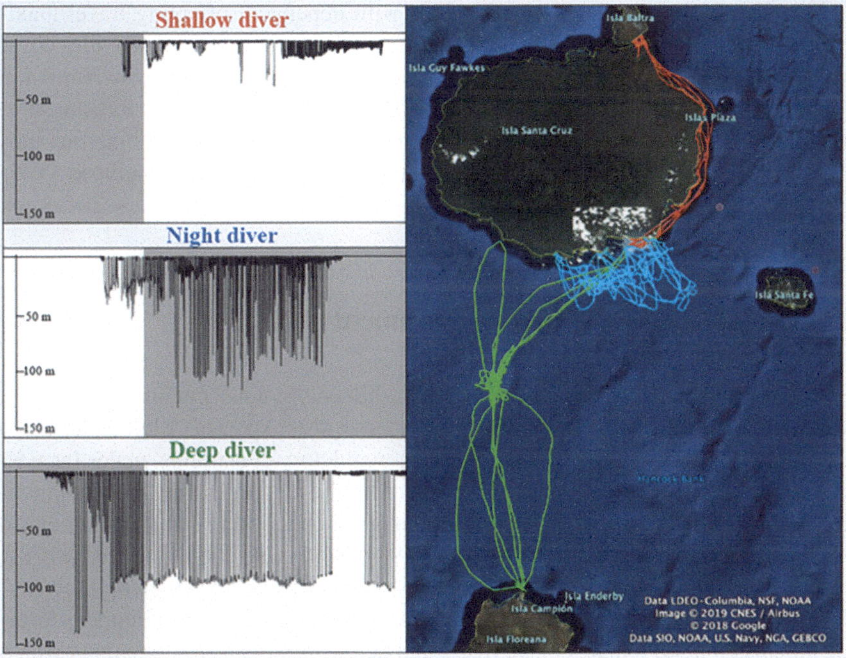

Fig. 8.6 Diving strategies of three female GSL from the Caamaño rookery. Diving profiles (18 hours, grey background represents night hours), and GPS tracks (two weeks) show the variability in dive depth and duration along with foraging location between the strategies. Schwarz et al., unpublished data

behavioral patterns are consistent throughout life or change with age, reproductive status, or ecological conditions, but in light of the documented juvenile personalities, it is tantalizing to suggest that these consistent individual differences in pup behavior have long-lasting consequences in other contexts. For example, are more explorative pups better able to exploit highly variable foraging grounds later in life, and thus have better access to a wider variety of prey? Or does the role of learning in early life depend on the personality and hence affect rates of cultural transmission?

GSL feed predominantly on small benthic and pelagic fish from approximately 12–33 different species and, in some rookeries, on cephalopods. A few fish species from the families Engraulidae, Carangidae, Serranidae, and Myctophidae dominate the diet (Dellinger and Trillmich 1999; Salazar and Bustamante 2003; Páez-Rosas and Aurioles-Gamboa 2010). There are dietary differences between GSL in the center of the archipelago that feed on a large variety of different fish species and western GSL that mainly feed on sardines and myctophids (Dellinger and Trillmich 1999; Páez-Rosas and Aurioles-Gamboa 2014), suggesting ecological differences in line with the genetic differentiation found (Jeglinski et al. 2015; Wolf et al. 2008). In the west, GSL live in sympatry with Galápagos fur seals, but both species exploit different foraging niches (Dellinger and Trillmich 1999; Páez-Rosas et al. 2012; Jeglinski et al. 2013; Villegas-Amtmann et al. 2013).

Sea lions forage on the shelf platform of the archipelago and, in the western archipelago, along its edge. Adult females with dependent offspring travel maximum distances between 50 and 97 km away from their colony, while juveniles cover maximum distances of less than 15 km (Jeglinski et al. 2013; Villegas-Amtmann et al. 2008; Villegas-Amtmann and Costa 2010). These foraging sojourns are short in comparison to other sea lion species. Both age groups haul out on numerous sites apart from their home colony, including other sea lion colonies (Jeglinski 2013; Villegas-Amtmann et al. 2008; Villegas-Amtmann and Costa 2010).

Reproductive Biology and Development

Galápagos sea lions display the typical otariid polygynous mating system with some slight distinctions from their more temperate sister species. Adult males try to monopolize access to females. Due to thermoregulatory constraints, males (as well as females) need direct access to water. Accordingly, the largest and most competitive males defend semiaquatic territories along beaches where females haul out to nurse their young (Trillmich and Trillmich 1984; Wolf et al. 2005). Indeed, only the most competitive males were able to establish territories in areas preferentially visited by females (Meise et al. 2014). In contrast to other polygynous pinnipeds forming harems on land, females move freely and define the social structure upon which male territories are superimposed (Wolf et al. 2007a; Wolf and Trillmich 2008).

Females come into estrus approximately 4 weeks after parturition (Heath 1989). The long reproductive season (September–January) and the predominance of aquatic mating make it impossible for males to monopolize access to receptive females. In contrast to other otariid species, they display and appear to depend on being chosen by a female, rather than copulating with each female that comes into estrus on their territory. Accordingly, reproductive success of Galápagos sea lions within a given season is not highly skewed toward territorial males (Pörschmann et al. 2010). Indeed, non-territorial males sire more than 50% of the annual number of pups where paternity could be assigned (Fig. 8.7; Pörschmann et al. 2010). Hence, for non-territorial males, roaming across the rookery proves to be an adequate strategy to gain reproductive success. The duration of attendance at the breeding rookery best predicts male reproductive success (Fig. 8.8; Pörschmann et al. 2010; Meise et al. 2014). Accordingly, large male body size might be selected for to allow extended fasting rather than to increase fighting abilities. The small rookery sizes and low density within Galápagos sea lion rookeries further suggest comparatively low pay-offs for territoriality. However, males establish territories in areas with highest female densities, thus increasing their chance to encounter receptive females which seem to be able to exercise female choice for indirect benefits (Lenz et al. 2013). Further, returning to territories where they previously gained matings increases a territorial male's lifetime reproductive success (Meise et al. 2014). For non-territorial males, the probability to encounter estrus females and mate successfully correlates positively with the size of their home range within a given season and thus with their ability to sneak into foreign territories (Wolf et al. 2005; Meise et al. 2014).

Galápagos sea lions can reproduce annually as do other otariids, but the breeding season is unusually long, extending over a period of more than 5 months. On Caamaño, it may start in early September, with a peak of birth in early November (Mueller et al. 2011). As the onset of the breeding season varies slightly among

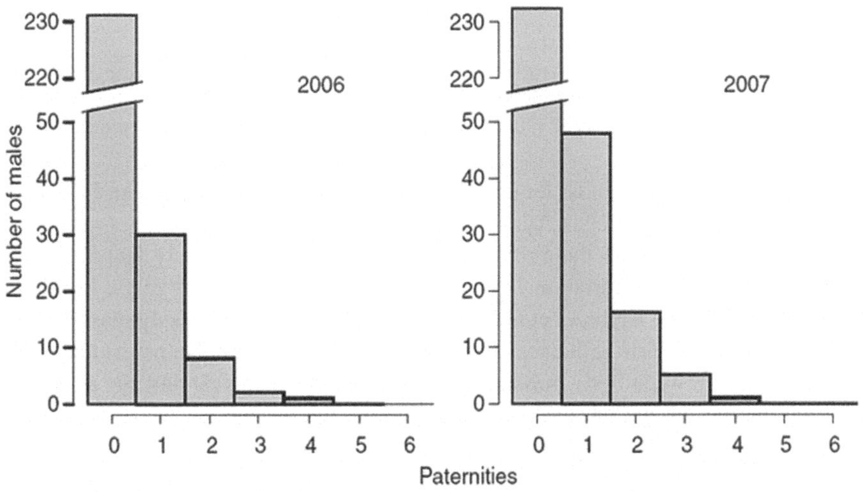

Fig. 8.7 Distribution of assigned paternities over all candidate males in the reproductive season 2006 (n = 272) and the reproductive season 2007 (n = 302). After Pörschmann et al. (2010)

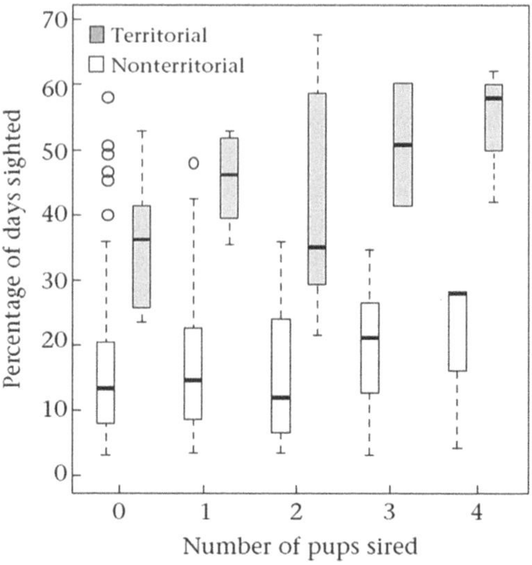

Fig. 8.8 Male reproductive success in relation to their attendance in the colony. Box plots show medians, interquartile ranges (IQR), 1.5 * IQR and outliers (circles). After Meise et al. (2014)

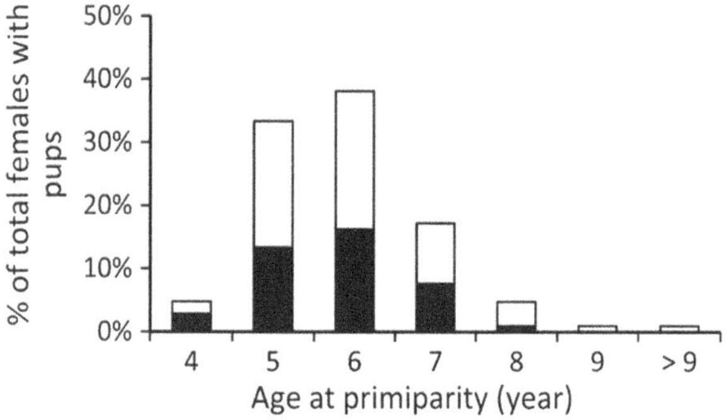

Fig. 8.9 Age at primiparity of female GSL born 2001 or later and seen every year during the breeding season from age of 4 until first pup (N = 105); pup sex: white, females (N = 62); black, males (N = 43). After Kalberer et al. (2018) and reproduced with permission from Springer Nature

colonies on different islands, females in various reproductive stages can be found throughout the year (Villegas-Amtmann et al. 2009).

Adult females initiate their reproductive life at a comparatively late age of 6 years with substantial variation between individual females (4–9 years, Fig. 8.9; Kalberer et al. 2018; Kalberer et al. 2019). Neither birth year nor body mass within the first week after birth influences age at primiparity. There is also no evidence for a relationship between body mass of females at the age of 1 year and age at primiparity (Kalberer et al. 2018). Hence, why some females start reproduction earlier than others is currently not understood.

Despite the annual breeding season, individual females reproduce every 2–3 years, resulting in a median offspring production of 0.4 pups per adult female per year (Kalberer et al. 2018). This low reproductive rate, compared to other otariid species, links to the variable productivity of the environment: Increased SST during the first 3 months after mating decreases female pupping probability (Mueller et al. 2011). Non-nursing, pregnant females return ashore 1–2 days prior to parturition (Trillmich 1986). Immediately after birth, females behave highly aggressively and defend the area around the newborn, calling frequently to it. The female stays on land with the newborn pup during a 4–7-day perinatal period. This period establishes an exclusive, mutual bond between mother and offspring (Trillmich 1981). Galápagos sea lion females, as income breeders, then resume a foraging cycle strategy, during which they continuously cycle between foraging trips at sea that last between 5 h and 4 days, in general increasing with offspring age (Trillmich 1986; Villegas-Amtmann et al. 2008; Jeglinski et al. 2012) and suckling bouts on land. Galápagos sea lion females lactate year-round. Also, given the long dependency periods of juvenile Galápagos sea lions, an annual reproduction would necessitate the simultaneous support of an older offspring, a newborn pup and self-maintenance, high costs that females rarely seem able to pay (Trillmich and Wolf 2008; Mueller et al. 2011). Therefore, inter-birth intervals of 1 and 2 years are most frequently observed (Kalberer et al. 2018). Sometimes (around 20%), females give birth to a pup while still suckling a dependent older offspring (Trillmich and Wolf 2008). This situation leads to sibling competition and can heavily impact the survival of the newborn pup when the older offspring is still around for suckling. Rarely, adult females manage to successfully support both the older offspring and the newborn, leading to the formation of a so-called trio (Trillmich and Wolf 2008). If a dependent offspring dies, lactation will end, but otherwise females will be essentially nursing all their reproductive lives, as long as the pup or juvenile offspring survives.

Demography

Galápagos sea lion females mature as early as in their 4th year, but most females exhibit an age of primiparity of 6 years of age (Kalberer et al. 2018). Females can produce one pup a year, because they always copulate about a week after parturition. Due to the slow development of their offspring, most females successfully rear a pup only every 2nd or 3rd year (Kalberer et al. 2018). Longevity is still unknown, but the oldest animals from the Caamaño population are at least 21 years old (unpublished data). Males presumably mature at a similar age to females.

Over the last 15 years, the study on Caamaño has compiled enough individual life histories to estimate female survival (Kalberer et al. 2019). As expected, pups had the lowest (0.688 ± 0.071 s.e.), juveniles intermediate (0.759 ± 0.033 s.e.), and adults the highest (0.825 ± 0.047 s.e.) average survival rate. Survival of all age classes varied among years (pups, 0.345–0.872; juveniles, 0.484–0.939; adults, 0.665–0.979), but overall showed a significant decrease since 2003 and particularly marked since 2008 (Fig. 8.10, Kalberer et al. 2019). Cold sea surface temperature and large birth mass positively affected survival in all age classes, particularly in pups. The 2015 El Niño reduced survival rates in all age classes. With these survival

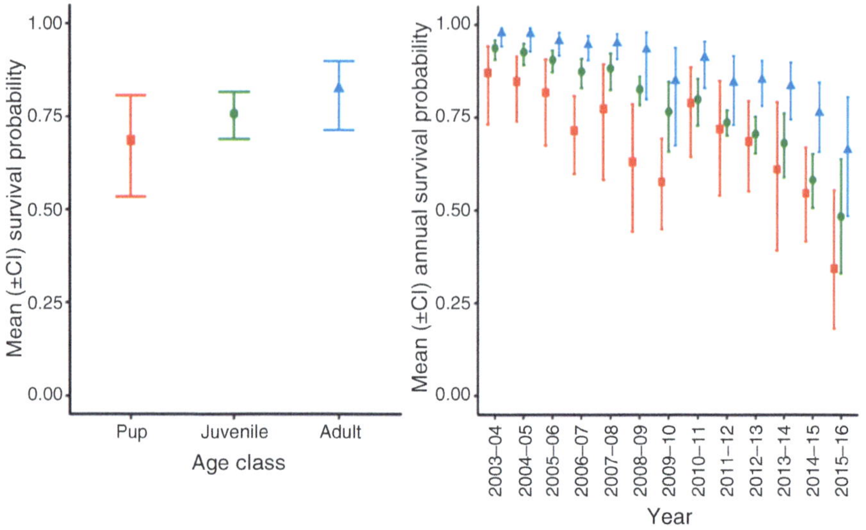

Fig. 8.10 Average female survival probability (± 95% confidence interval) (**a**) by age class and (**b**) over time ■; pup ●; juveniles ▲; ≥ 4 years old. After Kalberer (2019)

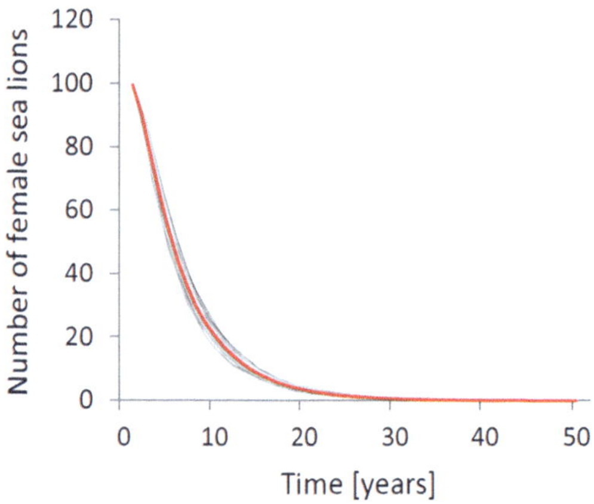

Fig. 8.11 Population projection over the next 50 years of a stochastic matrix model incorporating the pupping rate as of Kalberer et al. (2018), the average survival rates as depicted in Fig. 8.10, and the negative relationship between SST and survival rates as documented in Kalberer (2019). The average population trajectory is shown in red, and the thin grey lines depict the 100 individual simulations

estimates and the documented pupping rates, it was possible for the first time to construct matrix models and estimate population growth from vital rates. The estimated population growth rate for the Caamaño rookery in the next 50 years was predicted to be rather negative ($\lambda = 0.897$, Fig. 8.11). This growth rate translates into a population decline of 10% per annum. Prognoses worsened further when environmental stochasticity, represented by recent years' SST variation, was included ($\lambda = 0.840$) and even more so when the increase of 0.9 °C in mean SST predicted by the IPCC was considered ($\lambda = 0.829$). These results imply that over the next 50 years, the sea lion population on Caamaño is most likely facing a severe danger of extinction. Estimated population growth rate was most influenced by adult survival, as an elasticity analysis showed that 62% of variation in population growth rate was due to variation in adult survival rate. It is clear that during strong El Niño events (like 1982–1983, 1997–1998, and 2015–2016), when food availability for marine foragers is drastically reduced, numbers may substantially decline. During such events, pups of the year and adult territorial males were most affected (Trillmich and Limberger 1985; Trillmich and Dellinger 1991; Salazar and Bustamante 2003).

Conservation Status and Management

The GSL is currently classified as *Endangered* by the IUCN (Trillmich 2015). The Special Law for Galápagos (Ley Orgánica de Régimen Especial de la Provincia de Galápagos; Oficio No. SAN-2015-0989, Quito, 10 de junio de 2015) regulates that no endemic species of the Galápagos archipelago is subject to any trade, thereby providing complete protection. In addition, the sea lion population lives in an area that is protected since 1998 as the Galápagos Marine Reserve (GMR), covering an area of 138,000 km^2 and recognized in 2001 by the UNESCO as a World Heritage Site (Heylings et al. 2002). The GMR is defined as a 40-mile (74 km) zone around a baseline connecting the farthest points of the islands and therefore covers the main foraging area of the GSL. This almost excludes, but not totally, the danger of interaction with fisheries, since the international fishing activity around the perimeter of the GMR is intense and sometimes illegally encroaches into the reserve. Ghost nets and drifting long lines still endanger marine life within this region (Ventura et al. 2019).

Artisanal fishing by licensed local boats also interacts with sea lions because major sections of the coastline where sea lion rookeries exist are also used by local fishermen. Nevertheless, the increased tourism and loss or drifting of garbage led to a substantial increase in plastic pollution within the region (Alava et al. 2014). In particular, drifting nets have been documented to cause substantial mortality through entanglement (Poeta et al. 2017). In recent years, the catch per unit effort of the local legal fishery has declined. This implies a risk that in the future it may target lower trophic-level species and competition with GSL would increase or that perceived competition would lead to poaching.

Domestic animals, in particular rats, dogs, and cats in the settlements on Galápagos, pose a potential threat to the health of the GSL. All of these species must be considered a reservoir of infectious diseases like toxoplasmosis, leptospirosis, and even distemper virus (Levy et al. 2008; Alava et al. 2014). Sea lions are in intense contact with domestic animals as they often haul out in human settlements, most obvious on San Cristóbal and to a lesser extend Santa Cruz where they often rest on the dock area (Denkinger et al. 2015). Therefore, it is quite likely that exposure to introduced species in developed areas may impact immunity and overall health and fitness for specific sea lion colonies (Brock et al. 2013).

Connected to the increase in local human population and tourism, marine traffic has been increasing over the years, and ships have become much faster with the introduction of high-powered fiber glass boats. This obviously increases the risk of collision with pinnipeds as well as with whales and dolphins, but no systematic monitoring data are available. In addition, increased marine traffic harbors the risk of pollution which can have dramatic effects on GSL (Salazar 2003).

Finally, the potential increase in the frequency and intensity of El Niño events due to global warming (Cai et al. 2014) may endanger the persistence of GSL in the future. However, it is the strikingly low and variable adult survival which makes the Galápagos sea lion vulnerable to extinction in its variable environment, potentially strongly affected by future climate change. As the detailed local effects are difficult to predict (Karnauskas et al. 2015), the concerns listed above make further regular population monitoring as done over the last years by Galápagos National Park Directorate an urgent need, with a strong focus on interagency planning and management to achieve these goals (Páez-Rosas and Guevara 2017).

Acknowledgments We would like to thank the Galápagos National Park (GNP) for granting the permits that allowed our work on the Galápagos sea lion at Caamaño since 2003 and the Charles Darwin Research Station for their continued support. This work would have been impossible without the support of the German Research Foundation through grants TR 105/19-1 and KR 2089/10-1, the Volkswagen Stiftung, Friends of the Galápagos, and the Max Planck Society. Jochen Wolf, Birte Müller, and Ulo Pörschmann helped to establish the study on Caamaño, and field assistants too numerous to name them all provided enormous, much appreciated help. We are grateful to GNP rangers for their help in collecting data exposed in Fig. 8.2. Finally, we are also very grateful to Judith Denkinger and Jochen Wolf who provided many helpful comments which substantially improved this chapter.

References

Alava JJ, Salazar S (2006) Status and conservation of Otariids in Ecuador and the Galápagos Islands. In: Trites AW, Atkinson SK, DeMaster DP et al (eds) Sea lions of the world. Alaska Sea Grant College Program, Fairbanks, pp 495–520

Alava JJ, Palomera C, Bendell L et al (2014) Pollution as an emerging threat for the conservation of the Galápagos marine reserve: environmental impacts and management perspectives. In: Denkinger J, Vinueza L (eds) The Galápagos marine reserve. Springer Science, New York, pp 247–283. https://doi.org/10.1007/978-3-319-02769-2_12

Arntz W, Pearcy WG, Trillmich F (1991) Biological consequences of the 1982-83 El Niño in the eastern Pacific. In: Trillmich F, Ono KA (eds) Pinnipeds and El Niño. Responses to environmental Stress. Springer Verlag, Berlin, pp 22–42

Brock PM, Hall AJ, Goodman SJ et al (2013) Applying the tools of ecological immunology to conservation: a test case in the Galápagos sea lion. Anim Conserv 16:19–31

Cai W, Borlace S, Lengaigne M et al (2014) Increasing frequency of extreme El Niño events due to greenhouse warming. Nat Clim Chang 4:111. https://doi.org/10.1038/NCLIMATE2100

Capella JJ, Flórez-González L, Falk-Fernández P et al (2002) Regular appearance of otariid pinnipeds along the Colombian coast. Aquat Mamm 28:67–72

Ceballos G, Pompa S, Espinoza E et al (2010) Extralimital distribution of Galapagos (*Zalophus wollebaeki*) and northern (*Eumetopias jubatus*) sea lions in Mexico. Aquat Mammal 36(2):188–194

Collins M, An S-I, Cai W et al (2010) The impact of global warming on the tropical Pacific Ocean and El Niño. Nat Geosci 3:391–397. https://doi.org/10.1038/NGEO868

Dellinger T, Trillmich F (1999) Fish prey of the sympatric Galápagos fur seals and sea lions: seasonal variation and niche separation. Can J Zool 77:1204–1216

Denkinger J, Gordillo L, Montero-Serra I et al (2015) Urban life of Galápagos sea lions (*Zalophus wollebaeki*) on San Cristobal island, Ecuador: colony trends and threats. J Sea Res 105:10–14. https://doi.org/10.1016/j.seares.2015.07.004

DeRango E, Kalberer S, Schwarz J et al (2019) Intrinsic and maternal traits influence personality during early life in Galápagos sea lion (*Zalophus wollebaeki*) pups. Anim Behav 154:111–120

Fernández-Álamo MA, Färber-Lorda J (2006) Zooplankton and the oceanography of the eastern tropical Pacific: a review. Prog Oceanogr 69:318–359. https://doi.org/10.1016/j.pocean.2006.03.003

Heath CB (1989) The behavioral ecology of the California sea lion. PhD Dissertation, University of California, Santa Cruz, CA, Zalophus californianus

Heylings P, Bensted-Smith R, Altamirano M (2002) Zonificación e historia de la reserva Marina de Galápagos. In: Danulat E, Edgar GJ (eds) Reserva Marina de Galápagos. Línea base de la biodiversidad. Charles Darwin Foundation Press, Galápagos

Higdon JW, Bininda-Emonds ORF, Beck RMD et al (2007) Phylogeny and divergence of the pinnipeds (Carnivora: Mammalia) assessed using a multigene dataset. BMC Evol Biol 7:216

Jeglinski JWE (2013) Ecological differentiation in the Galápagos Sea: the role of foraging ontogeny and competition. Department of Animal Behaviour, Bielefeld University, PhD Dissertation

Jeglinski JWE, Robinson PW, Werner C et al (2012) Age, body mass and environmental variation shape the foraging ontogeny of Galápagos sea lions. Mar Ecol Prog Ser 453:279–296

Jeglinski JWE, Goetz KT, Werner C et al (2013) Same size – same niche? Foraging niche separation between sympatric juvenile Galápagos sea lions and adult Galápagos fur seals. J Anim Ecol 82:694–706. https://doi.org/10.1111/1365-2656.12019

Jeglinski JWE, Wolf JBW, Werner C et al (2015) Differences in foraging ecology align with genetically divergent ecotypes of a highly mobile marine top predator. Oecologia 179:1041–1052

Kalberer S (2019) Life history strategies and population dynamics of Galapagos sea lions: from individuals to population. PhD Dissertation, Department of Animal Behaviour, Bielefeld University

Kalberer S, Meise K, Trillmich F et al (2018) Reproductive performance of a tropical apex predator in an unpredictable habitat. Behav Ecol Sociobiol 72:108

Kalberer S, DeRango E, Trillmich F et al (2019) Evolution of age at primiparity in pinnipeds in the absence of the quality-quantity trade-off in reproduction. Ecol Evol 9:5450–5456

Karnauskas KB, Jenouvrier S, Brown CW et al (2015) Strong sea surface cooling in the eastern tropical Pacific and implications for Galápagos Penguin conservation. Geophys Res Lett 42:6432–6437. https://doi.org/10.1002/2015GL064456

Kraus C, Mueller B, Meise K et al (2013) Mama's boy: sex differences in juvenile survival in a highly dimorphic large mammal, the Galápagos sea lion. Oecologia 171:893–903

Krüger O, Wolf JBW, Jonker RM et al (2014) Disentangling the contribution of sexual selection and ecology to the evolution of size dimorphism in pinnipeds. Evolution 68:1485–1496

Lenz T, Mueller B, Trillmich F, Wolf JBW (2013) Divergent allele advantage at MHC-DRB through direct and maternal genotypic effects and its consequences for allele pool composition and mating. Proc R Soc B 280:20130714. https://doi.org/10.1098/rspb.2013.0714

Levy JK, Crawford PC, Lappin MR et al (2008) Infectious diseases of dogs and cats on Isabela island, Galápagos. J Vet Intern Med 22:60–65

Meise K, Piedrahita P, Krüger O et al (2014) Being on time: size-dependent attendance patterns affect male reproductive success. Anim Behav 93:77–86

Montero-Serra I, Páez-Rosas D, Murillo JC et al (2014) Environment-driven changes in terrestrial habitat use and distribution of the Galapagos sea lion. Endanger Species Res 24:9–19

Mueller B, Pörschmann U, Wolf JBW et al (2011) Growth under uncertainty: the influence of marine variability on early development of Galápagos sea lions. Mar Mamm Sci 27:350–365

Páez-Rosas D, Aurioles-Gamboa D (2010) Alimentary niche partitioning in the Galapagos sea lion, *Zalophus wollebaeki*. Mar Biol 157:2769–2781

Páez-Rosas D, Aurioles-Gamboa D (2014) Spatial variation in the foraging behaviour of the Galapagos sea lions (*Zalophus wollebaeki*) assessed using scat collections and stable isotope analysis. J Mar Biol Assoc UK 94:1099–1107

Páez-Rosas D, Guevara N (2017) Management strategies and conservation status of Galapagos sea lion populations at San Cristóbal Island, Galápagos, Ecuador. In: Alava JJ (ed) Tropical pinnipeds: bio-ecology, threats and conservation. CRC Press, Abingdon, pp 159–173

Páez-Rosas D, Aurioles-Gamboa D, Alava JJ et al (2012) Stable isotopes indicate differing foraging strategies in two sympatric otariids of the Galápagos islands. J Exp Mar Biol Ecol 424:44–52. https://doi.org/10.1016/j.jembe.2012.05.001

Páez-Rosas D, Villegas-Amtmann S, Costa D (2017) Intraspecific variation in feeding strategies of Galápagos sea lions: a case of trophic specialization. PLoS One 12:e0185165

Páez-Rosas D (2018) Lobo marino de Galápagos. In: Atlas de Galápagos, Ecuador: Especies Nativas e Invasoras. FCD y WWF-Ecuador, Quito, pp 158–159

Pennington JT, Mahoney KL, Kuwahara VS et al (2006) Primary production in the eastern tropical Pacific: a review. Prog Oceanogr 69:285–317. https://doi.org/10.1016/j.pocean.2006.03.012

Piedrahita P, Meise K, Werner C et al (2014) Lazy sons, self-sufficient daughters: are sons more demanding? Anim Behav 98:69–78

Poeta G, Staffieri E, Acosta ATR et al (2017) Ecological effects of anthropogenic litter on marine mammals: a global review with a "black list" of impacted taxa. Hystric, Italian J Mamm 28:253–264. https://doi.org/10.4404/hystrix-00003-2017

Pörschmann U, Trillmich F, Mueller B et al (2010) Male reproductive success and its behavioural correlates in a polygynous mammal, the Galápagos sea lion (*Zalophus wollebaeki*). Mol Ecol 19:2574–2586

Riofrío-Lazo M, Arreguín-Sánchez F, Páez-Rosas D (2017) Population abundance of the endangered Galapagos sea lion *Zalophus wollebaeki* in the southeastern Galápagos archipelago. PLoS One 12:e0168829

Robinson G, del Pino EM (1985) El Niño en las Islas Galápagos. El evento de 1982–1983. Fundación Charles Darwin para las Islas Galápagos, Quito

Salazar SK (2003) The Jessica oil spill effects on the Galapagos sea lion (*Zalophus wollebaeki*) population. Mar Pollut Bull 42:313–318

Salazar S, Bustamante RH (2003) Effects of the 1997–98 El Niño on population size and diet of the Galápagos sea lion (*Zalophus wollebaeki*). Noticias de Galápagos 62:40–45

Schramm Y, Mesnick SL, de la Rosa J et al (2009) Phylogeography of California and Galápagos sea lions and population structure within the California sea lion. Mar Biol 156:1375–1387

Sivertsen (1953) A new species of sea lion, *Zalophus wollebaeki*, from the Galapagos islands. K Norske Vidensk Selsk Forh 26:1–3

Stoffel MA, Humble E, Paijmans AJ et al (2018) Demographic histories and genetic diversity across pinnipeds are shaped by human exploitation, ecology and life-history. Nature Comms 9:4836

Timmermann A, Oberhuber J, Bacher A et al (1999) Increased El Niño frequency in a climate model forced by future greenhouse warming. Nature 398:694–697

Trillmich F (1979) Galápagos sea lions and fur seals. Noticias de Galápagos 29:8–14

Trillmich F (1981) Mutual mother-pup recognition in Galápagos fur seals and sea lions: cues used and functional significance. Behaviour 78:21–42

Trillmich F (1986) Attendance behaviour of Galápagos fur seals. In: Gentry RL, Kooyman GL (eds) Fur seals: maternal strategies on land and at sea. Princeton University Press, Princeton, pp 168–185

Trillmich F (2015) *Zalophus wollebaeki*. The IUCN red list of threatened species 2015: e.T41668A45230540. dx.doi.org/10.2305/IUCN.UK.2015-2.RLTS.T41668A45230540.en

Trillmich F, Dellinger T (1991) The effects of El Niño on Galápagos pinnipeds. In: Trillmich F, Ono KA (eds) Pinnipeds and El Niño. Springer, Berlin, pp 66–74

Trillmich F, Limberger D (1985) Drastic effects of El Niño on Galápagos pinnipeds. Oecologia 67:19–22

Trillmich F, Trillmich KGK (1984) The mating system of pinnipeds and marine iguanas: convergent evolution of polygyny. Biol J Linn Soc 21:209–216

Trillmich F, Wolf JBW (2008) Parent-offspring and sibling conflict in Galápagos fur seals and sea lions. Behav Ecol Sociobiol 62:363–375. https://doi.org/10.1007/s00265-007-0423-1

Trillmich F, Rea L, Castellini M et al (2008) Age-related changes in hematocrit in the Galápagos sea lion (*Zalophus wollebaeki*) and the Weddell seal (*Leptonychotes weddellii*). Mar Mamm Sci 24:303–314

Trillmich F, Jeglinski JWE, Meise K et al (2014) The Galápagos sea lion: adaptation to spatial and temporal diversity of marine resources within the archipelago. In: Denkinger J, Vinueza L (eds) The Galápagos marine reserve, social and ecological interactions in the Galápagos islands. Springer Science, New York, pp 61–70. https://doi.org/10.1007/978-3-319-02769-2_3

Trillmich F, Meise K, Kalberer S et al (2016) On the challenge of interpreting census data: insights from a study of an endangered pinniped. PLoS One 11:e0154588. https://doi.org/10.1371/journal.pone.0154588ne

Ventura F, Matthiopoulos J, Jeglinski JEW (2019) Minimal overlap between areas of high conservation priority for endangered Galápagos pinnipeds and the conservation zone of the Galápagos Marine Reserve. Aquat Conserv 29:115–126

Villegas-Amtmann S, Costa DP (2010) Oxygen stores plasticity linked to foraging behaviour and pregnancy in a diving predator, the Galápagos sea lion. Funct Ecol 24:785–795

Villegas-Amtmann S, Costa DP, Tremblay Y et al (2008) Multiple foraging strategies in a marine apex predator, the Galápagos sea lion *Zalophus wollebaeki*. Mar Ecol Prog Ser 363:299–309

Villegas-Amtmann S, Atkinson S, Costa DP (2009) Low synchrony in the breeding cycle of Galápagos sea lions revealed by seasonal progesterone concentrations. J Mammal 90:1232–1237

Villegas-Amtmann S, Jeglinski JWE, Costa DP et al (2013) Individual foraging strategies reveal niche overlap between endangered Galápagos pinnipeds. PLoS One 8:e70748. https://doi.org/10.1371/journal.pone.0070748

Wolf JBW, Trillmich F (2007) Beyond habitat requirements: individual fine-scale site fidelity in a colony of the Galapagos sea lion (*Zalophus wollebaeki*) creates conditions for social structuring. Oecologia 152:553

Wolf JBW, Trillmich F (2008) Kin in space. Social viscosity in a spatially and genetically substructured network. Proc R Soc B 275:2063–2069

Wolf JBW, Kauermann G, Trillmich F (2005) Males in the shade: habitat use and sexual segregation in the Galápagos sea lion (*Zalophus californianus wollebaeki*). Behav Ecol Sociobiol 59:293–302

Wolf JBW, Mawdsley D, Trillmich F et al (2007a) Social structure in a colonial mammal: unravelling hidden structural layers and their foundations by network analysis. Anim Behav 74:1293–1302

Wolf JBW, Tautz D, Trillmich F (2007b) Galápagos and Californian sea lions are separate species: genetic analysis of the genus *Zalophus* and its implications for conservation management. Front Zool 4:20. https://doi.org/10.1186/1742-9994-4-20

Wolf JBW, Harrod C, Brunner S et al (2008) Tracing early stages of species differentiation: ecological, morphological and genetic divergence of Galápagos sea lion populations. BMC Evol Biol 8:150

Chapter 9
Southern Elephant Seal (*Mirounga leonina*, Linnaeus 1758)

Mirtha Lewis and Elena Eder

Abstract Among the five seal species inhabiting the southern hemisphere and distributed around and south of the Antarctic Convergence, the southern elephant seal (*Mirounga leonina*, Pinnipedia, Phocidae) has the northernmost distribution, reaching sites surrounded by temperate waters. During the annual cycle, seals spend most of their time at sea, alternating with two highly synchronized shore-based haul-out periods for breeding and molting. On land, the distribution area is restricted to locations on islands or mainland, while at sea they travel distances of thousands of km^2. In the global distribution, there are four genetically identified stocks; two are settled in Latin America (Georgias del Sur-South Georgia and Península Valdés). In this chapter, we focus on the colonies of Argentina, Chile, and the Antarctic Peninsula that have resident animals for breeding and molting and from where tagged seal migration covers a wide latitudinal range from the equator to Antarctica. Long-term and systematic studies at the Península Valdés colony facilitated the comparison with other colonies further south. Annual cycle, demographic, and foraging aspects are described in this chapter emphasizing migrations at sea and distribution. The chapter closes with a consideration of conservation status and threats.

Keywords Southern elephant seal · *Mirounga leonina* · Latin American colonies · Seal distribution

---

M. Lewis (✉) · E. Eder
Centro para el Estudio de Sistemas Marinos (CESIMAR-CONICET), Puerto Madryn, Chubut, Argentina
e-mail: lewis@cenpat-conicet.gob.ar; eder@cenpat-conicet.gob.ar

G. Heckel, Y. Schramm (eds.), *Ecology and Conservation of Pinnipeds in Latin America*, https://doi.org/10.1007/978-3-030-63177-2_9

General Features

Elephant seals are the largest of the true seals and one of the marine mammals with the highest sexual dimorphism. The southern elephant seal (*Mirounga leonina*) is very similar in appearance and movements to its northern congener (*Mirounga angustirostris*), in terms of their fusiform body without externalization of the penis, testicles, and mammary glands, the lack of visible ears, the inability to move the hind flippers forward, and having a caterpillar movement on land (Le Boeuf and Laws 1994; Berta et al. 2015). In both species, adult males show noticeable secondary sexual characteristics to compete for females during breeding, like an elongated erectile proboscis which has a resonance function to vocal challenges (Sanvito et al. 2007), a thick cornified skin on the neck, and large canine teeth (Berta et al. 2015). The proboscis of the northern male is larger and the neck skin and the chest are more developed in the southern male. Females of the two species do not differ significantly in body mass; in contrast, males of the southern species become longer and heavier than the northern species (Le Boeuf and Laws 1994; Fig. 9.1).

Southern elephant seals exhibit a marked sexual dimorphism in adulthood: adult males up to 4.5 m average length, from nose to tail, while females less than 3.5 m. Adult males can weigh up to 3 tons (Le Boeuf and Laws 1994), while weight in females can reach 700 kg (Table 9.1). Maximum weight recorded in a pregnant female was 974.5 kg in the Antarctic Península area (Negrete pers. comm. in Eder et al. 2019a). Pups are born with 44 kg and a mean size of 1.32 m from nose to tail

Fig. 9.1 Southern elephant seals during the breeding season. The dominant male is displaying a threat vocalization to a competitor and a female is beside with a pup of an age close to weaning.

Table 9.1 Features in morphology and behavior used to classify southern elephant seals by age and social behavior at the time of the annual cycle in which they are most frequently seen. Offspring age is in days. Standard length is measured from nose to tail. The data belong to the Península Valdés population (Campagna et al. 1992, 2007, Lewis et al. 2001, Lewis and Almonacid (unpublished and validated from Le Boeuf and Laws 1994))

Sex and category		Age	Length (m)	Body mass (kg)	Pelage	Distinctive features	Time ashore and social behavior
Offspring	Pup	0–23 d	1.25	43.7	Black fur	Natal lanugo at about 1 month old	Breeding in the harem associated to female
	Weanling	24–70 d	1.4	~131	Soft hair, silver gray on the back and lighter on the belly	Bumbling movements on land	Post-breeding in groups of similar category
Juvenile	Yearling	1	1.8	183	Thick hair and light brown or yellowish brown	Fusiform and streamlined body, typical of phocids and can be confused with other seal species	Autumn-winter haul-out/molting mostly solitary
	Juvenile	2–3	2	~ 250	Light yellowish brown and short dense gray after molting		Autumn-winter haul-out/molting. Vagrant
Female	Female adult	3–12+	2.5–3.5	400–700	Short, dense, thick, and stiff hair. Color between brown, light yellowish beige, and deep gray after molting	Some scars around the neck. Along the age, the morphology is between older juvenile and subadult 1	In breeding/molting colonies and synchronized haul out ashore during the annual cycle.

(continued)

Table 9.1 (continued)

Sex and category		Age	Length (m)	Body mass (kg)	Pelage	Distinctive features	Time ashore and social behavior
Males	Subadult 1	4–5	2.8(2.5–3)	400–700	Thick and stiff hair in brown, beige yellowish and changing to dark grey after molting	Similar size of female with big and square head. Two incipient canine teeth	Molt in wide range of distribution. Vagrant in colonies at the end of breeding
	Subadult 2	6–7	3.5(3.2–3.7)	800–1000		Thickened skin around the neck and presence of scars. Incipient proboscis	In colonies at the end of breeding. Interactions between them are more to play than aggressiveness purpose. Vagrant long distances
	Subadult 3	8	4(3.8–4.5)	1000–1500		Distinctive nose of the species, crossed by two transversal grooves. Scars are present around the neck skin	Breeding/molting colonies. Fight in breeding season. Bachelor and a few reach the dominance in small harems
	Subadult 4 Adult	9+	>4.5	> 2000		Skin on the neck with grooves that give wrinkled appearance and many scars. Full developed proboscis falling in front of the face.	Breeding/molting. Compete for hierarchical dominance in harems

without evident differences between sexes at birth or weaning (Campagna et al. 1992). During the nursing period, they quickly modify their elongated appearance, becoming more round and robust. Juveniles have no bulbous nostrils and the sexual dimorphism is not apparent in weight and length. Changes between sexes can be visually differentiated starting at age 5, when male canine teeth come out, and they experience an accelerated growth in their body mass, achieving body sizes much larger than females at the same age (Berta et al. 2015). Males reach sexual maturity (puberty) when 5 years old; however, they begin to reproduce around 3–5 years later, after gaining significant size and social maturity (Le Boeuf and Laws 1994; Carlini et al. 2006). Females are sexually mature and begin to breed at the age of 3–4 years and may live over 20 years. Longevity is shorter in males, which rarely survive more than 15 years (Bastida and Rodríguez 2010). At Península Valdés, the oldest breeding female was re-sighted when it was 19 years old. Although pupping intervals appear to be 1 year, females may often skip breeding to achieve higher lifetime reproductive output (Shaughnessy 1999; Desprez et al. 2018; Griffen 2018).

Another distinctive feature of the species is its behavior at sea. Despite being an air-breathing mammal, elephant seals annually spend more than 90% of their time feeding at sea, diving continuously at depths between 200 and 700 m (up to 2000 m+), for periods of 20–30 min (up to 120 min), on a regular basis of 50–80 dives per day. These skills are the result of specialized anatomical and physiological adaptations to marine life, making them colossal deep divers (Kooyman 2006; Berta et al. 2015). They spend most of their time deep diving, returning briefly to the surface to replenish their oxygen stores between consecutive long dives (Hindell et al. 2016). The diving pattern of southern elephant seals is the result of oceanographic context and their foraging strategies, which allow them to cope with fasting on land. After the annual breeding and molting terrestrial periods, seals may have lost a significant amount of their body mass, mostly in lipid stores (up to 50% in breeding females, Carlini et al. 1999).

Annual Cycle

The annual cycle of elephant seals is characterized by two pelagic phases interrupted by two periods ashore (Fig. 9.2). The seals forage exclusively at sea, spending most of their time feeding (80%) in order to increase energy reserves that will be used to support terrestrial haul-out periods in complete fasting (Hindell and Perrin 2009). Adults usually spend 10 months at sea, distributed into a short post-breeding phase (2–3 months) and a longer post-molt foraging trip (7–8 months; Le Boeuf and Laws 1994). Immature individuals alternate with adults their terrestrial phases to molt and to rest haul-out (Hindell and Burton 1988; Lewis et al. 2004). The annual cycle of each species of elephant seal, *M. angustirostris* and *M. leonina*, occurs on a shifted time scale but with a similar haul-out pattern to breed, molt, and rest (juveniles haul-out). There are slight shifts in peak events, such as the maximum number of females during the austral breeding season (Fig. 9.2). Southern

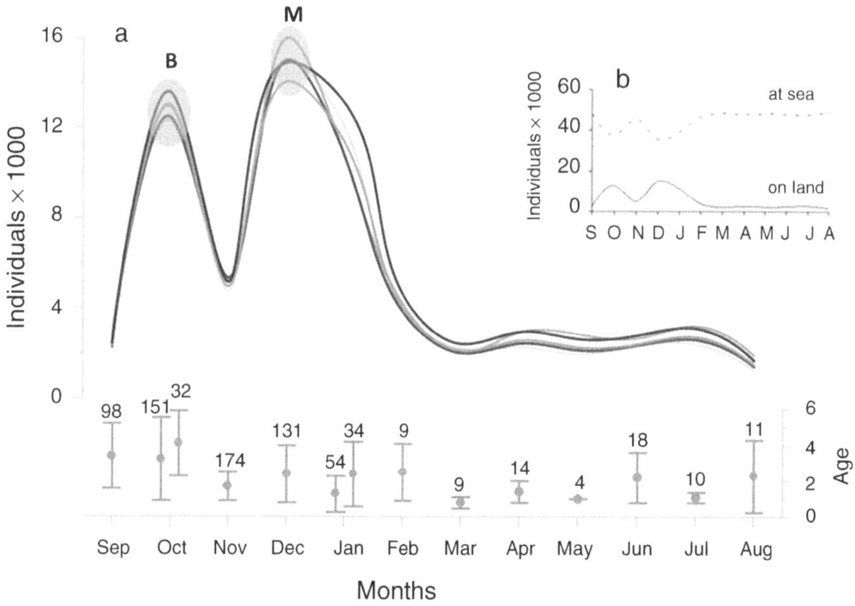

Fig. 9.2 (**a**) Schematic representation of five annual cycles of southern elephant seals showing the synchronism of relative number of animals ashore throughout the year at Península Valdés. B, breeding; M, molt. Below, age distribution of known-age individuals re-sighted during counts. (**b**) Smoothed curve was derived to estimate the number of animals at sea

elephant seals breed during the spring, from August through November, and their rookeries are widely distributed across a latitudinal gradient of 20°, from Península Valdés 42°50'S to Livingston Island 62°37'S. Along the breeding season, the maximum number of females ashore is delayed in correlation to latitude increase; the northernmost colony reaches its peak by October 3, while Antarctica's colonies reach it by October 22/31 (Campagna et al. 1993; Galimberti and Boitani 1999; Carlini et al. 2006). The effect of the daylight length was proposed as a determinant cause of the seasonal timing and sequence in adult's annual cycle (Campagna et al. 1993). Blastocyst implantation occurs during the molting season (Crocker et al. 2001), and it is affected by daylight dynamics; as a consequence, embryonic development differs by 3 weeks between colonies. This timing difference affects the whole annual cycle calendar along the latitudinal range, in such a way that early molting at Península Valdés prevents breeding seals from being highly exposed to thermal stress due to warmer weather (max range 23°–26°C) in late October and November (Campagna et al. 1993).

During the breeding season, the bulk of both sexes' mature population are present at breeding sites, with females spending nearly a month while males may stay one to 3 months ashore. The breeding season in Latin American colonies begins when the first adult males haul out in August and continues up to mid-November when the last males ashore return to sea after all the females have been mated (Campagna et al. 1993). At the end of the breeding season, most adults return to sea, and weaned pups stay on the beach developing their physiological and aerobic capacities for their first pelagic phase (Falabella et al. 1999). Concurrently with weanlings that venture to their first feeding trip, yearlings and juveniles of both sexes begin to haul out for their annual molt at the end of spring. Molting season continues up to the ending of the austral summer (December–March), when different sex and age categories alternate their presence at haul-out sites and older juveniles may take a shortly facultative haul-out during the following months of the austral autumn and winter (Lewis et al. 2004). Although the purposes and duration of these haul-outs remain unclear, changes in buoyancy while foraging at sea appear to influence juveniles' decisions to return to land (Orgeret et al. 2019).

At Península Valdés, the molting season begins in early-mid November with the first youngest juveniles arriving ashore. Adult females haul out to molt by December, a time that coincides with the largest number of animals ashore in the year (Fig. 9.2). Molting continues with the arrival of males on the second half of January up to March and until the end of the austral summer (Lewis et al. 2004). Cycles described show virtually the same pattern at all colonies. However, at the southernmost colonies, the maximum number of animals ashore is related to the breeding season. Molt process duration varies with seal body size (e.g., bigger individuals would take longer time to molt); although there is no regular order, the head, back, and flippers appear to be the last to shed (Le Boeuf and Laws 1994); this process usually takes one month of fasting ashore. During molting, elephant seals settle in large and crowded groups; this behavior helps reduce high metabolic costs of fasting and molting. At this time, seals are likely to be more sensitive to heat losses because of their active peripheral blood circulation that supplies nutrients for epidermis renewal and hair growth (Boyd et al. 1993; Chaise et al. 2019).

The number and age category of individuals seen at haul-out sites varies along the year; however, individuals settle in groups of homogeneous age categories (Fig. 9.2). Owing to their site fidelity, seals of both sexes and different ages return to their natal sites or roughly to the same site from year to year (Fabiani et al. 2006). Most of the adult females marked at Península Valdés (86%) were re-sighted within 0–3 km from where they had been previously seen. At least 15 females marked as weaned pups reproduced within 1 km from the place where they were born (2 overlapped with their mothers), and some of them were re-sighted during 8 seasons (Lewis et al. 1996).

Worldwide and Latin American Distribution and Abundance (Including Prior to Hunting)

The distribution of elephant seals has extreme contrasts. The southern species breeds in isolated groups from sub-Antarctic islands to temperate locations such as Península Valdés, and its northern counterpart has a comparatively limited breeding range that reaches subtropical islands with extensive interchange between colonies.

The southern elephant seal has populations separated in four stocks with limited interbreeding, over circumpolar range in the Atlantic, Indian, and Pacific Oceans, roughly 90°, 107°, and 163° longitudes apart (Le Boeuf and Laws 1994, Table 9.2,

Table 9.2 Abundance of southern elephant seal populations in 2000s/2010s compared with earlier data

Breeding groups		Pup production	Population size 1990s	Population size 2000s/2010s	Population trend	References
Atlantic Ocean	South Georgia Island	–	357,000	397,054	Stable	Hindell et al. (2016)
	Malvinas-Falkland Islands	532 (2007)[b]	1,827	1,872[a]	Stable	Galimberti et al. (2001), SCAR-EGS (2008)
	25 de Mayo/ King George Island	290–400 (2003)[b]	–	1,015[a]–1,400	Stable	SCAR-EGS (2008)
	Livingston Island	1510	2,528	5,530	Increasing	Torres (1981), Gil-Delgado et al. (2013)
	Nelson Island	135 (2001)[b]	–	–	Uncertain	Carlini et al. (2003)
	South Orkney Island	<100 (1985)[b]	–	–	Uncertain	McCann (1985)
	Gough Island	18	105	63	Decreasing	Bester et al. (2001)
	Bouvet Island	89	–	308	Uncertain	Kirkman et al. (2001)
	Península Valdés and Patagonian coast (Argentina)	16,200 (2015)[b]	33,726	56,700[a]	Increasing	Campagna and Lewis unpublished
Pacific Ocean	Fueguin Fjords (Chile)	–	–	247 (2012)[b]	Uncertain	Acevedo et al. (2016)
	Macquarie Islands stock	–	78,212	60,561	Decreasing/ stable?	Hindell et al. (2016) and references therein
Indian Ocean	Kerguelen Islands stock	–	188,669	219,157	Stable	Hindell et al. (2016) and references therein

[a]Total number of seals > 1 year old is estimated by converting pup production by a raising factor of 3.5 (McCann 1985). [b]Last year of the available data

Fig. 9.3). Each stock integrates subpopulations of variable size. The South Georgia stock is the largest one and has colonies at Georgias del Sur-South Georgia, South Orkney, South Shetland, South Sandwich (in the Scotia arc), Gough, Bouvet, and Malvinas-Falkland Islands. Genetic flow exists among these colonies, particularly between Georgias del Sur-South Georgia, Malvinas-Falkland, and South Shetland

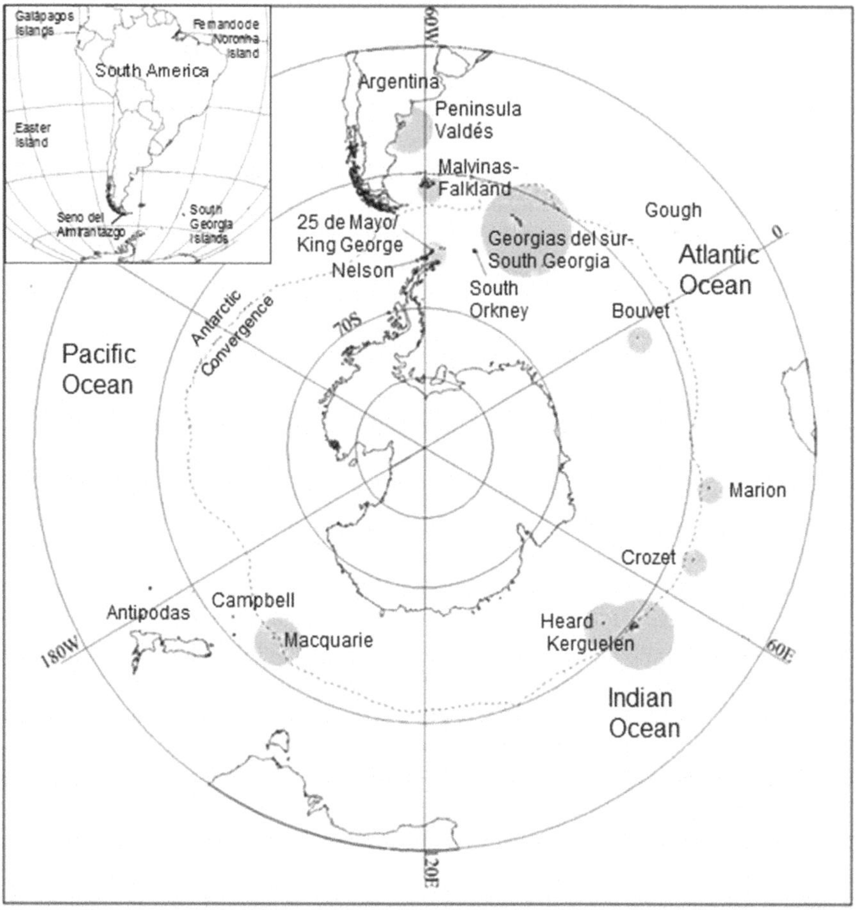

Fig. 9.3 Global distribution of southern elephant seal breeding colonies. Size circles are proportional to the estimated population sizes presented in Table 9.2

Islands (Fabiani et al. 2006; Lanusse et al. 2018). These locations share the Atlantic region with the population of Península Valdés as a genetically differentiated group, though Malvinas-Falkland could have a possible role as a gene flow conduit between the Georgias del Sur-South Georgia and the Península Valdés populations (Galimberti et al. 2001; Hoelzel et al. 2001). The Kerguelen stock is the second in population size and includes the subpopulations of Kerguelen, Heard, Marion, Prince Edward, Crozet, and Possession Islands. The Macquarie stock includes Macquarie, Campbell, Auckland, and Antipodes Islands (Le Boeuf and Laws 1994). In Latin America, observations of tagged and tracked seals reported on more widespread locations within the region (Campagna et al. 2006, 2007; Lewis et al. 2006a), which suggests movements between Península Valdés, Malvinas-Falkland Islands,

Seno del Almirantazgo, and Georgias del Sur-South Georgia elephant seal colonies. However, the geographic distance does not support the gene flow and divergence between the Península Valdés and Georgias del Sur-South Georgia colonies. There is almost an order of magnitude of greater divergence between the two populations, and although they are in the same oceanic region, Península Valdés may be the remainder of a population distributed primarily on the southern edge of the continental masses of Australia, Africa, and South America (Slade et al. 1998). Some seals from Península Valdés and the Malvinas-Falkland Islands were identified on an incipient breeding group in southern Chile, and mtDNA analysis suggested a greater genetic relationship with the second one (Acevedo et al. 2016).

Site fidelity and living in colonies during breeding and molting seasons contrast with the dispersion of vagrants observed in a broad range of latitudes throughout the south Atlantic and Pacific Ocean. Haul-out site selection is generally characterized by gradual slope, sandy beaches, next on pebbles or rocky shores of mainland coast, oceanic islands, and freshwater rivers (Lewis et al. 2006a; de Moura et al. 2011; Acevedo et al. 2016; Páez-Rosas et al. 2018, Table 9.3). The distribution and migrations of the species are influenced by marine currents where physical and oceanographic characteristics impact on prey availability and expand foraging areas from the Southern Ocean to temperate waters (Páez-Rosas et al. 2018). In Ecuador and Panama, vagrants were identified as southern elephant seals regarding their size and external morphology; most of the re-sights were subadult males and juveniles (Lewis et al. 2006a; Páez-Rosas et al. 2018; Redwood and Félix 2018).

Estimation of southern elephant seal population abundance is based on censuses of the main colonies during the breeding season. Global estimation of pups born is over 200,000; more than 50% of these births occur at Georgias del Sur-South Georgia Islands and Península Valdés (Table 9.2). The abundance of the main breeding groups has been stable over the past three generations (Forum for the Conservation of the Patagonian Sea 2019; Hofmeyr 2015), although there is a lack of current estimates throughout the entire population (Eder et al. 2019a, Table 9.2).

During the eighteenth and nineteenth centuries, an intensive commercial harvest of elephant seals took place at Georgias del Sur-South Georgia (Le Boeuf and Laws 1994). The indiscriminate slaughter of adult males was recurrent, being the main interest of the activity the oil derived from blubber of the largest individuals (more than 300 liters per individual, Bastida and Rodriguez 2010). After 1900, several ordinances were issued by the Malvinas-Falkland Islands government to manage commercial seal hunting sustainably at surrounding areas, until the activity became unsustainable, ending in 1964 (Le Boeuf and Laws 1994). The decrease in adult male population was further supported between 1948 and 1951 by observations of several key demographic parameters like the sex ratio on the breeding beaches, breeding group sizes, and low average age of males on the breeding beaches (Laws 1956). More than 250,000 animals were presumably hunted at Georgias del Sur-South Georgia during the twentieth century (Bastida and Rodriguez 2010). In 1951, Georgias del Sur-South Georgia's total population was about 338,000 seals; 20 years later, the population size was adjusted by the construction of life tables for each sex, resulting in 357,000 individuals (McCann 1985). It has remained stable since then (Hindell et al. 2016).

Estimation of the total population (individuals older than 1 year) in other groups was made by converting the number of pup counts at the end of the breeding season by a factor of 3.5 (McCann 1985). The worldwide population was estimated to be 650,000–750,000 in the mid-1990s (Le Boeuf and Laws 1994). Some populations have remained stable like Georgias del Sur-South Georgia, Kerguelen, and Heard Islands; others, such as Península Valdés, have increased, and some have continued decreasing (Macquarie and Marion Islands, McMahon et al. 2005) (Table 9.2). The reasons for these declines are still not fully understood; albeit most of the studies suggest they might be related to variations in food availability due to changes in ocean dynamics (Hindell 1991; Le Boeuf and Laws 1994; McMahon et al. 2005; van den Hoff et al. 2007, 2014). The large decreases in population size throughout most of the circumpolar distribution appear to be over (McMahon et al. 2005; Pistorius et al. 2011), and new breeding groups were identified along the Latin American continent in Chile and Argentina. Although there is no recent estimation of South Atlantic high-latitude island colony abundance, the total size might be approaching 460,000 individuals (Table 9.3 and references therein).

Atlantic Colonies

Elephant seals have been recorded along the coast of Argentina since the early 1800s; however, the first reports of breeding seals date back to the 1940s (Carrara 1952). Empirical models of population growth rate backward suggest that the Península Valdés colony could have been established early in the twentieth century, with a rate of 100 pups born per year (Ferrari et al. 2013).

The large breeding group settled at Península Valdés, Chubut, Argentina, is the only one among the globally identified stocks that has shown positive increments during the last decades (Campagna and Lewis 1992). Currently, it is considered stable with an annual increase rate <1% between 2001 and 2015 and 16,200 born pups per year (Ferrari et al. 2013; Campagna and Lewis unpublished). Moreover, during the last three decades, the colony has expanded southward outside the limits of the peninsula, a process that might have been promoted by the migration of subadult males in order to occupy peripheral expansion areas (Bonfili et al. 2019). In recent years, new breeding areas were registered in Chubut and Santa Cruz provinces, hundred kilometers south of the original area (Campagna, unpublished). There is no published data to substantiate this expansion as a process of recolonization, since no archaeological remains of this species were found in the historical records of the area (González 2020). Moreover, elephant seals have never been subjected to intensive commercial harvest in Patagonia; only a few males were hunted during the sea lion's commercial harvest in the mid-nineteenth century on the Argentinean coast (Carrara 1952).

Table 9.3 Latin American distribution of southern elephant seals for (B) breeding, (M) molting, and occasional haul-outs. Occupation sites by country where the seals were seen on the mainland and inshore with habitat descriptions of occupied sites. The sub-Antarctic Islands included were those locations in the distribution range of tagged animals recovered or tracked. Animals ashore described as R, resident; V, vagrant

Location	Area	Seasonality	Occurrence	Habitat	References
Brazil	Eight coastal states and Fernando de Noronha Island	Autumn-winter haul-out	V	Sandy beaches of the subtropical region	De Moura et al. (2011) and Lodi and Siciliano (1989)
Uruguay	Shoreline of Río de la Plata	Autumn-winter haul-out	V	Grassland and sandy beaches – freshwater coasts	Lewis et al. (2006a)
	Two coastal departments (Rocha-Maldonado)	Summer	V	Sandy beaches	Daneri (2009)
Argentina	Península Valdés	B and M	R	Sandy and pebblestone beaches, rocky shelves with tide pools	Lewis et al. and Retana et al. (2013)
	New breeding sites southward of Chubut	B and M autumn-winter haul-out	R	Pebblestone beaches and shelves with tide pools	Campagna and Lewis unpublished
Chile	Fueguin Fjords	B and M	R	Rugged coastline with few grassland and sandy beaches	Acevedo et al. (2016)
	Mainland and oceanic islands	M and autumn-winter haul-out	V	Sandy beaches and irregular geographic profile	Acevedo et al. (2016)
Peru	Punta San Juan	M	V	Rocky shore sharing with fur seals	Lewis et al. (2006a) and Acevedo et al. (2016)
Ecuador	Guayaquil Gulf and rivers therein	M	V	Shoreline and estuarine-riverine	Páez-Rosas et al. (2018)
Panama	Panama Gulf and islands	M	V	Sandy beaches	Redwood and Félix (2018)
Malvinas-Falkland Islands	Sea Lion Island	B and M	R	Sandy beaches	Sanvito and Boitani (2000)

(continued)

Table 9.3 (continued)

Location	Area	Seasonality	Occurrence	Habitat	References
Shetland Islands	Livingston, 25 de Mayo/King George, Nelson and Elephant Islands	B and M	R	Sandy and stony beaches with moss beds and flat rocks	Cruwyis and Davis (1994) and Vergani and Stanganelli (1990)
South Orkney Islands	Signy Island	B and M	R	Ice-free, sandy beaches with presence of grass tundra, mosses, and small ponds	Laws (1956)

Subadult males from the nearest colony of Malvinas-Falkland Islands were sighted at Península Valdés and on Patagonian coasts (Lewis et al. 2006a). On the other hand, studies of satellite tracking have shown young juveniles (1–3 years old) from Península Valdés migrating to high latitudes, to Malvinas Falkland Islands, to Georgias del Sur-South Georgia Island, and to the Pacific Ocean crossing the Drake Passage and the Magellan Strait after the molting season (Campagna et al. 2006, 2007; Eder et al. 2010). Migration between colonies can be related to the fact that both colonies forage over the Patagonian shelf and its break (Campagna et al. 2006). This shelf is characterized for having an area of 1×10^6 km^2 with a depth of less than 150 m and a width between 300 and 700 km. Seals from both colonies have to travel long distances in order to reach the foraging areas (Falabella et al. 2009), around 200 km from the Malvinas-Falkland colony and more than 500 km from the Península Valdés colony.

In the case of other colonies, highly productive shelf waters are localized within 9–100 km away offshore (Campagna et al. 2007). Georgias del Sur-South Georgia elephant seals prefer sub-Antarctic waters, and only few adventure to the Patagonian shelf and its slope or to waters surrounding the Malvinas-Falkland Islands (McConnell and Fedak 1996). This might be the reason why sub-Antarctic migrants have never been seen on the Patagonian coast.

Pacific Colonies

According to historical records, elephant seals were distributed in the eastern South Pacific on the oceanic islands of the Juan Fernández Archipelago and on Chile's mainland (from 37°S to Cape Horn Archipelago, Acevedo et al. 2016 and references therein). The Robinson Crusoe Island, at Juan Fernández Archipelago, was the main breeding group in Chile but disappeared before the twentieth century due to commercial hunt (Andrews 1914). However, since the 1970s, seals have been observed again mainly on the southern tip of Chile (Acevedo et al. 2016) as a recolonization process of the traditional sites (Torres 1981). The species has settled south of 51°S in Tierra del Fuego Island (Gibbons and Miranda 2001; Cáceres 2013) and on the mainland coast (52°14′S, Acevedo et al. 2016). Records suggest this is a resident group with immigration of juveniles and subadult males from Península Valdés and Malvinas-Falkland Islands (Torres et al. 1979). Residents are identified for their site fidelity; tagged subadult males were seen 1 year after being tagged. Satellite tracking of these males has shown that they travelled along the channels and fjords of southern Chile and the open Pacific Ocean in order to return to the area (Acevedo et al. 2016). One of these males travelled 33,336 km (Wildlife Conservation Society 2011), and one female less than 1 year old crossed the Magellan Strait and several channels to reach the open waters of the Pacific Ocean (Hückstädt et al. 2008).

Breeding Biology

Southern elephant seals have one of the most highly polygynous mating systems among mammals (Riedman 1990; Berta et al. 2015). Adult females gather in aggregations known as harems variable in size and in distribution. Coastal distribution depends on site topography and tidal amplitude, and the distribution is unique for each colony. At Península Valdés, the shape of the breeding assemblages and the wide female dispersion promote low-density conditions for reproduction, even during extreme high tides (Baldi et al. 1996). The average size of the harem there is of 22 females and can reach a size of 150 females. In contrast, harems at Georgias del Sur-South Georgia have an average size of 70 females with a maximum of 250 females per harem (Baldi et al. 1996).

Each harem is monopolized by a single large male, which hierarchically dominates the access to sexually receptive females using impressive behaviors like straightening upon its own lumbar waist and vocalizing to challenge and fight against competitor subordinate males. Vocal signaling is an important component of male behavior during the breeding season, as it is used to settle agonistic encounters. It is characterized by low-pitched sounds, made up of pulse trains, low-frequency modulation, and high sound level (Sanvito and Galimberti 2000). The breeding success is attained by less than 30% of the males (Fabiani et al. 2004), and few adults manage to inseminate more than 100 females per season (Bastida and Rodriguez 2010). Before males can compete successfully for hierarchical dominance of a harem, they must achieve a full body size (Table 9.1). Male size is strongly related to mating success, achievement of a higher social rank, and winning fights, although fighting experience may also influence. During the breeding season, males fast around 60 days, losing more than 12 kg/day (Le Boeuf and Laws 1994).

Adult females prefer to breed in large harems, which appears to be a strategy to avoid male harassment (Cassini 2000; Galimberti et al. 2000; Fabiani et al. 2006). Although this behavior is less frequent in phocids than in otariids, it is common in elephant seals at two moments: when females arrive at the breeding site (before joining harems) and when they return to sea (Galimberti and Boitani 1999); the latter is the most common. Pregnant females start hauling out to breed in early

September and stay during 31–35 days. They appear to arrive with high tide in order to reduce the distance between the shoreline and the harems to avoid harassment (Galimberti and Boitani 1999). About a week after arriving, they give birth to a single pup and nurse it for 23 days (southern species) to 28 days (northern species) (Campagna et al. 1993; Le Boeuf and Laws 1994). While nursing the pup, 18 days' postpartum, females become receptive to mate again, and they copulate more than once, until they return to the sea after abruptly weaning their pups.

Parturition appears to be more frequent at dusk or at night (Campagna et al. 1993), and twin births occur infrequently (McMahon and Hindell 2003). Females are extreme capital breeders because they rely entirely upon their stored energy reserves to nurse and wean their pups successfully (Pistorius et al. 2004), losing around 8 kg/day during lactation (Le Boeuf and Laws 1994).

During nursing, pup body mass increases at a rate of 4 kg/day and reaches a body weight 200% higher than their birth weight at the end of the lactation period (Campagna et al. 1992). Pup growth rate is directly proportional to its mother's size and is a good proxy for the female's foraging success (Ducatez et al. 2008). Once pups are weaned, their energy reserves will fuel a fasting period of 1–3 months. During this time, weanlings remain ashore improving their foraging skills and diving ability, losing between 0.80 and 1 kg/day (Campagna et al. 1992; Lewis et al. 2001). Diving skills in addition to their ability to avoid predation pressures will be crucial factors for survival and assure that the yearlings return to the coast (Orgeret et al. 2019). Mortality during the first trip is around 40% (Pistorius et al. 2011) compared to a low mortality rate of 2–5% during the nursing period (Campagna et al. 1992).

Foraging Ecology

Bio-logging technologies and satellite tracking have been crucial for the study of the elephant seal's pelagic phase (Hindell et al. 1991, 2016; Le Boeuf and Laws 1994; Campagna et al. 1995, 1998, 1999, 2006, 2007; McConnell and Fedak 1996; Slip 1997a, b; Jonker and Bester 1998; Bornemann et al. 2000; Bailleul et al. 2007; Tosh et al. 2009; McIntyre et al. 2011, 2012; Muelbert et al. 2013; O'Toole et al. 2014; McGovern et al. 2019). Southern elephant seals are extraordinarily wide-ranging, pelagic, and long, deep, continuous divers. While they forage at sea, they travel across a wide range of oceanographic regions and spend significant time in highly productive water masses, like fronts, currents, and marginal pack ice zones (Campagna et al. 2006, 2007; Bailleul et al. 2007; Biuw et al. 2010; McIntyre et al. 2011, 2012; Tosh et al. 2015; Paez-Rosas et al. 2018). There is a sexual segregation in the foraging areas; subadult males forage at continental margins (Hindell et al. 1991; Bailleul et al. 2007; Labrousse et al. 2017; Malpress et al. 2017), while adult females do so in deep waters at ocean frontal systems or within the marginal sea ice over the Antarctic shelf or shelf break of the Western Antarctic Peninsula (Bailleul et al. 2007; Costa et al. 2010; Hückstädt et al. 2012; Muelbert et al. 2013; van den

Hoff et al. 2014; Hindell et al. 2016). Young seals can travel as far as adults; although sexual dimorphism may not be apparent yet, spatial distribution of young males and females are similar to adults of the same sex (Field et al. 2005; Eder et al. 2011; Orgeret et al. 2019).

Seal dives have different characteristics depending on the activity being performed, such as transit between foraging sites, foraging, resting, and/or food digestion (Le Boeuf and Laws 1994; Sala et al. 2011). Foraging dives are usually associated with higher resident times in restricted areas (Campagna et al. 2006, 2007; McIntyre et al. 2011; Hindell et al. 2016) and characterized by long and deep dives (except over shallow shelf), meander movements with bursts of speed, steep angles both to descend and ascend along the water column, and vertical movements of the seal's head during prey capture (Campagna et al. 2006, 2007; Sala et al. 2011; McGovern et al. 2019). In contrast, transit dives to move between foraging locations are shallower, more linear, faster with higher flipper stroke frequency, involve farther straight-line distances, and are shorter, likely because oxygen stores are depleted more rapidly, reducing the aerobic dive limit (McGovern et al. 2019). Resting dives are shallower, longer, and slower and have lower stroking rates and higher variation in pitch and roll angle of body when descending (McGovern et al. 2019; Sala et al. 2011). Seals minimize swimming effort and save energetic costs of foraging, gliding at slower speeds or sinking passively when descending at any dive type (Richard et al. 2014), and maintaining a nearly constant swimming speed when ascending as their buoyancy increases. Seals exhibit a diel pattern with vertical migration during foraging dives, which may also persist in other types of diving, highlighting the plasticity of foraging behavior and strategies among populations of southern elephant seals (Biuw et al. 2010; McIntyre et al. 2011; Hindell et al. 2016; McGovern et al. 2019). Dive types and their behavior have been extensively studied in most of the populations, and it has been possible to establish a relation with the oceanographic dynamics of the water column.

At the Península Valdés colony, most of the animals need to cross the continental shelf to reach deeper waters at the shelf break and the Argentine Basin (Campagna et al. 1999, 2007, Fig. 9.4). However, at least 10% of the females, some juveniles, and subadult males remain foraging over shallow waters (less than 150 m deep). These differences in the bathymetry context are correlated to higher contribution of demersal, pelagic, bentho-pelagic, and benthic species in the diet of individuals that used the continental shelves (Green and Burton 1993; Eder et al. 2010). Differences in the trophic levels between sexes and age groups suggest resource partitioning, which is supported by individual patterns of tracking animals at sea (Lewis et al. 2006b; Newland et al. 2009, 2011).

There is no unique pattern to describe feeding habits in southern elephant seals due to their wide-ranging and diving behavior and the influence of many variables such as variations related to sex, age, reproductive status, season, and prey distribution and abundance (Rodhouse et al. 1992; Le Boeuf and Laws 1994; Green et al. 1998; Bowen and Siniff 1999; Daneri et al. 2000; Piatkowski et al. 2002; Bradshaw et al. 2003; van den Hoff et al. 2003; Field et al. 2007; Burdman et al. 2015). The traditional techniques to study diet (stomach content analysis and hard parts

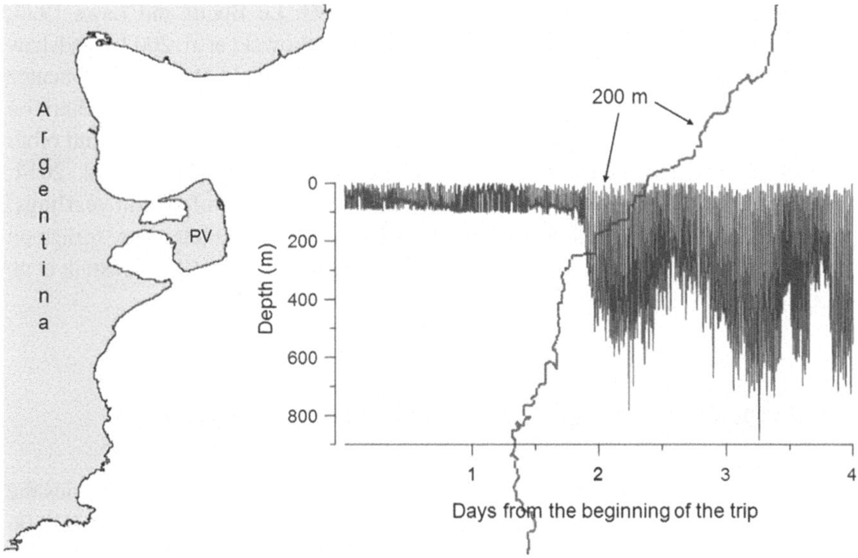

Fig. 9.4 Diving behavior of an adult female tracked on and off the continental shelf. The profile of each dive, depth, and duration suggests the activity carried out by the individual

recovered in the feces) indicate a predominant contribution (75–98%) of cephalopods (especially squid) and a lesser input of fish (18–25%) (Kastelein et al. 1991; Rodhouse et al. 1992; Daneri et al. 2000, 2015; Daneri and Carlini 2002). Diet may often include crustaceans, bivalves, gastropods, and polychaete worms (Clarke and MacLeod 1982; Rodhouse et al. 1992; Green and Burton 1993; Daneri et al. 2000; Daneri and Carlini 2002; Field et al. 2007). Pelagic prey appeared exclusively in female diet, while males often included demersal/pelagic and benthic prey (Rodhouse et al. 1992; Le Boeuf and Laws 1994; Daneri et al. 2000). More recently, indirect methods include the analysis of stable isotopes of carbon and nitrogen (Cherel et al. 2008; Ducatez et al. 2008; Lewis et al. 2006b; Eder et al. 2010, 2019b; Hückstädt et al. 2012; Orgeret et al. 2019), quantitative fatty acid signature analysis (Brown et al. 1999; Bradshaw et al. 2003; Banks et al. 2014), and the use of animal-borne video cameras (McGovern et al. 2019). Stable isotope values of carbon ($^{13}C/^{12}C$, or $\delta^{13}C$) and nitrogen ($^{15}N/^{14}N$, or $\delta^{15}N$) in the tissues of predators reflect those from assimilated prey (Kelly 2000) and are applied to study individual niche width and foraging strategies, since large-scale variations in $\delta^{13}C$ and $\delta^{15}N$ values are useful to reconstruct migratory movements and to determine foraging habitats (Newsome et al. 2006; Cherel and Hobson 2007; Hückstädt et al. 2012; Rita et al. 2017). All these methods (both traditional and more recently used) have biases associated with the characteristics of each method (Bowen and Iverson 2012). However, overall results indicate that southern elephant seals have a generalist diet with intraspecific variations between populations and seasons and show plasticity in prey selection (Green and Williams 1986; Clarke and MacLeod 1982; Kastelein et al.

1991; Rodhouse et al. 1992; Green and Burton 1993; Le Boeuf and Laws 1994; Daneri et al. 2000, 2015; Daneri and Carlini 2002; Piatkowski et al. 2002; Bradshaw et al. 2003; van den Hoff et al. 2003; Lewis et al. 2006b; Field et al. 2007; Ducatez et al. 2008; Eder et al. 2010; Newland et al. 2009). They feed in deep waters on mesopelagic prey of low trophic levels, mainly on squid and myctophids and other mesopelagic fish (Bradshaw et al. 2003; Cherel et al. 2008; Banks et al. 2014; McGovern et al. 2019). Some reports highlight the importance of krill in yearlings' diet (Walters et al. 2014; Lübcker et al. 2017). In the neritic habitats, they forage on demersal and benthic prey of higher trophic levels (Eder et al. 2010; Hückstädt et al. 2012; Orgeret et al. 2019).

Conservation

The International Union for Conservation of Nature (IUCN) currently classifies the species as *Least Concern*, both at global (Hofmeyr 2015) and regional scale (DS 06 MMA 2017; Forum for the Conservation of the Patagonian Sea 2019; Eder et al. 2019a) and *Vulnerable* (VU) in South American continental Chile (DS 06 MMA 2017). Haul-out sites located south of 60°S, where the species breeds and occurs, are under regulation of the Convention for the Conservation of Antarctic Marine Living Resources. The Falkland Islands Dependencies Conservation Ordinance provides protection for southern elephant seals in Georgias del Sur-South Georgia and South Sandwich Islands.

At the current time, the species has an economic value for tourism as cultural and recreation ecosystem service at Península Valdés, Malvinas-Falkland Islands (Galimberti et al. 2001; González et al. 2019), and Seno del Almirantazgo, Tierra del Fuego, in Chile (Miranda 2013), although the latter may be a source of stress, as the groups are very small (Gibbons and Miranda 2001; Cáceres 2013; Le Boeuf and Campagna 2013).

Península Valdés in Argentina is a provincial coastal protected area included in the UNESCO World Heritage sites list since 1999 (Lewis and Campagna 2008). However, outside protected areas, seals are exposed to unregulated visits and interaction with recreational activities like sport fishing and off-road driving and domestic animals which may lead to a variety of diseases (Lavigne and Schmitz 1990; Forum for the Conservation of the Patagonian Sea 2019).

Southern elephant seals face few threats and conflicts today. However, intensive fishing with important depletion of prey stocks (Hanchet et al. 2003), pollution, and climate change could have possible effects on the populations. Pollution mainly of plastics, fishing boxes, fishing nets and hooks, and buoys is reported on the beaches where seals breed and molt (Campagna et al. 2007; Esteves 2008; Lewis and Campagna 2008; Falabella et al. 2009). Interaction with squid fisheries is evident as entangled animals are relatively common (Campagna et al. 2007). Lately, few animals were also found entangled in long-line fishing gear; mortality rates of these interactions are unknown (Forum for the Conservation of the Patagonian Sea 2019).

Effects of global climate change on this species are not well known, although there may be an impact on foraging areas. Ocean warming and the reduction of ice extent in certain areas of Antarctica during the last decades (Stammerjohn et al. 2008) affect life cycle, distribution, and abundance of key species like krill (*Euphausia superba)* in the food web of the Southern Ocean ecosystem (Flores et al. 2012; Hindell et al. 2017).

References

Acevedo J, Aguayo-Lobo A, Brito J et al (2016) Review of the current distribution of southern elephant seals in the eastern South Pacific. N Z J Mar Freshwater Res 50:240–258. https://doi.org/10.1080/00288330.2015.1132746

Andrews RC (1914) Notes on the sea elephant (*Mirounga leonina*). Science 36:656–656. https://doi.org/10.1126/science.39.1009.656

Bailleul F, Charrasin J-B, Monestiez P et al (2007) Successful foraging zones of southern elephant seals from the Kerguelen Islands in relation to oceanographic conditions. Philos T R Soc B 362:2169–2181. https://doi.org/10.1098/rstb.2007.2109

Baldi R, Campagna C, Pedraza S et al (1996) Social effects of space availability on the breeding behaviour of elephant seals in Patagonia. Anim Behav 51:717–725. https://doi.org/10.1006/anbe.1996.0075

Banks J, Lea M-A, Wall S et al (2014) Combining bio-logging and fatty acid signature analysis indicates spatio-temporal variation in the diet of the southern elephant seal, *Mirounga leonina*. J Exp Mar Biol Ecol 450:79–90. https://doi.org/10.1016/j.jembe.2013.10.024

Bastida R, Rodriguez D (2010) Mamíferos marinos de la Patagonia y Antártida. Vazquez Mazzini, Buenos Aires

Berta A, Sumich JL, Kovacs KM (2015) Marine mammals: evolutionary biology. Elsevier, San Diego

Bester MN, Möller H, Wium J et al (2001) An update on the status of southern elephant seals at Gough Island. Afr J Wildl Res 31:68–71

Biuw M, Nøst OA, Stien A et al (2010) Effects of hydrographic variability on the spatial, seasonal and diel diving patterns of southern elephant seals in the eastern Weddell Sea. PLoS One 5:e13816. https://doi.org/10.1371/journal.pone.0013816

Bonfili P, Eder E, Ferrari M (2019) Un modelo de dos sexos con migración. Paper presented at VII Congreso de Matemática Aplicada, Computacional e Industrial (MACI), Córdoba, Argentina, 8–10 May 2019

Bornemann H, Kreyscher M, Ramdohr S et al (2000) Southern elephant seal movements and Antarctic sea ice. Antarct Sci 12:3–15. https://doi.org/10.1017/s095410200000002x

Bowen WD, Iverson SJ (2012) Methods of estimating marine mammal diets: a review of validation experiments and sources of bias and uncertainty. Mar Mamm Sci 29:719–754. https://doi.org/10.1111/j.1748-7692.2012.00604.x

Bowen WD, Siniff DB (1999) Distribution, population biology, and feeding ecology of marine mammals. In: Reynolds JE III, Rommel SA (eds) Biology of marine mammals. Smithsonian Press, Washington, DC, pp 423–484

Boyd I, Arnbom T, Fedak M (1993) Water flux, body composition, and metabolic rate during molt in female southern elephant seals (*Mirounga leonina*). Physiol Zool 66:43–60. https://doi.org/10.1086/physzool.66.1.30158286

Bradshaw CJA, Hindell MA, Best NJ et al (2003) You are what you eat: describing the foraging ecology of southern elephant seals (*Mirounga leonina*) using blubber fatty acids. Proc R Soc Lond B 270:1283–1292. https://doi.org/10.1098/rspb.2003.2371

Brown DJ, Boyd IL, Cripps GC et al (1999) Fatty acid signature analysis from the milk of Antarctic fur seals and Southern elephant seals from South Georgia: implications for diet determination. Mar Ecol Prog Ser 187:251–263. https://doi.org/10.3354/meps187251

Burdman L, Daneri G, Negrete J et al (2015) Cephalopoda as prey of juvenile Southern elephant seals at Isla 25 de Mayo/King George, South Shetland Islands. Iheringia Série Zoologia 105:12–19. https://doi.org/10.1590/1678-476620151051l219

Cáceres B (2013) Comportamiento reproductivo de la foca elefante (*Mirounga leonina*, Linneaus 1758) en Bahía Ainsworth, Tierra del Fuego, Chile y los potenciales efectos del turismo. B.Sc. Thesis, Universidad Austral de Chile

Campagna C, Lewis M (1992) Growth and distribution of a southern elephant seal colony. Mar Mamm Sci 8:387–396. https://doi.org/10.1111/j.1748-7692.1992.tb00053.x

Campagna C, Le Boeuf B, Lewis M et al (1992) Equal investment in male and female offspring in southern elephant seals. J Zool 226:551–561. https://doi.org/10.1111/j.1469-7998.1992.tb07498.x

Campagna C, Lewis M, Baldi R (1993) Breeding biology of southern elephant seals in Patagonia. Mar Mamm Sci 9:34–47. https://doi.org/10.1111/j.1748-7692.1993.tb00424.x

Campagna C, Le Boeuf B, Blackwell SB et al (1995) Diving behaviour and foraging location of female southern elephant seals from Patagonia. J Zool 236:55–71. https://doi.org/10.1111/j.1469-7998.1995.tb01784.x

Campagna C, Quintana F, Le Boeuf BJ et al (1998) Diving behaviour and foraging ecology of female southern elephant seals from Patagonia. Aquat Mamm 4:1–11

Campagna C, Fedak MA, McConnell BJ (1999) Post-breeding distribution and diving behavior of adult male southern elephant seals from Patagonia. J Mammal 80:1341–1352. https://doi.org/10.2307/1383185

Campagna C, Piola A, Marin MR et al (2006) Southern elephant seal trajectories, fronts and eddies in the Brazil/Malvinas Confluence. Deep-Sea Res I Oceanogr Res Pap 53:1907–1924. https://doi.org/10.1016/j.dsr.2006.08.015

Campagna C, Piola A, Marin MR et al (2007) Deep divers in shallow seas: southern elephant seals on the Patagonian shelf. Deep-Sea Res I Oceanogr Res Pap 54:1792–1814. https://doi.org/10.1016/j.dsr.2007.06.006

Carlini AR, Marquez MEI, Daneri GA et al (1999) Mass changes during their annual cycle in females of southern elephant seals at King George Island. Polar Biol 21:234–239. https://doi.org/10.1007/s003000050358

Carlini AR, Poljak S, Casaux R et al (2003) Southern elephant seals breeding at Nelson Island, South Shetland Islands. Polish Polar Res 24:143–147

Carlini AR, Poljak S, Daneri G et al (2006) The dynamics of male harem dominance in southern elephant seals (*Mirounga leonina*) at the South Shetland Islands. Polar Biol 29:796–805. https://doi.org/10.1007/s00300-006-0117-6

Carrara IS (1952) Lobos marinos, pingüinos y guaneras de las costas del litoral marítimo e islas adyacentes de la República Argentina. Ph.D. Dissertation, Facultad de Ciencias Veterinarias, Universidad Nacional de La Plata, Argentina

Cassini MH (2000) A model on female breeding dispersion and the reproductive systems of pinnipeds. Behav Processes 51:93–99. https://doi.org/10.1016/s0376-6357(00)00121-2

Chaise LL, McCafferty DJ, Krellenstein A et al (2019) Environmental and physiological determinants of huddling behavior of molting female southern elephant seals (*Mirounga leonina*). Physiol Behav 199:182–190. https://doi.org/10.1016/j.physbeh.2018.10.016

Cherel Y, Hobson KA (2007) Geographical variation in carbon stable isotope signatures of marine predators: a tool to investigate their foraging areas in the Southern Ocean. Mar Ecol Prog Ser 329:281–287. https://doi.org/10.3354/meps329281

Cherel Y, Ducatez S, Fontaine C et al (2008) Stable isotopes reveal the trophic position and mesopelagic fish diet of female southern elephant seals breeding on the Kerguelen Islands. Mar Ecol Prog Ser 370:239–247. https://doi.org/10.3354/meps07673

Clarke MR, MacLeod N (1982) Cephalopods in the diet of elephant seals at Signy Island, South Orkney Islands. British Antarctic Survey Bulletin 57:22–31

Costa DP, Hückstädt LA, Crocker DE et al (2010) Approaches to studying climatic change and its role on the habitat selection of Antarctic pinnipeds. Integr Comp Biol 50:1018–1030. https://doi.org/10.1093/icb/icq054

Crocker DE, Williams JD, Costa DP et al (2001) Maternal traits and reproductive effort in northern elephant seals. Ecology 82:541–3555. https://doi.org/10.1890/0012-9658(2001)082[3541,mtarei]2.0.co;2

Cruwyis E, Davis PB (1994) Southern elephant seal numbers during moult on Livingston Island, South Shetland Islands. Polar Rec 30:313–314. https://doi.org/10.1017/S0032247400024591

Daneri GA (2009) Two records of male southern elephant seals (*Mirounga leonina*) on the Atlantic coast of Uruguay. Lat Am J Aquat Mamm 7:101–102. https://doi.org/10.5597/lajam00144

Daneri G, Carlini A (2002) Fish prey of southern elephant seals, *Mirounga leonina*, at King George Island. Polar Biol 25:739–743. https://doi.org/10.1007/s00300-002-0408-5

Daneri GA, Carlini AR, Rodhouse PGK (2000) Cephalopod diet of the southern elephant seal, *Mirounga leonina*, at King George Island, South Shetland Islands. Antarct Sci 12:16–19. https://doi.org/10.1017/s0954102000000031

Daneri GA, Carlini AR, Marschoff ER et al (2015) The feeding habits of the Southern elephant seal, *Mirounga leonina*, at Isla 25 de Mayo/King George Island, South Shetland Islands. Polar Biol 38:665–676. https://doi.org/10.1007/s00300-014-1629-0

de Moura J, Di Dario B, Moreira Lima L et al (2011) Southern elephant seals (*Mirounga leonina*) along the Brazilian coast: review and additional records. Mar Biodivers Rec 3:e18. https://doi.org/10.1017/S1755267209991138

Desprez M, Gimenez O, McMahon CR et al (2018) Optimizing lifetime reproductive output: intermittent breeding as a tactic for females in a long-lived, multiparous mammal. J Anim Ecol 87:199–211. https://doi.org/10.1111/1365-2656.12775

DS 06 MMA (2017) Ministerio de Medio Ambiente de Chile. Ficha de antecedentes de especie. *Mirounga leonina*. Available via http://www.mma.gob.cl/clasificacionespecies/fichas13proceso/fichas-inicio/Mirounga_leonina_INICIO_13RCE.pdf

Ducatez S, Dalloyau S, Richard P et al (2008) Stable isotopes document winter trophic ecology and maternal investment of adult female southern elephant seals (*Mirounga leonina*) breeding at the Kerguelen Islands. Mar Biol 155:413–420. https://doi.org/10.1007/s00227-008-1039-3

Eder EB, Lewis M, Campagna C et al (2010) Evidence of demersal foraging from stable isotope analysis of juvenile elephant seals from Patagonia. Mar Mamm Sci 26:430–442. https://doi.org/10.1111/j.1748-7692.2009.00347.x

Eder EB, Lewis MN, Marin MR et al (2011) On- and off-shelf diving effort of juvenile elephant seals from Península Valdés determined by light loggers. J Mammal 92:811–818. https://doi.org/10.1644/10-mamm-a-292.1

Eder EB, Negrete J, Gribaudo C et al (2019a) *Mirounga leonina*. In: SAyDS, SAREM (eds) Categorización 2019 de los mamíferos de Argentina según su riesgo de extinción. Lista Roja de los mamíferos de Argentina. Available via http://cma.sarem.org.ar. https://doi.org/10.31687/saremlr.19.159. Accessed 13 Nov 2019

Eder EB, Muelbert MMC, Hindell MA et al (2019b) Foraging strategies of female elephant seals from Península Valdés, Patagonia, inferred from whisker stable isotope signatures of their pups. Aquat Mamm 54:1–13. https://doi.org/10.1578/am.45.1.2019.1

Esteves JL (2008) Contaminación. In: Estado de Conservación del Mar Patagónico y Áreas de Influencia. Available via http://www.marpatagonico.org

Fabiani A, Galimberti F, Sanvito S et al (2004) Extreme polygyny among southern elephant seals on Sea Lion Island, Falkland Islands. Behav Ecol 15:961–969. https://doi.org/10.1093/beheco/arh112

Fabiani A, Galimberti F, Sanvito S et al (2006) Relatedness and site fidelity at the southern elephant seal, *Mirounga leonina*, breeding colony in the Falkland Islands. Anim Behav 72:617–626. https://doi.org/10.1016/j.anbehav.2005.11.024

Falabella V, Lewis M, Campagna C (1999) Development of cardiorespiratory patterns associated with terrestrial apneas in free-ranging southern elephant seals. Physiol Biochem Zool 72:64–70. https://doi.org/10.1086/316637

Falabella V, Campagna C, Croxall J (2009) Atlas del Mar Patagónico. Especies y espacios. Wildlife Conservation Society, BirdLife International, Buenos Aires

Ferrari MA, Campagna C, Condit R et al (2013) The founding of a southern elephant seal colony. Mar Mamm Sci 29:407–423. https://doi.org/10.1111/j.1748-7692.2012.00585.x

Field IC, Bradshaw CJA, Burton HR et al (2005) Resource partitioning through oceanic segregation of foraging juvenile southern elephant seals. Oecologia 142:127–135. https://doi.org/10.1007/s00442-004-1704-2

Field IC, Bradshaw CJA, van den Hoff J et al (2007) Age-related shifts in the diet composition of southern elephant seals expand overall foraging niche. Mar Biol 150:1441–1452. https://doi.org/10.1007/s00227-006-0417-y

Flores H, Atkinson A, Kawaguchi S et al (2012) Impact of climate change on Antarctic krill. Mar Ecol Prog Ser 458:1–19. https://doi.org/10.3354/meps09831

Forum for the Conservation of the Patagonian Sea (2019) Marine mammals. In: Falabella V, Campagna C (eds) Report of the IUCN Regional Red List First Workshop for Species of the Patagonian Sea – 2016. Available via http://marpatagonico.org/en/publica/report-of-the-iucn-regional-red-list-first-workshop-for-species-of-the-patagonian-sea-marine-mammals/

Galimberti F, Boitani L (1999) Demography and breeding biology of a small, localized population of southern elephant seals (*Mirounga leonina*). Mar Mamm Sci 15:159–178. https://doi.org/10.1111/j.1748-7692.1999.tb00787.x

Galimberti F, Boitani L, Marzetti I (2000) Female strategies of harassment reduction in southern elephant seals. Ethol Ecol Evol 4:367–388. https://doi.org/10.1080/08927014.2000.9522793

Galimberti F, Sanvito S, Boitani L et al (2001) Viability of the southern elephant seal population of the Falkland Islands. Anim Conserv 4:81–88. https://doi.org/10.1017/s1367943001001093

Gibbons J, Miranda C (2001) Presencia del elefante marino del sur (*Mirounga leonina*) (Phocidae) en seno Almirantazgo, Tierra del Fuego. Anales del Instituto de la Patagonia 29:157–179

Gil-Delgado JA, Villaescusa JA, Diazmacip ME, et al (2013). Minimum population size estimates demonstrate an increase in southern elephant seals (Mirounga leonina) on Livingston Island, maritime Antarctica. Pol Biol 36:607–610. https://doi.org/10.1007/s00300-012-1280-6

González MB (2020) Distribución espacial de elefantes marinos en Península Valdés y sus implicancias para su manejo como recurso turístico Ph.D. Dissertation. Universidad Nacional de la Patagonia San Juan Bosco, Argentina

González MB, Sapoznikow A, Lewis MN (2019) Avistaje de elefantes marinos en Península Valdés, Argentina: Importancia y satisfacción desde la perspectiva de guías de turismo y turistas. Rev Interam Amb Tur 15:46–59. https://doi.org/10.4067/s0718-235x2019000100046

Green K, Burton HR (1993) Comparison of the stomach contents of southern elephant seals, *Mirounga leonina*, at Macquarie and Heard Islands. Mar Mamm Sci 9:10–22. https://doi.org/10.1111/j.1748-7692.1993.tb00422.x

Green K, Williams R (1986) Observations on food remains in faeces of elephant, leopard and crabeater seals. Polar Biol 6:43–45. https://doi.org/10.1007/bf00446239

Green K, Slip DJ, Moore GJ (1998) The take of fish species by seabirds and marine mammals in the Australian fisheries zone around heard island: the potential for competition with a commercial fishery. Polar Biol 20:273–280. https://doi.org/10.1007/s003000050303

Griffen BD (2018) Reproductive skipping as an optimal life history strategy in the southern elephant seal. Ecol Evol 8:9158–9170. https://doi.org/10.1002/ece3.4408

Hanchet S, Horn P, Stevenson M (2003) Fishing in the ice: is it sustainable? Water & Atmosphere 11:24–25

Hindell MA (1991) Some life-history parameters of a declining population of southern elephant seals, *Mirounga leonina*. J Anim Ecol 60:119–134. https://doi.org/10.2307/5449

Hindell MA, Burton HR (1988) Seasonal haul-out patterns of the southern elephant seal (*Mirounga leonina* L.), at Macquarie Island. J Mammal 69:81–88. https://doi.org/10.2307/1381750

Hindell MA, Perrin WF (2009) Elephant seals *Mirounga angustirostris* and *M. leonina*. In: Perrin WF, Würsig B, Thewissen GM (eds) Encyclopedia of marine mammals. Academic Press, San Diego, pp 364–368. https://doi.org/10.1016/b978-0-12-373553-9.00088-2

Hindell MA, Burton HR, Slip DJ (1991) Foraging areas of southern elephant seals, *Mirounga leonina*, as inferred from water temperature data. Mar Freshw Res 42:115–128. https://doi.org/10.1071/mf9910115

Hindell MA, McMahon CR, Bester MN et al (2016) Circumpolar habitat use in the southern elephant seal: implications for foraging success and population trajectories. Ecosphere 7:e01213. https://doi.org/10.1002/ecs2.1213

Hindell MA, Sumner M, Bestley S et al (2017) Decadal changes in habitat characteristics influence population trajectories of southern elephant seals. Glob Chang Biol 23:5136–5150. https://doi.org/10.1111/gcb.13776

Hoelzel R, Campagna C, Arnbom T et al (2001) Differentiation between island and mainland southern elephant seal populations. Proc Biol Sci 268:325–332. https://doi.org/10.1098/rspb.2000.1375

Hofmeyr GJG (2015) *Mirounga leonina*. The IUCN red list of threatened species 2015: e.T13583A45227247. https://doi.org/10.2305/IUCN.UK.2015-4.RLTS.T13583A45227247.e

Hückstädt LA, Hassrick J, Olavarría C et al (2008) Patrón de movimiento y buceo durante el primer viaje de una cría de elefante marino del sur (*Mirounga leonina*). Paper presented at the XIII Reunión de Trabajo de Especialistas en Mamíferos Acuáticos de América del Sur y 7° Congreso SOLAMAC: 'Estatus de los Mamíferos Acuáticos en Latinoamérica'. Montevideo, Uruguay, October 13–17 2008

Hückstädt LA, Koch PL, McDonald BI et al (2012) Stable isotope analyses reveal individual variability in the trophic ecology of a top marine predator, the southern elephant seal. Oecologia 169:395–406. https://doi.org/10.1007/s00442-011-2202-y

Jonker FC, Bester MN (1998) Seasonal movements and foraging areas of adult southern female elephant seals, *Mirounga leonina*, from Marion Island. Antarct Sci 10:21–30. https://doi.org/10.1017/s0954102098000042

Kastelein RA, Kershaw J, Wiepkema PR (1991) The food consumption of southern elephant seals (*Mirounga leonina*). Aquat Mamm 17:76–87

Kelly JF (2000) Stable isotopes of carbon and nitrogen in the study of avian and mammalian trophic ecology. Can J Zool 78:1–27. https://doi.org/10.1139/z99-165

Kirkman SP, Hofmeyr GJG, Bester MN et al (2001) Counts of southern elephant seals, *Mirounga leonina*, at Bouvet Island. Polar Biol 24:62–65. https://doi.org/10.1007/s003000000180

Kooyman GL (2006) Mysteries of adaptation to hypoxia and pressure in marine mammals. The Kenneth S. Norris lifetime achievement award lecture. Mar Mamm Sci 22:507–526. https://doi.org/10.1111/j.1748-7692.2006.00069.x

Labrousse S, Sallée JB, Fraser AD et al (2017) Under the sea ice: exploring the relationship between sea ice and the foraging behaviour of southern elephant seals in East Antarctica. Prog Oceanogr 156:17–40. https://doi.org/10.1016/j.pocean.2017.05.014

Lanusse L, Sánchez J, Negrete J et al (2018) El elefante marino del sur (*Mirounga leonina*) en Península Potter, Antártida: variabilidad genética de la colonia reproductiva y flujo génico con otras colonias. Paper presented at the XXXI Jornadas Argentinas de Mastozoología, La Rioja, Argentina, 23–26 October 2018

Lavigne DM, Schmitz OJ (1990) Global warming and increasing population densities: a prescription for seal plagues. Mar Pollut Bull 21:280–284. https://doi.org/10.1016/0025-326x(90)90590-5

Laws RM (1956) The elephant seal (*Mirounga leonina* Linn.) II. General, social and reproductive behaviour. Falkland Island Dependencies Survey Scientific Reports 13:1–88

Le Boeuf BJ, Campagna C (2013) Wildlife viewing spectacles: Best practices from elephant seal (*Mirounga* sp.) colonies. Aquat Mamm 39:132–146. https://doi.org/10.1578/AM.39.2.2013.132

Le Boeuf BJ, Laws RM (1994) Elephant seals: population ecology, behavior and physiology. University of California Press, Berkeley

Lewis M, Campagna C (2008) Mamíferos marinos. In: Estado de Conservación del Mar Patagónico y Áreas de Influencia. Available via http://www.marpatagonico.org

Lewis M, Campagna C, Quintana F (1996) Site fidelity and dispersion of southern elephant seals from Patagonia. Mar Mamm Sci 12:138–147. https://doi.org/10.1111/j.1748-7692.1996.tb00312.x

Lewis M, Campagna C, Uhart M et al (2001) Ontogenetic and seasonal variation in blood parameters in southern elephant seals. Mar Mamm Sci 17:862–872. https://doi.org/10.1111/j.1748-7692.2001.tb01302.x

Lewis M, Campagna C, Zavatti J (2004) Annual cycle and inter-annual variation in the haul-out pattern of an increasing southern elephant seal colony. Antarct Sci 16:219–226. https://doi.org/10.1017/s0954102004002020

Lewis M, Campagna C, Marin MR et al (2006a) Southern elephant seals north of the Antarctic Polar Front. Antarct Sci 18:213–221. https://doi.org/10.1017/s0954102006000253

Lewis R, O'Connell TC, Lewis MN et al (2006b) Sex-specific foraging strategies and resource partitioning in the southern elephant seal (*Mirounga leonina*). P Roy Soc B-Biol Sci 273:2901–2907. https://doi.org/10.1098/rspb.2006.3642

Lodi L, Siciliano S (1989) A southern elephant seal in Brazil. Mar Mamm Sci 5:313

Lübcker N, Reisinger R, Oosthuizen WC et al (2017) Low trophic level diet of juvenile southern elephant seals *Mirounga leonina* from Marion Island: a stable isotope investigation using vibrissal regrowths. Mar Ecol Prog Ser 577:237–250. https://doi.org/10.3354/meps12240

Malpress V, Bestley S, Corney S et al (2017) Bio-physical characterisation of polynyas as a key foraging habitat for juvenile male southern elephant seals (*Mirounga leonina*) in Prydz Bay, East Antarctica. PLoS One 12:e0184536. https://doi.org/10.1371/journal.pone.0184536

McCann TS (1985) Size, status and demography of southern elephant seal (*Mirounga leonina*) populations. In: Ling LK, Bryden MM (eds) Sea mammals of south latitudes. Proceedings of a Symposium of the 52nd ANZAAS Congress, Sydney

McConnell BJ, Fedak MA (1996) Movements of southern elephant seals. Can J Zool 74:1485–1496. https://doi.org/10.1139/z96-163

McGovern KA, Rodríguez DH, Lewis M et al (2019) Classification and behavior of free-ranging female southern elephant seal dives based on three-dimensional movements and video-recorded observations. Mar Ecol Prog Ser 620:215–232. https://doi.org/10.3354/meps12936

McIntyre T, Bornemann H, Plötz J et al (2011) Water column use and forage strategies of female southern elephant seals from Marion Island. Mar Biol 158:2125–2139. doi:https://doi.org/10.1007/s00227-011-1719-2

McIntyre T, Bornemann H, Plötz J et al (2012) Deep divers in even deeper seas: habitat use of male southern elephant seals from Marion Island. Antarct Sci 24:561–570. https://doi.org/10.1017/s0954102012000570

McMahon C, Hindell M (2003) Twinning in southern elephant seals: the implications of resource allocation by mothers. Wildl Res 30(1):35–39. https://doi.org/10.1071/WR01069

McMahon CR, Hindell MA, Burton HR (2005) Comparison of southern elephant seal populations, and observations of a population on a demographic knife-edge. Mar Ecol Prog Ser 288:273–283. https://doi.org/10.3354/meps288273

Miranda C (2013) El turismo como una herramienta para la conservación del elefante marino del sur (*Mirounga leonina*), en Tierra del Fuego, Chile. M.Sc. Thesis, Universidad de Magallanes, Chile

Muelbert MMC, de Souza RB, Lewis MN et al (2013) Foraging habitats of southern elephant seals, *Mirounga leonina*, from the Northern Antarctic Peninsula. Deep Sea Res II 88-89:47–60. https://doi.org/10.1016/j.dsr2.2012.07.009

Newland C, Field IC, Nichols PD et al (2009) Blubber fatty acid profiles indicate dietary resource partitioning between adult and juvenile southern elephant seals. Mar Ecol Prog Ser 384:303–312. https://doi.org/10.3354/meps08010

Newland C, Fiel IC, Cherel Y et al (2011) Diet of juvenile southern elephant seals reappraised by stable isotopes in whiskers. Mar Ecol Prog Ser 424:247–258. https://doi.org/10.3354/meps08769

Newsome SD, Koch PL, Etnier MA et al (2006) Using carbon and nitrogen isotope values to investigate maternal strategies in Northeast Pacific otariids. Mar Mamm Sci 22:556–572. https://doi.org/10.1111/j.1748-7692.2006.00043.x

O'Toole M, Hindell MA, Charrassin J-B et al (2014) Foraging behaviour of southern elephant seals over the Kerguelen Plateau. Mar Ecol Prog Ser 502:281–294. https://doi.org/10.3354/meps10709

Orgeret F, Weimerskirch H, Guinet C et al (2019) Body condition influences ontogeny of foraging behavior in juvenile southern elephant seals. Ecol Evol 9:223–236. https://doi.org/10.1002/ece3.4717

Páez-Rosas D, Riofrío-Lazo M, Ortega J et al (2018) Southern elephant seal vagrants in Ecuador: a symptom of La Niña events? Mar Biodivers Rec 11:13. https://doi.org/10.1186/s41200-018-0149

Piatkowski U, Vergani DF, Stanganelli ZB (2002) Changes in the cephalopod diet of southern elephant seal females at King George Island, during El Niño-La Niña events. J Mar Biol Assoc UK 82:913–916. https://doi.org/10.1017/s0025315402006343

Pistorius PA, Bester MN, Lewis MN et al (2004) Adult female survival, population trend, and the implications of early primiparity in a capital breeder, the southern elephant seal (*Mirounga leonina*). J Zool 263:107–119. https://doi.org/10.1017/s0952836904004984

Pistorius PA, de Bruyn N, Bester MN (2011) Population dynamics of southern elephant seals: a synthesis of three decades of demographic research at Marion Island. Afr J Mar Sci 33:523–534. https://doi.org/10.2989/1814232x.2011.637357

Redwood SD, Félix F (2018) The most northerly record of a southern elephant seal (*Mirounga leonina*) in the Pacific Ocean at the Island of Taboga. Gulf of Panama, Panama Aquat Mamm 44:13–18. https://doi.org/10.1578/AM.44.1.2018.13

Retana MV, Guzmán J, Lewis MN (2013). Fine-Scale Breeding Habitat Preferences of Southern Elephant Seals (Mirounga leonina) on Península Valdés. Aquat Mamm 39:233–240. https://doi.org/10.1578/AM.39.3.2013.233

Richard G, Vacquié-Garcia J, Jouma'a J et al (2014) Variation in body condition during the post-moult foraging trip of southern elephant seals and its consequences on diving behavior. J Exp Biol 217:2609–2619

Riedman M (1990) The pinnipeds: seals, sea lions, and salruses. Univ of California Press, Berkeley

Rita D, Drago M, Galimberti F et al (2017) Temporal consistency of individual trophic specialization in southern elephant seals *Mirounga leonina*. Mar Ecol Prog Ser 585:229–242. https://doi.org/10.3354/meps12411

Rodhouse PG, Arbom T, Fedak MA et al (1992) Cephalopod prey of the southern elephant seal, *Mirounga leonina* L. Can J Zool 70:1007–1015. https://doi.org/10.1139/z92-143

Sala JE, Quintana F, Wilson R et al (2011) Pitching a new angle on elephant seal dive patterns. Polar Biol 34:1197–1209. https://doi.org/10.1007/s00300-011-0981-6

Sanvito S, Galimberti F (2000) Bioacoustics of southern elephant seals. I. Acoustic structure of male aggressive vocalizations. Bioacoustics 10:259–285. https://doi.org/10.1080/09524622.2000.9753438

Sanvito S, Galimberti F, Miller EH (2007) Vocal signalling of male southern elephant seals is honest but imprecise. Anim Behav 73:287–299. https://doi.org/10.1016/j.anbehav.2006.08.005

SCAR-EGS (2008) Scientific committee for Antarctic research – expert group on seals report. Available via http://www.seals.scar.org/pdf/statusofstocs.pdf

Shaughnessy PD (1999) The action plan for Australian seals. CSIRO Wildlife and Ecology, Canberra

Slade W, Moritz C, Hoelzel R et al (1998) Molecular population genetics of the southern elephant seal *Mirounga leonina*. Genetics 149:1945–1957

Slip DJ (1997a) Foraging ecology of southern elephant seals from Heard Island. Ph.D. Dissertation. University of Tasmania, Australia

Slip DJ (1997b) Diving and foraging behavior of juvenile southern elephant seals from Heard Island. In: Hindell M, Kemper C (eds) Marine mammal research in the Southern Hemisphere, vol 1. Status, ecology and medicine. Surrey Beatty and Sons, Chipping Norton, pp 114–124

Stammerjohn SE, Martinson DG, Smith RC et al (2008) Trends in Antarctic annual sea ice retreat and advance and their relation to El Niño–Southern Oscillation and Southern Annular Mode variability. J Geophys Res 113:1–20. https://doi.org/10.1029/2007jc004269

Torres D (1981) Notas sobre el elefante marino del sur, *Mirounga leonina* (Linn., 1758) y hallazgo de un ejemplar en Chañaral, III Región, Chile. (Pinnipedia: Phocidae). Boletín Antártico Chileno 1:11–14

Torres D, Yañez J, Cattan PE (1979) Mamíferos marinos de Chile: Antecedentes y situación actual. Biología Pesquera 11:49–81

Tosh CA, Bornemann H, Ramdohr S et al (2009) Adult male southern elephant seals from King George Island utilize the Weddell Sea. Antarct Sci 21:113–121. https://doi.org/10.1017/s0954102008001557

Tosh CA, de Bruyn N, Steyn J et al (2015) The importance of seasonal sea surface height anomalies for foraging juvenile southern elephant seals. Mar Biol 162:2131–2140. https://doi.org/10.1007/s00227-015-2743-4

van den Hoff J, Burton H, Davies R (2003) Diet of male southern elephant seals (*Mirounga leonina* L.) hauled out at Vincennes Bay, East Antarctica. Polar Biol 26:27–31. https://doi.org/10.1007/s00300-002-0447-y

van den Hoff J, Burton H, Raymond B (2007) The population trend of southern elephant seals (*Mirounga leonina* L.) at Macquarie Island (1952–2004). Polar Biol 30:1275–1283. https://doi.org/10.1007/s00300-007-0288-9

van den Hoff J, McMahon CR, Simpkins GR et al (2014) Bottom-up regulation of a pole-ward migratory predator population. P Roy Soc B-Biol Sci 281:20132842. https://doi.org/10.1098/rspb.2013.2842

Vergani DF, Stanganelli ZB (1990) Fluctuations in breeding populations of elephant seals, *Mirounga leonina*, at Stranger Point, King George Island, 1980–1988. In: Kerry K, Hempel G (eds) Antarctic ecosystems, ecological change and conservation. Springer Verlag, Berlin, pp 241–245

Walters A, Lea MA, van den Hoff J et al (2014) Spatially explicit estimates of prey consumption reveal a new krill predator in the Southern Ocean. PLoS One 9:e86452. https://doi.org/10.1371/journal.pone.0086452

Wildlife Conservation Society (2011) Elephant seal travels 18,000 miles. Available via https://phys.org/news/2011-12-elephant-miles.html

Chapter 10
Northern Elephant Seal (*Mirounga angustirostris*, Gill 1866)

María C. García-Aguilar

Abstract The northern elephant seal (*Mirounga angustirostris*) is the largest carnivore in the Northern Hemisphere. They are sexually dimorphic with adult males weighing up to 2000 kg and measuring up to 4.5 m and adult females weighing up to 700 kg and measuring up to 3 m. In addition to the differences in weight and length, adult males have a facial proboscis, which gives the common name "elephant." It is a deep-diving marine mammal that can dive to depths over 1500 m. The species forages within the cold waters of the North Pacific Ocean. Breeding sites are located mainly on sandy beaches of temperate islands and a few mainland coastal areas of California, USA, and Baja California, Mexico. Despite being nearly driven to extinction due to the intense commercial harvesting in the nineteenth century, the northern elephant seal has recolonized its historical range and as of 2010 had an estimated population size of 201,000 individuals. However, due to low genetic diversity, a consequence of the population bottleneck, the main threats to the species appear to be disease outbreaks and environmental changes.

Keywords Phocids · Northeastern Pacific Ocean · Deep diving · Sexual dimorphism

General Features

The northern elephant seal (*Mirounga angustirostris*) is a member of the family Phocidae ("true seals"), subfamily Monachinae ("southern seals"), which includes three groups: Lobodontini, *Mirounga*, and *Monachus* (Fyler et al. 2005). While the North Atlantic origin of Monachinae seals is widely accepted, the origin of *Mirounga* is not clear due to the lack of fossil record (Deméré et al. 2003; Fyler et al. 2005). The known phylogenetic history of *Mirounga* begins with *Callophoca*, a lineage

M. C. García-Aguilar (✉)
Departamento de Oceanología Biológica, Centro de Investigación Científica y de Educación Superior de Ensenada, Ensenada, Baja California, Mexico
e-mail: gaguilar@cicese.mx

G. Heckel, Y. Schramm (eds.), *Ecology and Conservation of Pinnipeds in Latin America*, https://doi.org/10.1007/978-3-030-63177-2_10

that may have evolved in the eastern North Atlantic, more than 4.5 Ma, and dispersed westward through the North Equatorial Current to the coasts of North America (Ray 1976; de Muizon 1982). Some members of this lineage could have traveled in the late Miocene or early Pliocene through the Central American Seaway, open at those times, and settled in the eastern South Pacific (Deméré et al. 2003). Speciation resulted in the southern species (*M. leonina*), while the northern species (*M. angustirostris*) could have resulted from a transequatorial event and allopatric speciation, possibly during the Pleistocene (Deméré et al. 2003). Currently, the southern elephant seal is distributed in the Southern Ocean and the northern elephant seal in the Northeastern Pacific Ocean (Hindell and Perrin 2008).

Northern elephant seals are extremely sexually dimorphic in adulthood (Fig. 10.1) with males weighing as much as 2000 kg and measuring up to 4.5 m in length while females weigh between 400 and 700 kg and measure about 3 m in length (Deutsch et al. 1990, 1994). At birth there are no weight and length differences between the sexes, but when weaned, males are slightly heavier (4%) and longer (1%) than females (Salogni et al. 2018). Along with the difference in body size, adult males develop a fleshy proboscis on their head and have a wide hairless neck and a reddish dermal shield on the chest (Le Boeuf and Laws 1994). In both sexes, the pelage is short and rough and replaced to the epidermis once a year when they haul out to molt (Le Boeuf and Laws 1994). Pelage in adult males is dark brown, and in females, it is light brown. Pups are born black and then molt between the fifth and sixth week

Fig. 10.1 Northern elephant seal male and group of adult females with pups. (Photo: M.C. García-Aguilar)

Fig. 10.2 Group of weaned pups showing pelage of molted and unmolted pup. (Photo: M.C. García-Aguilar)

to a silvery color (Fig. 10.2) (Le Boeuf and Laws 1994). Northern elephant seals have vibrissae on their snout and above their large, round, and black eyes. The hind flippers are emarginated and hairy and lack nails. The adult dentition is i 2/1, c 1/1, pc 5/5, total 30, and the canines are sexually dimorphic in size and shape (Stewart and Huber 1993).

Females can live up to 21 years, while males live up to 15 years (Condit et al. 2013). The age of first reproduction for females ranges from 3 to 8 years, with an average of 3–4 years (Reiter et al. 1981; Sydeman et al. 1991; Sydeman and Nur 1994). The age of first reproduction for males varies between 5 and 10 years (average 8 years), being more successful breeders after 12 years of age (Clinton and Le Boeuf 1993).

Northern elephant seals are physiologically adapted to cold water, but heat dissipation can be a problem when they are on land. Elephant seals do not pant or sweat, and heat is dissipated through thermal windows (Khamas et al. 2012; Codde et al. 2016), which are body surfaces that function as a temporary heat dissipater under conditions of thermal stress (Mauck et al. 2003). Locomotion on land is energetically very expensive given their enormous body size and the fact that they do not use the hind flippers to move (Tennett et al. 2018). Their haul-out period has four phases (Fig. 10.3): (1) the winter breeding season, (2) the spring molt of adult females and juveniles (i.e., sexually immature individuals of both sexes), (3) the summer molt of subadult and adult males, and (4) the autumn haul-out of juveniles (Le Boeuf and Laws 1994). The preferred sites to breed, molt, and haul out are soft sloping sandy beaches, but they are sometimes found on rocky substrates.

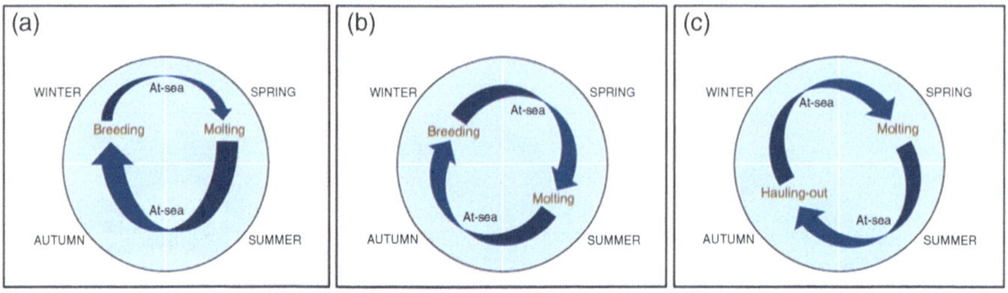

Fig. 10.3 The annual cycle of northern elephant seals. (**a**) Adult females, (**b**) subadult and adult males, and (**c**) juveniles

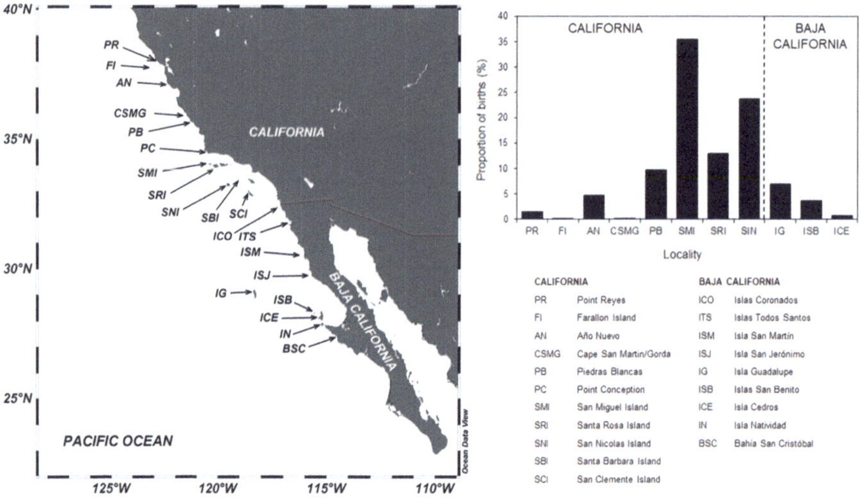

Fig. 10.4 Breeding and haul-out sites in California, USA, and Baja California, Mexico, with 2009–2010 proportion (%) of births in the main colonies. Based on Lowry et al. (2014) and García-Aguilar et al. (2018)

Distribution and Abundance

Archaeological evidence shows that northern elephant seals were established in southern California 15,000 years ago and that they were an important resource for Native Americans (Le Boeuf and Laws 1994). Their breeding and haul-out sites are located mainly on islands of California, USA, and Baja California, Mexico, with some isolated continental mainland locations of the California coast (Le Boeuf and Laws 1994; Lowry et al. 2014) (Fig. 10.4).

The northern elephant seal was subjected to intense commercial harvest during the nineteenth century. Population size and distribution before hunting are unknown, but it is presumed that they were distributed from central California to Bahía San Cristóbal, in Baja California Sur, Mexico (Fig. 10.4). Scammon (1870) noted that elephant seals were scarce in the 1850s and were considered extinct by the end of the 1870s (Townsend 1885). Nevertheless, in the early 1880s, some elephant seals were found at Bahía San Cristobal and at Isla Guadalupe, Baja California; those animals were killed, and in 1884 the species was again declared as being extinct (Le Boeuf and Laws 1994). A few years later, in 1892, nine individuals were discovered at Isla Guadalupe, seven of which were killed for the Smithsonian Museum's collection, and again elephant seals were considered extinct (Le Boeuf and Laws 1994). At the beginning of the twentieth century, expeditions to Isla Guadalupe continued to hunt elephant seals. In 1911, the Mexican government banned the killing of elephant seals and in the 1920s conducted five scientific expeditions to Isla Guadalupe (Table 10.1). Finally, in 1928 the island was decreed as a protected area.

There is no certainty about the number of elephant seals during the population bottleneck that resulted from commercial and scientific harvests. Estimates for the 1890s range from 10 to 30 (Hoelzel et al. 1993) to a few hundred individuals (Bartholomew and Hubbs 1960). The first estimate of the population size was made by Anthony (1924), who obtained an estimate of 1250 individuals using the Isla Guadalupe count of 1922 (see Table 10.1). In the following years, elephant seals were sighted at Islas San Benito, Baja California, and San Miguel Island, California. In subsequent decades (the 1940s and the 1950s), northern elephant seals expanded their distribution to other islands of Baja California, as well as to the islands of southern and central California (Bartholomew and Hubbs 1960). In 1957 the total population size was estimated at around 13,000 individuals, of which 91% were at Isla Guadalupe, 8% at Islas San Benito, and 1% at the Channel Islands, including San Miguel, Santa Barbara, and San Nicolas Islands (Bartholomew and Hubbs 1960). Stewart et al. (1994) estimated the 1991 population size at 127,000 individuals, with California colonies contributing 75% of births. The 1991 estimate calculated an overall annual rate of increase of 6.3% between 1965 and 1991; however, during that period, the colonies of California increased at an accelerated rate, while those of Baja California seemed to be stable. By 2010, the abundance of elephant seals in California was estimated at 179,000 individuals, and the population was still increasing at an average annual rate of 3.8% between 1980 and 2010 (Lowry et al. 2014), whereas in Baja California, the abundance was estimated at 22,000

Table 10.1 Northern elephant seal counts from Isla Guadalupe during the early twentieth century

Date	AM	SAM	AF	J	P	W	Total	References
1907	–	–	–	–	–	–	40	Rothschild (1910)
March 2, 1911	"Some"	"Many"	0	"Numerous"	0	6	125	Townsend (1912)
July 12, 1922	259	0	0	5	0	0	264	Anthony (1924)
July 16, 1923	361	0	0	5	0	0	366	Huey (1924)
August 30, 1924	9	109	0	6	0	0	124	Huey (1925)
June 23, 1926	363	99	0	3	0	0	465	Huey (1927)
September 28, 1929	5	0	0	464	0	0	469	Huey (1930)

AM adult males, *SAM* subadult males, *AF* adult females, *J* juveniles, *P* newborn pups, *W* weaned pups

individuals for 2009, but the population was decreasing at an average annual rate of 0.7% for the period 1970–2009 (García-Aguilar et al. 2018).

Currently there are ten colonies in California and five in Baja California (Lowry et al. 2014; García-Aguilar et al. 2018) (Fig. 10.4). The main breeding sites in California are located at San Miguel Island, San Nicolas Island, Santa Rosa Island, Piedras Blancas, and Año Nuevo, followed by Point Reyes, Farallon Islands, Cape San Martin/Gorda, Santa Barbara Island, and San Clemente Island (Lowry et al. 2014). In Baja California, breeding sites are located at Isla Guadalupe, Islas San Benito, Isla Cedros, Islas Coronados, and Islas Todos Santos (Arias-del-Razo et al. 2017). A few pups are occasionally born in other localities of California, such as Point Conception (Lowry et al. 2014), and Baja California, such as Natividad, San Martín, and San Jerónimo islands (Le Boeuf and Mate 1978; Franco-Ortíz 2012), as well as in Oregon, Washington, and British Columbia (Hückstädt 2015).

Breeding Season

The reproduction of the northern elephant seal is restricted in space and time. The breeding season occurs in the northern winter, and during this season, the species is highly gregarious, congregating on breeding beaches by the hundreds or thousands (Le Boeuf 1978). The mating system is extreme polygyny with establishment of harems, whose size can vary between 2 and 100 females, depending on breeding site density, characteristics of the reproductive beaches, and males' dominance (Le Boeuf 1972).

The breeding season begins with the arrival of adult males to the reproductive areas in late November or early December, depending on location, followed a few days later by the first adult females (Le Boeuf 1978). Males remain on land, fasting throughout the breeding season (mean 91 days; Deutsch et al. 1994). In contrast, females are asynchronous and are present on land for 32–34 days (Fig. 10.5) (Reiter et al. 1981; García-Aguilar 2004). Therefore, at no time of the season are all the females present on land. Females give birth to only one pup, 5–6 days after arrival (Le Boeuf 1972). Estrus occurs an average 24 days after birth and lasts around 3 days (Le Boeuf et al. 1972; García-Aguilar 2004). The mating period begins in the

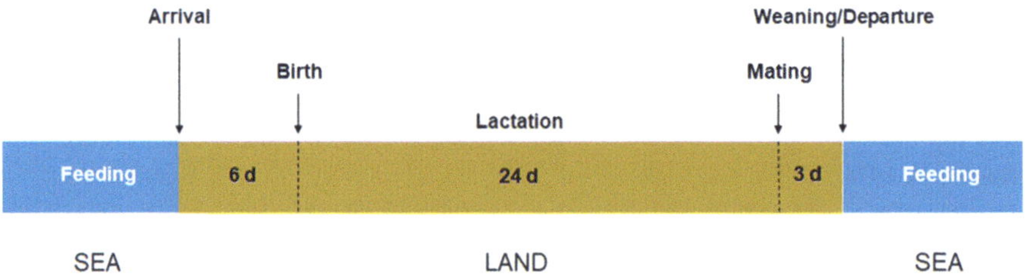

Fig. 10.5 Timeline of reproductive events during the stay on land of adult females

first week of January and ends in the first week of March, with a maximum occurring in mid-February. After mating, females leave the reproductive areas (Le Boeuf et al. 1972).

There are latitudinal differences in the occurrence of several reproductive parameters. At the Islas San Benito colony (~28°N), the first females' arrival occurs at the end of November, the first births happen in early December, and the peak of the season is in mid-January (García-Aguilar 2004). Further north, at the Año Nuevo colony (~37°N), the arrival of the first females occurs in mid-December, the first births happen around December 20, and the peak of the breeding season is at the end of January (Le Boeuf et al. 1972; Reiter et al. 1981). Despite these differences, births in both locations end at the beginning of February, and the breeding season culminates in early March. Other differences are related to natality rate, which seems to be lower at San Benito than at Año Nuevo, 74% and 84% in average, respectively (Robinson et al. 2012; Elorriaga-Verplancken and García-Aguilar 2018), and also to pre-weaning mortality rate, which was estimated at 8% in San Benito (García-Aguilar 2004) and 24% at Año Nuevo (Le Boeuf et al. 1994). Nevertheless, since these rates vary over time depending on various environmental and demographic factors, they should not be taken as fixed characteristics of the colonies.

Maternal Investment

In elephant seals, males do not provide postnatal care; thus the females' reproductive success is conditioned by their ability to obtain vital resources for themselves and their offspring. Since birth and rearing are on land, space availability is the main resource that determines the distribution of females during the breeding season (Fig. 10.6). Females prefer areas that offer protection to their pups from predators and storms and that contain resources for thermoregulation (i.e., access to water and

Fig. 10.6 Northern elephant seal female and pup. (Photo: M.C. García-Aguilar)

sandy beaches) (Le Boeuf 1991). Females form dense aggregations to prevent harassment by peripheral males, which can cause the separation of mothers from their pups, as well as serious injuries to females, and even their death (Cassini 1999).

Northern elephant seal females show embryonic diapause, or delayed implantation, which consists in keeping the embryonic development in a dormant state to delay the timing of implantation, and thus ensure that pups are born at the most favorable time for their survival (Riedman 1990). The gestation length is estimated at approximately 8 months (Le Boeuf et al. 1972; Deutsch et al. 1994). The northern elephant seal's maternal strategy is the "fasting strategy" where females arrive to reproductive areas with a large energy store in the form of blubber and fast throughout lactation (Boness and Bowen 1996). Lactation is short, averaging 27 days (Le Boeuf 1972), but the maternal investment is intense, and hence the growth rate of the pups is high. The body mass of the pups increases from 40 kg at birth to 130–140 kg at weaning, gaining on average 3.8 kg/day (Rea and Costa 1992; Crocker et al. 2001). Weaning is abrupt, and after that pups remain on land, fasting for up to 2.5 months, until they develop diving and swimming skills (Reiter et al. 1981; Le Boeuf et al. 1989). To cope with this long fasting period, the body mass of weaned pups is 42% fat and 11% protein (Crocker et al. 2001).

Elephant seal females produce nutrient-rich, energy-dense milk with fat contributing most of energy content. Since females fast throughout lactation, they use their stores for milk lipid synthesis, but they must also use the fatty acids for their own maintenance (Riedman 1979). Recent findings suggest that females preferentially use medium-chain monounsaturated fatty acids and saturated fatty acids to meet their energy requirements and long-chain monounsaturated fatty acids for milk production (Fowler et al. 2014). Milk composition changes throughout lactation as a result of both the physiological stress of the mother and the requirements of the pup: water declines from 75% after birth to 35% 2 weeks later, while fat content increases from 12% to 50% in the same period (Riedman 1979).

Adult females lose 36–37% of their body mass between their arrival and departure from the breeding areas, which represents an average of 180 kg or 7.2 kg/day (Deutsch et al. 1994; Crocker et al. 2001). Moreover, during the breeding fast, there may be an increase in oxidative damage to proteins (Sharick et al. 2015). Given the high energy cost of reproduction, it is expected that not all females will survive or breed successfully. In fact, both the probability of survival and the reproductive success of primiparous females seem to be lower than those of experienced females (Lee 2011).

Male Reproductive Effort

In all polygynous species, the variability in reproductive success is greater in males than in females (Clutton-Brock 1985), and the northern elephant seal is no exception. Male mating success is correlated with dominance rank, which in turn depends largely on the body size and condition (Haley et al. 1991). Overall, a male of high rank is able to monopolize a large number of females (up to more than 100) in a single breeding season, while males of lower rank may mate with a minimal fraction of females (Le Boeuf 1974).

Males arrive at the reproductive areas before females and remain on land without foraging for more than 90 days, and during that period, they compete to control access to females. Initially, the social status is determined by physical confrontations, which have an extremely high energetic costs (Norris et al. 2010) and risk of lethal injuries (Fig. 10.7). Once dominance ranks have been established, they are maintained mainly by ritualized displays, which include vocalizations and body posturing (Le Boeuf 1974; Sandegren 1976; Cox 1981). These signals are usually sufficient to control the movement of subordinate males, so the frequency of physical confrontations usually decreases during the course of the breeding season.

Despite the effort that lactation represents for adult females, the energy cost for reproduction in adult males seems to be higher. Although male fasting duration over the reproductive season is three times longer than that of females, the average mass loss is similar (~36%) (Deutsch et al. 1994), but total expended energy could be three times the standard metabolic rate (Crocker et al. 2012). Moreover, the prolonged fasting of males causes oxidative damage to lipids and DNA (Sharick et al. 2015).

Unlike females, males' reproductive cost is related to both age class and social status. Average rate of mass loss is greater in adult males than in subadult males, 7.1 and 4.6 kg per day, respectively, and the percentage of body mass loss is greater in high-ranking males than in subordinates males, 41% and 34%, respectively (Deutsch et al. 1990). Also, the occurrence of external injuries, which can seriously compromise survival, is up to 20 times greater in males than in females (Deutsch et al. 1994).

Fig. 10.7 Confrontation between adult males. (Photo: M.C. García-Aguilar)

At-Sea Behavior

Northern elephant seals are deep-diving mammals that spend more than 90% of their time at sea submerged (Le Boeuf et al. 1988). Based on time-depth profiles, their dives are classified into five categories (Le Boeuf et al. 1988; Crocker et al. 1997; Kuhn et al. 2009): type A dives represent transit dives, type C food processing or rest dives, type D pelagic foraging dives, type E benthic dives used for both traveling along the continental shelf and benthic foraging, and type I short and shallow dives (<100 m).

After the post-weaning fast, weaned pups perform their first foraging migration at sea, and although the routes are not well known, some observations suggest that they move northward (Condit and Le Boeuf 1984). The mean dive depths achieved at this early stage of life is 200 m and the maximum >500 m (Le Boeuf et al. 1996). As juveniles (up to 4 years of age), elephant seals perform two migrations: one in summer, after the molt, and the second one during winter, after their resting period on land. Both female and male juveniles start their migration moving north toward the coastal waters of Washington, and then their migratory paths diverge. Females migrate toward open oceanic waters, while males remain near the continental shelf. The average dive depth is ~400 m in both sexes, with maximum dives of >600 m (Le Boeuf et al. 1996). In general, females make more transit and pelagic foraging dives than males, while males perform more benthic dives (Le Boeuf et al. 1996).

As adults, northern elephant seals alternate their stays on land to breed and molt with two migrations per year (Le Boeuf 1994). The adult females' post-breeding migration lasts 2–2.5 months (short migration), and the post-molting migration lasts 7–8 months (long migration); for adult males, both migrations last around 4 months (Le Boeuf 1994). Although both sexes migrate toward the northeast Pacific, male and female routes and destinations differ. While females exhibit similar latitudinal movements during both migrations, with a strong preference between 40° and 45°N (Robinson et al. 2012), their longitudinal movements vary seasonally. During the short migration, females travel over 5000 km and remain east of 160°W, while in the long migration, they can travel near 180°E, covering a distance of ~10,000 km (Fig. 10.8) (Robinson et al. 2012). Despite this difference in the routes between migrations, their transit speed is similar, ranging from 40 to 65 km/day (Le Boeuf et al. 2000). All these findings come from satellite tracking studies conducted on animals from California, but isotopic data from adult females at the San Benito colony indicates they forage ~8° south of those from California (Aurioles et al. 2006). Males travel about 10,000 km during each migration, moving further north than females and foraging in the Gulf of Alaska and near the Aleutian Islands, between 42 and 58°N (Fig. 10.9) (Stewart and DeLong 1995; Le Boeuf et al. 2000). Their average transit speed is around 85 km/d (Le Boeuf et al. 2000).

Adult female mean dive depth is approximately 500 m, with maximum dives greater than 1500 m (Robinson et al. 2012; Hückstädt et al. 2018). Overall, females spend more time at sea on transit and foraging dives (types A and D), while benthic dives and shallow dives (types E and I) are usually uncommon (Le Boeuf et al.

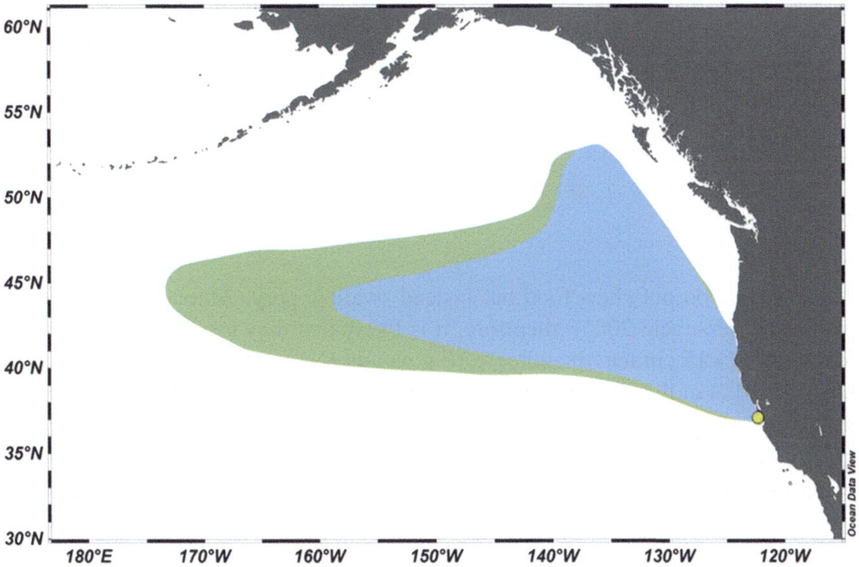

Fig. 10.8 Post-breeding (blue) and post-molting (green) migratory paths of northern elephant seal females from Año Nuevo, California (yellow dot). Based on Robinson et al. (2012)

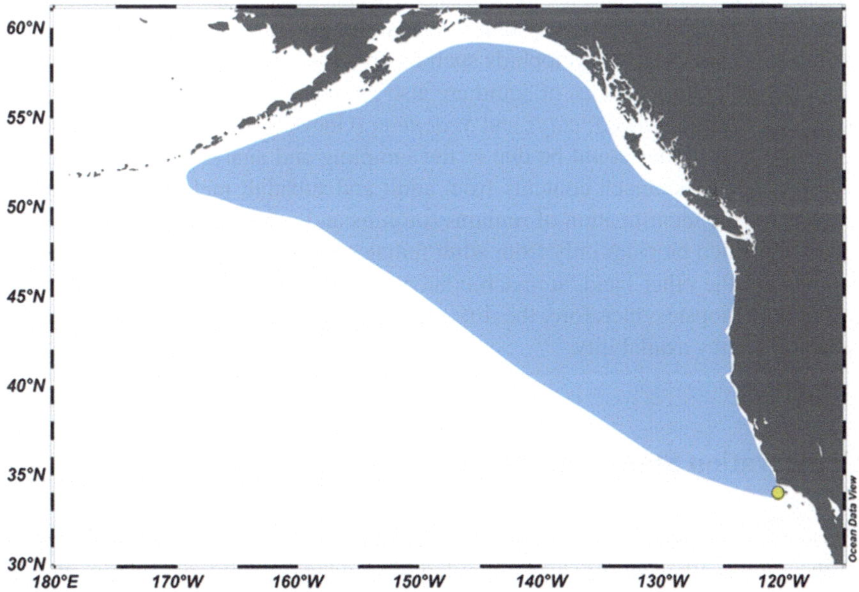

Fig. 10.9 Migratory paths of northern elephant seal males from San Miguel Island, California (yellow dot). Based on Stewart and DeLong (1994)

2000; Kuhn et al. 2009; Robinson et al. 2012). The dives of adult males are longer than those of adult females (~20%) but with a lower mean dive depth, which ranges from 350 to 450 m (DeLong and Stewart 1991). Moreover, type D dives are less frequent than in females, and type E dives are more frequent (Le Boeuf 1994).

Diet

Elephant seals do not chew food but instead swallow prey (Antonelis et al. 1987; Abbott and Verstraete 2005); therefore, it is likely that they use suction to feed on small prey of <15 cm length and/or < 70 g weight (Naito et al. 2013; Adachi et al. 2019). The diet of the northern elephant seal is not well understood, but diving patterns and isotopic data suggest that there are sexual differences both in feeding strategies and in the type of prey. That is, females seem to consume mainly epi- and mesopelagic prey of wide distribution, while males appear to preferably consume benthic prey in localized areas of the continental shelf (Le Boeuf et al. 2000; Riofrío-Lazo et al. 2012).

Elephant seals have traditionally been assumed to feed primarily on squid. Antonelis et al. (1994) identified more than 28 species of squid as part of the diet, being the most important prey *Octopoteuthis deletron*, *Histioteuthis heteropsis*, *Stigmatotheuthis dofleini*, and *Gonatopsis borealis*, although the Pacific hake, *Merluccius productus*, and the tuna crab, *Pleuroncodes planipes*, also contributed notably. However, a recent study indicates that the diet consists mainly of mesopelagic fishes, particularly myctophids such as *Tarletonbeania taylori* and *Electrona risso*; non-migratory species of squid are also an important component of the diet, especially *Chiroteuthis* c.f. *calyx* and *S. dofleini* (Goetsch 2018). The discrepancies between these studies could be due to the sampling and analyses. Antonelis et al. (1994) collected stomach contents from adult and subadult males and from adult females for the identification of remains (otoliths and squid beaks), while Goetsch (2018) collected biopsies only from adult females for the analysis of fatty acid signatures. On the other hand, stomach contents were collected around two decades earlier than biopsies; therefore, the difference in diet could be due also to temporal variation in prey availability.

Conservation Status and Threats

The northern elephant seal was almost driven to extinction by overhunting for its oil in the nineteenth century, leaving a few individuals who took refuge at Isla Guadalupe. After being protected by Mexican law in the early twentieth century, the species began to recover during the first half of the century and is currently located in all areas of its known historical distribution. At the beginning of the twenty-first century, the northern elephant seal population was estimated at approximately

201,000 individuals, of which 90% are in California, USA, and 10% in Baja California, Mexico. However, the numbers in California are increasing, while those in Baja California are decreasing (Lowry et al. 2014; García-Aguilar et al. 2018).

The northern elephant seal is currently listed as a species of *Least Concern* in the Red List of Threatened Species of the International Union for Conservation of Nature (IUCN) (Hückstädt 2015). In 1992 the species was removed from Appendix II of the Convention on International Trade in Endangered Species of Wild Fauna and Flora (CITES). In the United States, it is protected by the Marine Mammal Protection Act, but is not considered a strategic stock, and is not listed as *Endangered* or *Threatened* by the Endangered Species Act. In Mexico, elephant seals are protected under the Norma Oficial Mexicana NOM-059-ECOL-2010, and all their breeding and haul-out sites in Baja California are in protected areas decreed by the Mexican government. Although there are movements of individuals between the colonies of Baja California and California, there are no international agreements for the management of the species between the governments of Mexico and the United States.

Unlike many other pinniped species, incidental fishery mortality is very low, approaching a zero rate in US fisheries (Carretta et al. 2017), but incidental fishery mortality rate is not known in Mexico. In recent times, there has been an increase in the number of entangled animals (which can potentially die), presumably related to the "Great Pacific Garbage Patch" (Hückstädt 2015), located in waters between California and Hawaii (Lebreton et al. 2018).

Oceanographic anomalies, such as El Niño Southern Oscillation and recently the 2013–2016 warm water anomaly ("The Blob"; Kintisch 2015), impact pinniped populations. For the northern elephant seal, the most notable effects are increased pup mortality, reduced juvenile survival, low reproductive success of females, and increased foraging effort by adult females (Le Boeuf and Reiter 1991; Crocker et al. 2006). However, perhaps the greatest threat to the northern elephant seal is its low genetic diversity, a consequence of the population bottleneck they went through (Hoelzel et al. 2002; Abadía-Cardoso et al. 2017). The lack of genetic diversity makes the northern elephant seal population highly vulnerable to infectious diseases and environmental changes (Hückstädt 2015).

Although no disease outbreaks have been reported so far, several pathogens have been detected in both captive and free-ranging northern elephant seals. Bacteria such as *Leptospira* sp. (Colegrove et al. 2005; Serrano-de-la-Vega 2012), *Escherichia coli*, *Streptococcus* sp., and *Enterococcus* sp. (Spraker et al. 2014) have been isolated, as well as various viruses such as caliciviruses (Kennedy-Stoskopf 2001), gammaherpesvirus (Goldstein et al. 2006), and H1N1 influenza virus (Goldstein et al. 2013). Infections by the *Otostrongylus circumlitus* nematode have also been reported (Elson-Riggins et al. 2001).

Climate change appears as another major concern, because of the increased sea surface temperature (SST) resulting in nutritional stress (Le Boeuf and Crocker 2005) and disrupting the northern elephant seal's ability to thermoregulate while on land (Noren 2002). Climate projections predict that both the SST and air temperatures will increase above current and historical levels (IPCC 2014). Under those conditions, heat becomes an environmental threat, especially for northern elephant seal pups because of their diminished heat tolerance that makes them highly susceptible to hyperthermia (or heat stroke), potentially leading to death (Salogni et al. 2015).

References

Abadía-Cardoso A, Freimer NB, Deiner K et al (2017) Molecular population genetics of the northern elephant seal *Mirounga angustirostris*. J Hered 108(6):618–627

Abbott C, Verstraete RJM (2005) The dental pathology of northern elephant seals (*Mirounga angustirostris*). J Comp Pathol 132(2–3):169–178

Adachi T, Hückstädt LA, Tift MS et al (2019) Inferring prey size variation from mandible acceleration in northern elephant seals. Mar Mamm Sci 35(3):893–908

Anthony AW (1924) Notes on the present status of the northern elephant seal, *Mirounga angustirostris*. J Mammal 5(3):145–152

Antonelis GA, Lowry MS, Fiscus CH (1987) Assessing northern elephant seal feeding habits by stomach lavage. Mar Mamm Sci 3(4):308–322

Antonelis GA, Lowry MS, Fiscus CH et al (1994) Diet of the northern elephant seal. In: Le Boeuf BJ, Laws RM (eds) Elephant seals: population ecology, behavior and physiology. University of California Press, Berkeley/Los Angeles/Oxford, pp 211–223

Arias-Del-Razo A, Schramm Y, Heckel G et al (2017) Distribution of four pinnipeds (*Zalophus californianus*, *Arctocephalus philippii townsendi*, *Phoca vitulina richardii*, and *Mirounga angustirostris*) on islands off the west coast of the Baja California Peninsula, Mexico. Aquat Mamm 43(1):40–51

Aurioles D, Koch PL, Le Boeuf BJ (2006) Differences in foraging location of Mexican and California elephant seals: evidence from stable isotopes in pups. Mar Mamm Sci 22(2):326–338

Bartholomew GA, Hubbs CL (1960) Population growth and seasonal movements of the northern elephant seal, *Mirounga angustirostris*. Mammalia 24(3):313–324

Boness DJ, Bowen WD (1996) The evolution of maternal care in pinnipeds. Bioscience 46(9):645–654

Carretta, JV, Forney KA, Oleson E et al (2017) U.S. Pacific marine mammal stock assessments, 2016. NOAA Technical memorandum 577

Cassini MH (1999) The evolution of reproductive systems in pinnipeds. Behav Ecol 10(5):612–616

Clinton WL, Le Boeuf BJ (1993) Sexual selection's effects on male life history and the pattern of male mortality. Ecology 74(6):1884–1892

Clutton-Brock TH (1985) Selection in relation to sex. In: Bendall S (ed) Evolution from molecules to men. Cambridge University Press, Cambridge, pp 457–481

Codde SA, Allen SG, Houser DS, Crocker DE (2016) Effects of environmental variables on surface temperature of breeding female northern elephant seals, *Mirounga angustirostris*, and pups. J Therm Biol 61:98–105

Colegrove KM, Lowenstine LJ, Gulland FMD (2005) Leptospirosis in northern elephant seals (*Mirounga angustirostris*) stranded along the California coast. J Wildlife Dis 41(2):426–430

Condit R, Le Boeuf BJ (1984) Feeding habits and feeding grounds of the northern elephant seal. J Mammal 65(2):281–290

Condit R, Reiter J, Morris PA et al (2013) Lifetime survival rates and senescence in northern elephant seals. Mar Mamm Sci 30(1):122–138

Cox CR (1981) Agonistic encounters among male elephant seals: frequency, context, and the role of female preference. Am Zool 21:197–209

Crocker DE, Le Boeuf BJ, Costa DP (1997) Drift diving in female northern elephant seals: implications for food processing. Can J Zool 75:27–39

Crocker DE, Williams JD, Costa DP et al (2001) Maternal traits and reproductive effort in northern elephant seals. Ecology 82(12):3451–3555

Crocker DE, Costa DP, Le Boeuf BJ et al (2006) Impact of El Niño on the foraging behavior of female northern elephant seals. Mar Ecol Prog Ser 309:1–10

Crocker DE, Houser DS, Webb PM (2012) Impact of body reserves on energy expenditure, water flux, and mating success in breeding male northern elephant seals. Physiol Biochem Zool 85(1):11–20

De Muizon C (1982) Phocid phylogeny and dispersal. Ann S Afr Mus 89(2):175–213

DeLong RL, Stewart BS (1991) Diving patterns of northern elephant seal bulls. Mar Mamm Sci 7(4):369–384

Deméré TA, Berta A, Adam PJ (2003) Pinnipedimorph evolutionary biogeography. B Am Mus Nat Hist 279:32–76

Deutsch CJ, Haley MP, Le Boeuf BJ (1990) Reproductive effort of male northern elephant seals: estimates from mass loss. Can J Zool 68:2580–2593

Deutsch CJ, Crocker DE, Costa DP et al (1994) Sex- and age-related variation in reproductive effort of northern elephant seals. In: Le Boeuf BJ, Laws RM (eds) Elephant seals: population ecology, behavior, and physiology. University of California Press, Berkeley/Los Angeles/Oxford, pp 169–210

Elorriaga-Verplancken F, García-Aguilar MC (2018) Inter-annual (2002–2016) variation on the natality rate of the northern elephant seal (*Mirounga angustirostris*) at the San Benito colony, Baja California, Mexico. Mar Mamm Sci 34(3):823–828

Elson-Riggins JC, Al-Banna L, Platzer EG et al (2001) Characterization of *Otostrongylus circumlitus* from Pacific harbor and northern elephant seals. J Parasitol 87(1):73–78

Fowler MA, Debier C, Mignolet E et al (2014) Fatty acid mobilization and comparison to milk fatty acid content in northern elephant seals. J Comp Physiol B 184:284–294

Franco-Ortiz M (2012) Distribución y abundancia del elefante marino del norte, *Mirounga angustirostris*, en México. Dissertation, Universidad Autónoma de Baja California

Fyler CA, Reeder TW, Berta A et al (2005) Historical biogeography and phylogeny of monachine seals (Pinnipedia: Phocidae) based on mitochondrial and nuclear DNA data. J Biogeogr 32:1267–1279

García-Aguilar MC (2004) Breeding biology of the northern elephant seal (*Mirounga angustirostris*) at the Isla San Benito del Oeste, Eastern Pacific, Mexico. Aquat Mamm 30(2):289–295

García-Aguilar MC, Turrent C, Elorriaga-Verplancken F et al (2018) Climate change and the northern elephant seal (*Mirounga angustirostris*) population in Baja California, Mexico. PLoS One 13(2):e0193211

Gill T (1866) Prodrome of a monograph of the pinnipeds. Proc Essex Inst 5:3–13

Goetsch C (2018) Illuminating the twilight zone: diet and foraging strategies of a deep-sea predator, the northern elephant seal. Ph.D. Dissertation, University of California Santa Cruz

Goldstein T, Lowenstine LJ, Lipscomb TP et al (2006) Infection with a novel gammaherpesvirus in northern elephant seals (*Mirounga angustirostris*). J Wildlife Dis 42(4):830–835

Goldstein T, Mena I, Anthony SJ et al (2013) Pandemic H1N1 influenza isolated from free-ranging northern elephant seals in 2010 off the Central California coast. PLoS One 8(5):e62259

Haley MP, Deutsch CJ, Le Boeuf BJ (1991) A method for estimating mass of large pinnipeds. Mar Mamm Sci 7(2):157–164

Hindell MA, Perrin W (2008) Elephant seals, *Mirounga angustirostris* and *M. leonine*. In: Würsig B, Perrin W, Thewissen JGM (eds) Encyclopedia of marine mammals, 2nd edn. Academic Press, Burlington/San Diego/New York/London, pp 364–368

Hoelzel AR, Halley J, Campagna C et al (1993) Elephant seal genetic variation and the use of simulation models to investigate historical population bottlenecks. J Hered 84(6):443–449

Hoelzel AR, Fleischer RC, Campagna C et al (2002) Impact of a population bottleneck on symmetry and genetic diversity in the northern elephant seal. J Evol Biol 15(4):567–575

Hückstädt L (2015) Mirounga angustirostris. In: The IUCN red list of threatened species. https://doi.org/10.2305/IUCN.UK.2015-2.RLTS.T13581A45227116.en. Accessed 19 Nov 2019

Hückstädt LA, Holser RR, Tift MS et al (2018) The extra burden of motherhood: reduced dive duration associated with pregnancy status in a deep-diving mammal, the northern elephant seal. Biol Lett 14:20170722
Huey LM (1924) Recent observations on the northern elephant seal. J Mammal 5(4):237–242
Huey LM (1925) Late information on the Guadalupe Island elephant seal herd. J Mammal 6(2):126–127
Huey LM (1927) The latest northern elephant seal census. J Mammal 8(2):160–161
Huey LM (1930) Past and present status of the northern elephant seals with a note on the Guadalupe fur seal. J Mammal 11(2):188–194
Intergovernmental Panel on Climate Change (IPCC) (2014) Climate change 2014: synthesis report. Geneva
Kennedy-Stoskopf S (2001) Viral diseases. In: Dierauf LA, Gulland FMD (eds) CRC handbook of marine mammal medicine, 2nd edn. Taylor & Francis Group, Boca Raton, pp 285–308
Khamas WA, Smodlaka H, Leach-Robinson J et al (2012) Skin histology and its role in heat dissipation in three pinniped species. Acta Vet Scand 54(1):46
Kintisch E (2015) 'The blob' invades Pacific, flummoxing climate experts. Science 348:17–18
Kuhn CE, Crocker DE, Tremblay Y et al (2009) Time to eat: measurements of feeding behaviour in a large marine predator, the northern elephant seal *Mirounga angustirostris*. J Anim Ecol 78:515–523
Le Boeuf BJ (1972) Sexual behaviour in the northern elephant seal, *Mirounga angustirostris*. Behaviour 41:1):1–1)26
Le Boeuf BJ (1974) Male-male competition and reproductive success in elephant seals. Am Zool 14:163–176
Le Boeuf BJ (1978) Social behavior in some marine and terrestrial carnivores. In: Reese ES, Lighter FJ (eds) Contrast in behaviour. Wiley, Hoboken, pp 251–279
Le Boeuf BJ (1991) Pinniped mating systems on land, ice and the water: emphasis on the Phocidae. In: Renouf D (ed) Behavior of pinnipeds. Chapman and Hall, London, pp 45–65
Le Boeuf BJ (1994) Variation in the diving pattern of northern elephant seals with age, mass, sex, and reproductive condition. In: Le Boeuf BJ, Laws RM (eds) Elephant seals: population ecology, behavior and physiology. University of California Press, Berkeley/Los Angeles/Oxford, pp 237–252
Le Boeuf BJ, Crocker DE (2005) Ocean climate and seal condition. BMC Biol 3:9
Le Boeuf BJ, Laws RM (1994) Elephant seals: an introduction to the genus. In: Le Boeuf BJ, Laws RM (eds) Elephant seals: population ecology, behavior, and physiology. University of California Press, Berkeley/Los Angeles/Oxford, pp 1–26
Le Boeuf BJ, Mate BR (1978) Elephant seal colonize additional Mexican and California islands. J Mammal 59(3):621–622
Le Boeuf BJ, Reiter J (1991) Biological effects associated with El Niño southern oscillation, 1982-83, on northern elephant seals breeding at Año Nuevo, California. In: Trillmich F, Ono KA (eds) Pinnipeds and El Niño, ecological studies, vol 88. Springer, Berlin/Heidelberg, pp 206–218
Le Boeuf BJ, Whiting RJ, Gantt RF (1972) Perinatal behavior of northern elephant seal females and their young. Behaviour 43(1):121–156
Le Boeuf BJ, Costa DP, Huntley AC et al (1988) Continuous, deep diving in female northern elephant seals, *Mirounga angustirostris*. Can J Zool 66:446–458
Le Boeuf BJ, Condit R, Reiter J (1989) Parental investment and the secondary sex ratio in northern elephant seals. Behav Ecol Sociobiol 25(2):109–117
Le Boeuf BJ, Morris P, Reiter J (1994) Juvenile survivorship of northern elephant seals. In: Le Boeuf BJ, Laws RM (eds) Elephant seals: population ecology, behavior and physiology. University of California Press, Berkeley/Los Angeles/Oxford, pp 121–136
Le Boeuf BJ, Morris PA, Blackwell SB et al (1996) Diving behavior of juvenile northern elephant seals. Can J Zool 74:1632–1644

Le Boeuf BJ, Crocker DE, Costa DP et al (2000) Foraging ecology of northern elephant seals. Ecol Monogr 70(3):353–382

Lebreton L, Slat B, Ferrari F et al (2018) Evidence that the great Pacific garbage patch is rapidly accumulating plastic. Sci Rep UK 8:4666

Lee DE (2011) Effects of environmental variability and breeding experience on northern elephant seal demography. J Mammal 92(3):517–526

Lowry MS, Condit R, Hatfield B et al (2014) Abundance, distribution, and population growth of the northern elephant seal (*Mirounga angustirostris*) in the United States from 1991 to 2010. Aquat Mamm 40(1):20–31

Mauck B, Bilgmann K, Jones DD et al (2003) Thermal windows on the trunk of hauled seals: hot spots for thermoregulatory evaporation? J Exp Biol 202:1727–1738

Naito Y, Costa DP, Adachi T et al (2013) Unravelling the mysteries of a mesopelagic diet: a large apex predator specializes on small prey. Funct Ecol 27:710–717

Noren DP (2002) Thermoregulation of weaned northern elephant seal (*MIrounga angustirostris*) pups in air and water. Physiol Biochem Zool 75(5):513–523

Norris AL, Houser DS, Crocker DE (2010) Environment and activity affect skin temperature in breeding adult male elephant seals (*Mirounga angustirostris*). J Exp Biol 213:4205–4212

Ray CE (1976) Geography of phocid evolution. Syst Zool 25:391–406

Rea LD, Costa DP (1992) Changes in standard metabolism during long-term fasting in northern elephant seal pups (*Mirounga angustirostris*). Physiol Zool 65:97–111

Reiter J, Panken KJ, Le Boeuf BJ (1981) Female competition and reproductive success in northern elephant seals. Anim Behav 29:670–687

Riedman M (1979) Changes in milk composition during lactation in the northern elephant seal. Physiol Zool 52(2):240–249

Riedman M (1990) The pinnipeds: seals, sea lions, and walruses. University of California Press, Berkeley/Los Angeles/Oxford

Riofrío-Lazo M, Aurioles-Gamboa D, Le Boeuf BJ (2012) Ontogenetic changes in feeding habits of northern elephant seals revealed by $\delta^{15}N$ and $\delta^{13}C$ analysis of growth layers in teeth. Mar Ecol Progr Ser 450:229–224

Robinson PW, Costa DP, Crocker DE et al (2012) Foraging behavior and success of a mesopelagic predator in the Northeast Pacific Ocean: insights from a data-rich species, the northern elephant seal. PLoS One 7(5):e36728

Rothschild W (1910) Notes on sea elephants (*Mirounga*). Novitates Zoologicae 17(3):445–446

Salogni E, Sanvito S, Galimberti F (2015) Postmortem examination and causes of death of northern elephant seal (*Mirounga angustirostris*) pups at the San Benito Islands, Baja California, Mexico. Mar Mamm Sci 32(2):743–752

Salogni E, Galimberti F, Sanvito S et al (2018) Male and female pups of the highly sexually dimorphic northern elephant seal (*Mirounga angustirostris*) differ slightly in body size. Can J Zool 97:241–250

Sandegren FE (1976) Agonistic behavior in the male northern elephant seal. Behavior 57(1–2):136–158

Scammon CM (1870) Sea-elephant hunting. Overland Mon 4(2):112–117

Serrano-de-la-Vega MI (2012) Exposición a *Leptospira* sp. patógena del elefante marino, *Mirounga angustirostris*. Dissertation, Universidad Autónoma de Baja California

Sharick JT, Vasquez-Medina JP, Ortiz RM et al (2015) Oxidative stress is a potential cost of breeding in male and female northern elephant seals. Funct Ecol 29(3):367–376

Spraker TR, Lyons ET, Kuzmina TA et al (2014) Causes of death in preweaned northern elephant seal pups (*Mirounga angustirostris*, Gill, 1866), Año Nuevo State Reserve, California, 2012. J Vet Diagn Investig 26(2):320–326

Stewart BS, DeLong RL (1994) Postbreeding foraging migration of northern elephant seals. In: Le Boeuf BJ, Laws RM (eds) Elephant seals: population ecology, behavior and physiology. University of California Press, Berkeley/Los Angeles/Oxford, pp 291–309

Stewart BS, DeLong RL (1995) Double migrations of the northern elephant seal, *Mirounga angustirostris*. J Mammal 76(1):196–205

Stewart BS, Huber HR (1993) Mirounga angustirostris. Mamm Species 449:1–10

Stewart BS, Yochem PK, Huber HR et al (1994) History and present status of the northern elephant seals population. In: Le Boeuf BJ, Laws RM (eds) Elephant seals: population ecology, behavior and physiology. University of California Press, Berkeley/Los Angeles/Oxford, pp 29–48

Sydeman WJ, Nur N (1994) Life history strategies of female northern elephant seals. In: Le Boeuf BJ, Laws RM (eds) Elephant seals: population ecology, behavior and physiology. University of California Press, Berkeley/Los Angeles/Oxford, pp 137–153

Sydeman WJ, Huber HR, Emslie SD et al (1991) Age-specific weaning success of northern elephant seals in relation to previous breeding experience. Ecology 72(6):2204–2217

Tennett KA, Costa DP, Nicastro AJ et al (2018) Terrestrial locomotion of the northern elephant seal (*Mirounga angustirostris*): limitation of large aquatically adapted seals on land? J Exp Biol 221:jeb180117

Townsend CH (1885) An account of recent captures of the California Sea-elephant and statistics related to the present abundance of the species. Proc US Nat Mus 8:90–94

Townsend CH (1912) The northern elephant seal, *Macrorhinus angustirostris* Gill. Zoologica 1(8):159–173

Chapter 11
Pacific Harbor Seal (*Phoca vitulina richardii*, Gray 1864)

Yolanda Schramm and Gisela Heckel

Abstract The Pacific harbor seal, *Phoca vitulina richardii* (sometimes referred to as *P. v. richardsi*), occurs along the coast and islands from Japan to Alaska; the west coast of Canada and USA; and to the west-central coast of the Baja California Peninsula, Mexico. There are approximately 348,961 individuals, including 11,967 (95% Credibility Interval: 9,619–16,284) individuals in Mexico. The harbor seal's diet in Mexico has been documented using fecal analysis of prey hard parts and metagenomic studies and from stable isotope studies which identified habitat use. Genetic studies of seals from five colonies in Mexico indicated that haplotype diversity was two times lower and nucleotide diversity was four to nine times lower than in the rest of the Pacific. A latitudinal cline has been described for the timing of both pupping and molting. There is a delay in the mean pupping date from the southernmost colony, San Roque Island (27 January), to the northern colony on Todos Santos Island (16 February). Molting starts on San Roque on 27 February, while on Todos Santos, it occurs on 18 April. At one coastal colony (Punta Banda Estuary, Baja California), recreational activities affected important events such as nursing and molting. This colony was monitored for six pupping seasons, which showed that pup production was reduced with unfavorable oceanographic conditions, i.e., warm sea surface temperature and low chlorophyll concentration, proxies for food availability. Fishermen at four colonies west of the Baja California Peninsula considered that the harbor seal interferes little with their livelihood. The International Union for the Conservation of Nature (IUCN) Red List has harbor seals listed as *Least Concern* worldwide and occupy the lowest category (*Under Special Protection*) in the Mexican *List of Species at Risk*. New data collected on harbor seals in Mexico indicate that the Mexican listing may be changed to *Threatened* in the future.

Keywords Abundance · Distribution · Correction factor · Trophic ecology · Genetics · Conservation status · *Phoca vitulina richardsi*

Y. Schramm (✉)
Facultad de Ciencias Marinas, Universidad Autónoma de Baja California, Ensenada, Baja California, Mexico
e-mail: yschramm@uabc.edu.mx

G. Heckel
Departamento de Biología de la Conservación, Centro de Investigación Científica y de Educación Superior de Ensenada, Ensenada, Mexico

G. Heckel, Y. Schramm (eds.), *Ecology and Conservation of Pinnipeds in Latin America*, https://doi.org/10.1007/978-3-030-63177-2_11

Worldwide Distribution

Harbor seals (*Phoca vitulina*) are distributed along the North Atlantic and North Pacific regions (King 1983). Three subspecies are currently recognized (Committee on Taxonomy 2019): (1) *P. v. vitulina* in the Atlantic, from Portugal to Norway (including Svalbard) in the east and, in the west, from New Jersey, USA, to Hudson Bay, Canada, including part of Greenland and Iceland; (2) *P. v. richardii* (sometimes referred to as *P. v. richardsi*) in the Pacific, from Japan to Baja California (including the Commander Islands, the Kamchatka Peninsula, and the western Aleutian Islands); and (3) *P. v. mellonae* in freshwater lakes and rivers in eastern Canada, on the Ungava Peninsula (Reeves et al. 2002).

The southern distribution limit of the Pacific harbor seal is located along the west coast and islands of the western Baja California Peninsula, Mexico (Fig. 11.1). They are found at nine islands (listed from north to south): (1) Coronado, (2) Todos

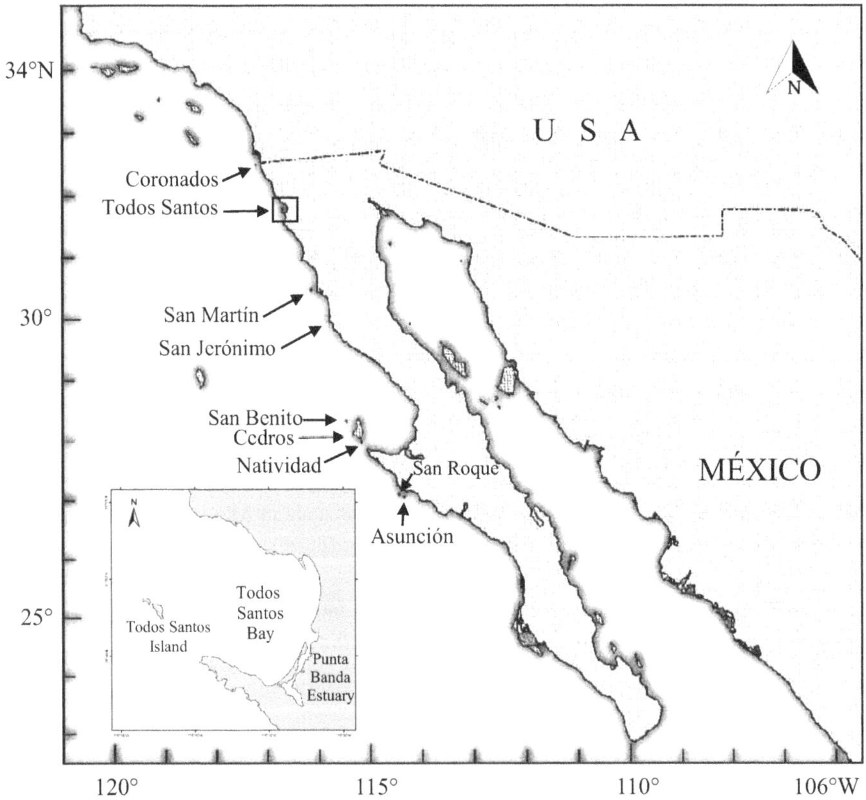

Fig. 11.1 Distribution of *Phoca vitulina richardii* in Mexico. Islands with reproductive colonies are shown. The inset is an enlargement of Todos Santos Bay, with the Todos Santos Island and Punta Banda Estuary colonies. There are also reproductive colonies along the Peninsular west coast, distributed from Todos Santos Bay to Asunción Bay

Santos, (3) San Martín, (4) San Jerónimo, (5) San Benito, (6) Cedros, (7) Natividad, (8) San Roque, and (9) Asunción (Gallo-Reynoso and Aurioles-Gamboa 1984; Arias-Del-Razo et al. 2016; 2017; Lubinsky-Jinich et al. 2017). Forty-four breeding colonies (confirmed by the presence of pups) have been located along the western Baja California Peninsula coast, between Todos Santos Bay and Asunción Bay (Lubinsky-Jinich et al. 2017).

Worldwide, the harbor seal's terrestrial habitat is very diverse; however, in Mexico it hauls out on a few substrate types. Although it prefers beaches, during the breeding season, the harbor seal gathers on three substrates: sandy beaches, cobblestone beaches, and rocky platforms (Arias-Del-Razo et al. 2016). During the non-breeding season, herds are smaller, and individuals are dispersed across the entire island or colony (Arias-Del-Razo et al. 2017).

General Characteristics

The harbor seal (*Phoca vitulina*) has two pelage colorations, a dark and a light color phase (or morph), with irregular patterns of black spots or clear rings (Shaughnessy and Fay 1977; Kelly 1980) (Fig. 11.2). The Pacific harbor seal is the only pinniped species in Mexico that is spotted. Along the coast and islands west of the Baja California Peninsula, 92% of harbor seals belong to the dark morph group (Shaughnessy and Fay 1977). During molting, the pelage turns into brown (Fig. 11.3). Newborn pups have the same pelage pattern and color as adults, as their

Fig. 11.2 Pelage colorations in harbor seals. Light and dark phases, although there is a higher proportion of dark seals in Mexico. Photo by Gisela Heckel, Todos Santos Island, 17 May 2006

Fig. 11.3 Molting harbor seals with old, brown pelage. Photo by Alejandra Baez-Flores, Cedros Island, 26 May 2009

long soft white lanugo hair molts in utero (Scheffer and Slipp 1944; Boulva and McLaren 1979). Premature pups, however, are born with lanugo (Boulva 1975).

Adult length varies, from 140 to 175 cm in females and 138 to 190 cm in males, with the largest individuals found in the western Pacific (McLaren 1993). In British Columbia, adult females (≥ 5 years) average 147.7 ± 2.4 cm (95% confidence limits) in length and weigh 64.8 ± 4.4 kg, and adult males (≥ 9 years) average 161.1 ± 4.9 cm in length and weigh 87.0 ± 6.6 kg (Bigg 1969). Pacific harbor seals are slightly sexually dimorphic. Fully grown males are approximately 9% longer and 34% heavier than fully grown females (Bigg 1969). Newborn harbor seal (*Phoca vitulina*) pups are 65–100 cm in length and weigh 8–12 kg. Pacific harbor seals (*P. v. richardii*) in British Columbia have been measured at an average length of 81.6 ± 6.2 cm and an average weight of 10.2 ± 1.5 kg (Bigg 1969). Males live for 20 years and females for 30 years (Bigg 1969), and the survival rate for both sexes is approximately 20% during the first 5 years of life. Afterward, the mortality rate in males is 29% and 15% in females (Bigg 1969). It should be noted that the information pertaining to British Columbia was generated more than 40 years ago, so some features may have changed since then.

Abundance

The Pacific harbor seal global abundance estimate is approximately 348,961 individuals; this total number includes estimates for different years and regions, from Japan to Mexico. In Alaska, from 1998 to 2007, 152,602 individuals were estimated

(SE = 2,703) (Allen and Angliss 2015), 105,000 in Canada from 2004 to 2008 (Olesiuk 2010); 24,732 (CV = 0.12) in Washington and Oregon in 1999, 13,692 in Washington inland waters in 1999, and 30,968 (CV = 0.157) in California in 2012 (Carretta et al. 2014). There are no current abundance estimates for Pacific harbor seals from Japan to the west of the Aleutian Islands, the Commander Islands, and the Kamchatka Peninsula, although some reports indicate that the population has decreased in some colonies since 1970 (Kobayashi et al. 2014) and that it could number under 10,000 individuals (Bjørge et al. 2010). The counts for all the estimates mentioned were carried out more than 8 years ago, so abundances may have changed since then.

In 2016, the abundance of the Pacific harbor seal in Mexico was 11,967 individuals (95% Credibility Interval: 9,619–16,284) (Lubinsky-Jinich 2019). This estimate was obtained using the 1.58 (95% Credibility Interval: 1.27–2.15) multiplier (correction factor for seals not counted because they were in the water during the surveys) specific to the Mexican population on land during the molting season (Lubinsky-Jinich 2019). In Mexico, the highest number of individuals on land was observed to haul out during the molting season (Fernández-Martín et al. 2016; Tapia-Harris et al. 2017; Lubinsky-Jinich et al. 2019). Previous studies on temporal variation of harbor seal abundance at two colonies in Mexico (Lubinsky-Jinich et al. 2019) indicated censuses should be conducted during the first days of June in the northern colonies (Coronado, Todos Santos, San Martín, and San Jerónimo islands) and the last days of April in the southern colonies (San Benito Archipelago, and Cedros, Natividad, San Roque, and Asunción islands) and along the coast of the Baja California Peninsula (Lubinsky-Jinich et al. 2019).

Movements

In general, harbor seals are considered to be a non-migratory species, with local variations in the distance that individuals travel from their colonies (Burns 2009) and whose movements are related to (1) local bathymetry (Frost et al. 2006), (2) stage of the annual life cycle, and (3) age or sex (Lowry et al. 2001, Peterson et al. 2012). The dispersal radius around the colonies related to foraging depends on the study area. In Scotland, it has been recorded at 25 (Cunningham et al. 2009) and 30 km (Tollit et al. 1998). In Alaska, a radius of 5–10 km in adults and 10–25 km in juveniles was reported, although long-range movements of up to 500 km may also occur (Lowry et al. 2001). In Washington, USA, males moved in average 104 km and females 23 km from the capture sites (Peterson et al. 2012). Despite variations between regions or colonies, the harbor seal's home range is small compared to other pinnipeds.

In Mexico, six satellite-tagged harbor seals from Todos Santos Island, west of Ensenada in Northern Baja California (Fig. 11.1), indicated that between May and September 2016, after the molting season, individuals traveled an average of 5–30 km from the tagging site (Lubinsky-Jinich 2019). Stable isotope analysis of harbor seal pup fur collected at Natividad Island in the central region of the Baja California Peninsula (Fig. 11.1) indicated that adult female harbor seals forage in areas near their colonies (Elorriaga-Verplancken et al. 2016). Further indirect evidence of the small dispersal radius of harbor seals in Mexico is seen in the differences in both prey type and relative consumption between two colonies separated by only 16 km, i.e., Todos Santos Island and the Punta Banda Estuary (Fig. 11.1, Alamán-De Regules 2014).

Pupping and Molting Phenology

While harbor seals, like all pinnipeds, spend most of the year at sea, they carry out important activities on land for pupping, nursing, breeding, and molting (Bigg 1969). Once the pup is weaned, the mother comes into estrus (Bigg 1969; Burns 1986); however, females without pups seem to ovulate earlier than mothers during the same season (Venables and Venables 1957; Bigg 1969; Harrison 1963, cited by Bigg 1969). As with other mammals, harbor seals have delayed implantation of the blastocyst (Fisher 1954) that in Vancouver, Canada has been noted to occur 2 months after fertilization (Bigg 1969), one and a half to 2 months in Alaska (Bishop 1968, cited by Bigg 1969), and 2–3 months in the northern Atlantic (Harrison 1963, cited by Bigg 1969).

Pupping occurs on almost precisely the same dates every year (Temte 1991). Blastocyst implantation is mainly regulated by photoperiod (Temte 1994), resulting in a northward latitudinal delay in pupping (Temte 1991). Therefore, a latitudinal cline occurs in which pupping starts at the end of December at the southern extreme of the subspecies' distribution in Baja California, Mexico (Fernández-Martín 2018), and at the end of May in Washington, USA (Temte 1991). Because molting follows the pupping season, the start dates for molting also present a latitudinal delay (Daniel et al. 2003).

Pupping and molting clines also have been observed for harbor seals in Mexico, where on San Roque Island, one of the southernmost colonies of its distribution, the pupping season occurs from the end of December through March (Table 11.1, Fernández-Martín 2018; Lubinsky-Jinich et al. 2019), and the molting season occurs from mid-February through June (Tapia-Harris et al. 2017; Lubinsky-Jinich et al. 2019). On Todos Santos Island, one of the northernmost harbor seal colonies in Mexico, the pupping season occurs from the end of January to the beginning of April (Fernández-Martín 2018; Lubinsky-Jinich et al. 2019), and the molting season occurs from mid-April to mid-July (Tapia-Harris et al. 2017; Lubinsky-Jinich et al. 2019). At the coastal colony located at Punta Banda Estuary, the pupping season occurs from mid-February to mid-April (Fernández-Martín 2018), and the molting season occurs from the end of April to mid-July (Table 11.1, Fernández-Martín et al. 2016; Lubinsky-Jinich et al. 2019). Even though Todos Santos Island is located on the same latitude as the Punta Banda Estuary, a 27-day delay in the mean

Table 11.1 Start and end of pupping and molting seasons at the distribution ends of *Phoca vitulina richardii* in Mexico

LOCATION	PUPPING SEASON						MOLTING SEASON							
	MONTHS						MONTHS							
	D	J	F	M	A	M	J	F	M	A	M	J	J	A
TODOS SANTOS ISLAND (31°47'37''N 116°46'45''W)		■	■	■	■					■	■	■		
PUNTA BANDA ESTUARY (31°46'04''N 116°37'07''W)			■	■	■					■	■	■	■	
SAN ROQUE ISLAND (27°08'53''N 114°22'45''W)	■	■	■	■				■	■	■	■	■		

Note that Todos Santos Island and Punta Banda Estuary are almost at the same N latitude, within Todos Santos Bay, Baja California

pupping date has been reported for the latter colony (Fernández-Martín 2018). This demonstrates that these two colonies respond to different photoperiods: 13.6 h light/day on Todos Santos Island (mid-February) and 14.1 h light/day at Punta Banda Estuary (mid-March), probably due to the genetic isolation of both these colonies (Fernández-Martín 2018). Monitoring undertaken at the small colony at Punta Banda Estuary for six breeding seasons has shown little variation in pupping dates (Fernández-Martín 2018).

Reproductive Biology: Social Organization and Breeding Strategies

Gathering information about the harbor seal's reproductive biology is not easy, mainly because of its reproductive behavior and mating system. The evasive behavior of the species, difficulty in accessing their haul-out sites, and their high response reaction to disturbance make the study of their behavior challenging. In addition, as mating occurs in the water, observations of this nature are not easy to accomplish.

Harbor seal herds present little social structure at their haul-out sites, which tend to be random assemblages of individuals exhibiting site fidelity, mainly females with pups (Godsell 1988). These groups usually contain animals of different sex and age classes that do not seem to stay together in the long term (Bigg 1969; Godsell 1988).

Female Reproductive Biology

Females are sexually mature after their first ovulation, which occurs at 2–5 years of age, with the majority (70%) reaching maturity at 2–3 years (Bigg 1969). They give birth to a single pup annually, which indicates that they ovulate only once per year (Bigg 1969), while the gestation period (from fertilization to birth) lasts 10.5 months (Bigg 1969). The high gestation rate observed at harbor seal colonies (≥80%) means that most of the reproductive females become pregnant and give birth every year (Bigg 1969; Thompson and Wheeler 2008). The gestation or fecundity rate is 70% for 2- to 7-year-old females and 97% for 8- to 28-year-old females (Bigg 1969). The only data of this kind that are specific to Mexico pertain to an indirect observation conducted on a small continental colony of approximately 100 individuals (Fernández-Martín et al. 2016) at Punta Banda Estuary, which counted approximately 40 pups in years with favorable environmental conditions (Fernández-Martín 2018).

Harbor seal births occur on land (Newby 1973a), with mothers giving birth while segregated away from the rest of the group, either individually or with other females with newborn pups (collectively known as a nursing herd) (Newby 1973a; Slater and Markowitz 1983; Kovacs et al. 1990; Wilson and Jones 2018). The mother and pup bond with each other within the first 5 min after birth (Lawson and Renouf 1987). Harbor seal mother-pup pairs stay together on land and in water, and this necessitates mutual recognition. The pups' age in days can be determined by the status of either the umbilical cord or the umbilicus (Boulva 1975; Dierauf et al. 1986). Pups are precocious and follow the mother into the water a few minutes after birth (Newby 1973a; Bowen et al. 1999; Skinner 2006; Wilson and Jones 2018), apparently avoiding harassment from birds that devour the placenta. Minutes later, both mother and pup return to land and start nursing.

Nursing takes place on land (Venables et al. 1955; Newby 1973a; Wilson 1974, 1978, 2001; Renouf et al. 1983) with the lactation period lasting between 2 and 6 weeks, depending on the region (Fisher 1954; Venables et al. 1955; Harrison 1963, cited by Bigg 1969). In Mexico, nursing has been reported to last for 4–5 weeks (Fernández-Martín et al. 2016; Ruiz-Mar 2016). Contrary to most phocids, female harbor seals' fasting is very short or non-existent (Bowen et al. 1992; Boness et al. 1994), probably because of their low capacity for energy storage due to their small size (Bowen et al. 1992). The pup accompanies its mother during her foraging trips, usually riding on her back while she swims (Fig. 11.4). In Mexico, pups have been observed floating asleep on the canopy of kelp (*Macrocystis pyrifera*) forests found around Natividad Island while their mothers dive in search of food nearby (F. Galimberti and S. Sanvito,[1] pers. comm.). When the weaning time approaches, the mother gradually rejects the pup's attempts to nurse (Lawson and Renouf 1987).

[1] Filippo Galimberti and Simona Sanvito, Elephant Seal Research Group, Sea Lion Island, Falkland Islands (www.eleseal.org)

Fig. 11.4 Pup riding on its mother's back. Photo by Fernando R. Elorriaga-Verplancken, San Benito Islands, winter 2013

Male Reproductive Biology

Male harbor seals attain sexual maturity at 3–6 years of age, although most mature at 5 years and produce sperm for 8–9 months per year (Bigg 1969). During the breeding season, reproductive males' fidelity to haul-out sites usually increases, although they remain isolated and do not interact with adult females (Walker and Bowen 1993), while overt fights between males occur rarely (Davis and Renouf 1987; Godsell 1988; Walker and Bowen 1993). However, agonistic behavior between males is assumed to occur in the water, as males present head, neck, tail, and hind flipper wounds that increase in frequency as more females become estrous (Boulva and McLaren 1979; Sullivan 1981, 1982; Davis and Renouf 1987; Godsell 1988; Walker and Bowen 1993), suggesting competition for reproductive females (Boulva and McLaren 1979; Sullivan 1981; Godsell 1988; Walker and Bowen 1993).

Evidence suggests patterns of dominance or hierarchies among reproductive males (Sullivan 1981, 1982; Nicholson 1997). Each reproductive male uses a small territory in which to vocalize (Hanggi and Schusterman 1994; Van-Parijs et al. 1997; Hayes et al. 2004a, b) and performs diving displays to attract estrous females, thus suggesting a lek-type mating system (Van-Parijs et al. 1997). This system is also defined by the males' fidelity to aquatic territories that they defend year after year and the free movements of females (Hayes et al. 2004a, b). Small territories are distributed across a wide geographic area around the haul-out site, with some found in foraging areas and others in corridors running between sites (Van-Parijs et al. 1997). During the mating season, males stay nearshore, reducing their foraging effort, increasing activities associated with competition for females, and reducing their distribution area (Boness et al. 2006). The lek-type system is also supported by molecular evidence: a small number of paternity assignments suggest the selection of males by females and a low to medium polygyny level (Boness et al. 2006; Hayes et al. 2006). Furthermore, there is evidence that females copulate with more than one male during the breeding season (Peterson 1998).

Although various interactions occurring in the water between adult seals have been described and interpreted as courtship and copulation (Scheffer and Slipp 1944; Venables and Venables 1957; Sullivan 1981, 1982), there remains insufficient information on the harbor seal's mating behavior.

Diet and Foraging Strategies

The Pacific harbor seal's diet and the seasonal and geographic variations in its foraging habits have been studied by identifying hard part remains (e.g., cephalopod beaks and fish otoliths) recovered from feces (Stewart and Yochem 1994; Lance and Jeffries 2007; Lance et al. 2012; Luxa and Acevedo-Gutiérrez 2013; Gibble and Harvey 2015), fatty acids (Bromaghin et al. 2013), stable isotopes (Bjorkland et al. 2015), and metabarcoding (Schwarz et al. 2018). Recent studies have generated information on the trophic ecology of the harbor seal in Mexico (Elorriaga-Verplancken et al. 2013; Alamán-De Regules 2014; Durazo-Rodríguez 2015; Elorriaga-Verplancken et al. 2016; Brassea-Pérez et al. 2019; Juárez-Rodríguez et al. 2020).

The harbor seal, like other marine mammals, is considered a tertiary consumer in the trophic chain (Pauly et al. 1998). In Mexico it is categorized with a trophic level of between 4.2 and 4.9 (Elorriaga-Verplancken et al. 2013; Alamán-De Regules 2014; Durazo-Rodríguez 2015), thus confirming it as a top predator. Its prey averaged between 15 cm and 25 cm in total length (Brown and Mate 1983; Harvey 1987; Bowen and Harrison 1996). Harbor seal diet is highly variable due to its wide latitudinal distribution where it exploits a wide variety of habitats, and preys are seasonally abundant and available. Therefore, it has been considered an opportunistic (Scheffer and Sperry 1931; Harvey and Torok 1994) and generalist consumer (Payne and Selzer 1989; Schwarz et al. 2018). However, Harvey and Torok (1994) state that only a low percentage of prey is consumed in high quantities, noting that, in San Francisco Bay, five species constituted more than 93% of the harbor seal's estimated dietary mass (Harvey and Torok 1994). Moreover, in the Salish Sea, the generalist diet of the species appears to be comprised by specialist individuals (Lance et al. 2012; Bromaghin et al. 2013, Bjorkland et al. 2015; Schwarz et al. 2018; Voelker 2018). In Mexico, in two neighboring colonies (Todos Santos Island and Punta Banda Estuary), a study based on hard part remains found that ≥70% of the harbor seal's diet was comprised of three to five species (Alamán-De Regules 2014). Moreover, another study with the same method conducted in Mexico on four distant colonies during the nursing, molting, and resting seasons found that four to ten prey species comprised 90% of the harbor seal's diet (Durazo-Rodríguez 2015). In addition, based on a trophic amplitude of 0.13–0.35 (Levin's index), the harbor seal is considered a specialist predator (Alamán-De Regules 2014; Durazo-Rodríguez 2015), which has been confirmed by stable isotope analyses (Elorriaga-Verplancken et al. 2013; Juárez-Rodríguez et al. 2020).

Harbor seals prefer to forage in shallow waters at a depth of between 5 and 100 m (Tollit et al. 1998; Eguchi and Harvey 2005). They are known to dive as deep as 481 m and can stay under water up to 32.25 min (Eguchi and Harvey 2005). Average dive duration during foraging is 0.5 min (SD = 0.3 min) to 3.3 min (SD = 1.3 min). Harbor seals spend a greater proportion of time diving at night than diving during daytime (Harvey and Torok 1994; Frost et al. 2001). Moreover, recent fine-scale analysis of diving behavior indicated that foraging dive bouts ≥20 m deep were predominant when compared to dive bouts <20 m. However, a high variability was observed relative to haul-out site, season, sex, and light (day/night) (Wilson et al. 2014).

Globally, harbor seals have been reported to prefer bony fish (Pitcher 1980; Brown and Mate 1983; Payne and Selzer 1989; Tollit et al. 1997; Lunneryd 2001; Luxa and Acevedo-Gutiérrez 2013). Harbor seals in California that were tagged with temperature and pressure sensors indicated they are benthic predators (Eguchi and Harvey 2005). However, they are also known to consume pelagic prey species during at least one season of the year (Harvey and Torok 1994; Gibble and Harvey 2015). In the Georgia Strait, Canada, differences between sexes were found: females foraged on demersal fish, while males foraged on pelagic species (salmonids) (Wilson and Jones 2018). Studies conducted in different regions of the subspecies' distribution have shown a high contribution to its diet by benthic and demersal fish, such as flounders (Paralichthyidae and Pleuronectidae), cod (Gadidae), rockfish (Sebastidae), sand lances (Ammodytidae), and sculpins (Cottidae) (Brown and Mate 1983; Payne and Selzer 1989; Tollit et al. 1997; Lunneryd 2001). At Mexican colonies, harbor seals prefer to consume benthic or demersal fish (Elorriaga-Verplancken et al. 2013; Alamán-De Regules 2014; Durazo-Rodríguez 2015; Brassea-Pérez et al. 2019; Juárez-Rodríguez et al. 2020). However, they are also known to consume octopus and squid frequently (Elorriaga-Verplancken et al. 2013; Alamán-De Regules 2014; Durazo-Rodríguez 2015), and in some seasons at certain colonies, cephalopods are more important than fish in the harbor seal diet (Alamán-De Regules 2014; Durazo-Rodríguez 2015).

At Todos Santos Bay, the most consumed prey (based on hard part remains) during the 2013 molting season were the California lizardfish (*Synodus lucioceps*), rockfish (*Sebastes* spp.), two-spotted octopus (*Octopus bimaculatus*), and market squid (*Doryteuthis opalescens*) (Alamán-De Regules 2014).

Also based on hard part remains, the harbor seal's diet on four island colonies (Todos Santos, San Jerónimo, Natividad, and San Roque) in Baja California, Mexico, showed no seasonal (nursing, molting, resting) changes in diet composition, although there were changes between common prey species consumed (Durazo-Rodríguez 2015). A total of 55 species were identified in the harbor seal's diet, of which 31 had a low frequency or low importance index (Durazo-Rodríguez

2015). The most important prey species in the diet of harbor seals from those islands were Pacific sanddab (*Citharichthys sordidus*), rockfish (*Sebastes* spp.), octopus (*Octopus* spp.), market squid (*Doryteuthis opalescens*), California lizardfish (*Synodus lucioceps*), plainfin midshipman (*Porichthys notatus*), spotted cusk eel (*Chilara taylori*), speckled sanddab (*C. stigmaeus*), and Dover sole (*Microstomus pacificus*) (Durazo-Rodríguez 2015). In the same study, crustacean remains were found in all samples, while, for the first time, the pelagic red crab (*Pleuroncodes planipes*) was also identified as harbor seal prey (Durazo-Rodríguez 2015).

Moreover, the same fecal samples from Durazo-Rodríguez (2015) were analyzed using a molecular technique (metabarcoding analysis) and confirmed the harbor seal's most often consumed prey reported by Durazo-Rodríguez (Brassea-Pérez et al. 2019). The following other preferred species were also identified (Brassea-Pérez et al. 2019): longfin sanddab (*Citharichthys xanthostigma*), white weakfish (*Atractoscion nobilis*), California skate (*Raja inornata*), and Pacific red octopus (*Octopus rubescens*) (Brassea-Pérez et al. 2019). Other species that do not leave hard remains were identified: bivalves, one cnidarian, one salp, one myxine, one ray, and six crustaceans, including the pelagic red crab (Brassea-Pérez et al. 2019). However, metabarcoding analysis does not convey the importance level of the species in the harbor seal's diet. Overall, 14 soft-bodied species were reported, while 11 species had not been identified via the hard remains analysis, representing almost 25% of the species identified by metabarcoding (Brassea-Pérez et al. 2019).

Genetics

Genetic information on the harbor seal (*Phoca vitulina*) based on molecular markers has facilitated the understanding about differences between populations in the Pacific and those in the Atlantic Oceans. The evolutionary and demographic history of Pacific harbor seal (*P. v. richardii*) colonies is more complex than the Atlantic harbor seal's (*P. v. vitulina*), because Pacific populations are more recent (Westlake and O'Corry-Crowe 2002), or because there has been recent genetic exchange (Burg et al. 1999). In addition, at least two colonizations occurred from the Atlantic to the Pacific; during the second colonization (about 380,000 years ago), colonies established in British Columbia, Washington, California, and Baja California (Burg et al. 1999). The Baja California colonies are the result of a founding event from California harbor seals, and no significant exchanges within these colonies seem to have occurred (Fernández-Martín 2018).

In general, genetic diversity is higher in Pacific colonies (Lamont et al. 1996; Kappe et al. 1997; Burg et al. 1999) than in Atlantic colonies (Stanley et al. 1996, Goodman 1998; Andersen et al. 2011). Diversity loss in Atlantic populations may have occurred during the last glaciation (Stanley et al. 1996; Kappe et al. 1997; Goodman 1998), followed by bottlenecks due to hunting (Bowen and Lidgard 2013), or epizootics (Härkönen et al. 2006), as well as founding events (Stanley et al. 1996; Goodman 1998). All these events may deplete genetic diversity, although they also have an effect on the genetic structure (population subdivision) since they cause genetic drift (Goodman 1998), i.e., allele frequencies (genetic diversity) in small populations are reduced from generation to generation. Therefore, the Atlantic harbor seal has a stronger genetic structure (higher values) (Stanley et al. 1996; Goodman 1998; Andersen et al. 2011) than the Pacific harbor seal (Lamont et al. 1996; Stanley et al. 1996; Burg et al. 1999; O'Corry-Crowe et al. 2003; Herreman et al. 2009; Huber et al. 2010, 2012).

Harbor seal colonies in Baja California, Mexico, have low genetic diversity, both in a mitochondrial marker (control region sequences) and in a nuclear marker (microsatellites) (Fernández-Martín 2018). Mitochondrial markers indicated that the haplotype diversity of the harbor seal in Mexico was two times lower than that found in the rest of the Pacific (Westlake and O'Corry-Crowe 2002; O'Corry-Crowe et al. 2003; Huber et al. 2010; Dishman 2011). The nucleotide diversity was four to nine times lower (Fernández-Martín 2018) than in other harbor seal populations (Westlake and O'Corry-Crowe 2002; Huber et al. 2010; Dishman 2011). Eighteen haplotypes were found, 13 of which were exclusive for seals in Mexico (Fernández-Martín 2018), and only 5 of those haplotypes were similar to colonies in California and Oregon, USA (Stanley et al. 1996; Lamont 2002; Dishman 2011), or Vancouver, Canada (Burg et al. 1999). Furthermore, the 15 variable sites that define harbor seal haplotypes in Mexico are much fewer than the 30 reported by Lamont et al. (1996) in Washington, Oregon, and California, the 52 reported by Huber et al. (2010) in Washington, the 60 reported by Burg et al. (1999) in Alaska and Canada, and the 89 reported by O'Corry-Crowe et al. (2003) in Alaska. The low mitochondrial diversity and the low difference among haplotypes of Mexican seals suggest that a founding event occurred rather than a genetic bottleneck (Fernández-Martín 2018). Similarly, no genetic bottleneck has been identified across the entire distribution of the Pacific harbor seal (Lamont 2002). The expected microsatellite heterozygosity shows, in general terms, 1.5 times lower diversity in harbor seals from Mexico (Fernández-Martín 2018) to what has been reported for them along the rest of the Pacific (Burg et al. 1999; Dishman 2011; Huber et al. 2012).

In terms of the harbor seal's genetic structure in Mexico, clear (high fixation indices) and statistically significant differences have been found, for both markers, between one continental colony (Punta Banda Estuary) and four island colonies (Todos Santos, San Jerónimo, Natividad, and San Roque) (Fernández-Martín 2018). This differentiation may be due to the small size of the Punta Banda Estuary colony (99 individuals; Fernández-Martín et al. 2016) and, therefore, its susceptibility to a genetic drift that favors differentiation even between this colony and the Todos Santos colony located only 16 km west. There is also evidence to show isolation, based on the trophic behavior observed at the Punta Banda Estuary colony, which indicates that seals from these two colonies forage at different sites (Alamán-De Regules 2014). In addition, even though the colonies are located on the same latitude, there is a lag in the mean pupping dates, and therefore they respond to different photoperiods (Fernández-Martín et al. 2016). The genetic difference between neighboring colonies has also been reported for coastal and inland water colonies in Washington, USA (Lamont et al. 1996; Huber et al. 2010, 2012).

A larger structure was found with the mitochondrial marker than with the nuclear marker, which suggests a gene flow between colonies due to greater movement by males than females (Fernández-Martín 2018). This coincides with results obtained by studies conducted at other harbor seal sites (Burg et al. 1999; O'Corry-Crowe et al. 2003; Herreman et al. 2009). Further to the differentiation observed at the Punta Banda Estuary colony, paired comparisons between colonies based on haplotypes have found differences between Todos Santos and the rest of the island colonies, as well as between San Jerónimo and Natividad, specifically (Fernández-Martín 2018). As to the nuclear marker, paired differences have been found between the Todos Santos and San Roque colonies, as well as between the San Jerónimo and Natividad colonies (Fernández-Martín 2018). These differences do not adhere to a distance-based isolation pattern, which may be because insufficient time has passed for the populations to diverge or because complex individual movements among colonies occur (Fernández-Martín 2018). However, the genetic structure found in Mexico for the harbor seal is strong compared to larger regions in the north Pacific, from Alaska to California. For example, seven populations were identified via mitochondrial markers (Burg et al. 1999; Lamont 2002; O'Corry-Crowe et al. 2003; Huber et al. 2010; Huber et al. 2012) and only two populations using the nuclear marker (Burg et al. 1999; Herreman et al. 2009). In addition, three populations have been identified in inland waters, based on both markers (Huber et al. 2010, 2012).

Threats: Global Change, Hunting, and Fisheries Interactions

The main threats faced by the Pacific harbor seal are directed hunts (Newby 1973b; Itoo and Shukunobe 1986; Olesiuk 2010) and incidental mortality by commercial fisheries (DeMaster et al. 1985; Barlow et al. 1994; Fraker 1996; DeMaster et al. 2001; Jamieson and Olesiuk 2001; Read et al. 2006; Wright et al. 2007; Moore et al. 2009). However, there are some reports in which the cause of the reduction of harbor seal populations is unknown (Pitcher 1990; Jemison et al. 2006; Mathews and Pendleton 2006). Also of concern are future effects due to climate change and associated ecosystem changes on harbor seals and other pinniped populations (Kovacs et al. 2012).

In Mexico, there are no records of harbor seals being hunted, and information of interactions with commercial fisheries is scarce. A study on the perception of fishermen working for cooperatives (fishing associations that have been granted a governmental concession to extract marine products from the islands and surrounding waters) found that the harbor seals are negligibly affected by those fisheries (Arias-Del-Razo 2016). This is uncommon compared to other countries where harbor seals occur and mortality associated with fishing operations can reach thousands of animals each year (Barlow et al. 1994; Lunneryd 2001; Thompson et al. 2007; Kauppinen et al. 2005; Read et al. 2006; Moore et al. 2009). The fishing cooperatives studied on four islands in Baja California, Mexico (San Jerónimo, Natividad, San Roque and Asunción), obtain their income mainly from abalone (*Haliotis* sp.)

and lobster (*Panulirus interruptus*) (Arias-Del-Razo 2016), which are not consumed by harbor seals (Brassea-Pérez et al. 2019). Therefore, no retaliation measures are exacted by the fishermen, and they do not use gill nets where harbor seals may be entangled (Arias-Del-Razo 2016). There is indirect evidence from examination of stranded harbor seals along the coast of Ensenada, Baja California, that no harbor seals have been known to have died from gunshots or knife wounds or entangled in gill nets or fish hooks during 1998–2001 (Bravo et al. 2005).

There are two documented cases of the effect of environmental changes on seal colonies in Mexico. For six pupping seasons, the colony at Punta Banda Estuary, Baja California was monitored (Fernández-Martín et al. 2016), finding an evident effect of changing oceanographic conditions (namely, sea surface temperature and chlorophyll concentration, proxies for food availability) on the number of pups. Around 40 pups were born in years with favorable conditions (low SST and high Chl *a*), while in years with oceanographic anomalies (high SST and low Chl *a*), pupping plummeted to 21 pups. Only 12 pups were born when unfavorable conditions persisted for 2 subsequent years (Fernández-Martín 2018). When food availability is low, females may be unable to complete gestation successfully (Jemison and Kelly 2001, Bowen et al. 2003). Moreover, Juárez-Rodríguez et al. (2020) evidenced isotopic ($\delta^{13}C$) values for harbor seals from Natividad Island in winter of 2015 that were significantly lower than those reported by Elorriaga-Verplancken et al. (2013) in the same colony during winter of 2013. This may be explained by more offshore foraging habits in 2015 (than 2013), related to ocean warming anomalies in the northeastern Pacific from the end of 2013 to 2015 (Cavole et al. 2016), which may have changed prey distribution, increasing the foraging effort of harbor seals (Juárez-Rodríguez et al. 2020).

Human disturbance is a documented threat in Mexico. The impact of different disturbance sources on seal behavior was studied at Punta Banda Estuary, finding that the immediate effect of the disturbance meant that seals flushed to the water, reducing their time on land and, probably, increasing their energy expenditure (Ruiz-Mar 2016). Disturbances were very frequent (0.5 disturbance events/hour; Ruiz-Mar 2016) which may be considered a high incidence, given seals remain on land at Punta Banda Estuary solely at low tide (when their haul-out sites are accessible), for the 6-hour period that elapses until high tide. The disturbances affected the seals' resting time and important events such as nursing and molting (Ruiz-Mar 2016). Another disturbance caused by human activity was the massive displacement of individuals from colonies along the west coast of the Baja California Peninsula. Lubinsky et al. (2017) reported 44 breeding colonies in 2009, while, in contrast, in 2016, only 6 colonies were observed during aerial surveys (Lubinsky-Jinich 2019). The colonies may have been disturbed by the extraction of pebble stones from the beaches, an activity observed during the same aerial surveys.

A potential threat for seal colonies in Mexico is the presence of different toxic substances in the marine environment coming from human activities. The presence of certain bacteria in the seals' gut makes it possible to deduce the existence of substances with industrial or agricultural origin in the environment. A study on the microbiome of harbor seals at five colonies in Baja California (Punta Banda Estuary,

Todos Santos, San Jerónimo, Natividad, and San Roque islands) revealed the presence of bacteria with metabolic pathways that are involved in the degradation of persistent pollutants (e.g., pesticides) and plastics. This is probably related to the distribution of the colonies near important agricultural settlements. In addition, this study reported pathways related to primary immunodeficiency and neurodegenerative diseases (Pacheco-Sandoval et al. 2019).

Conservation Status

The harbor seal is listed by the International Union for the Conservation of Nature (IUCN) Red List under the category *Least Concern* (Thompson et al. 2008), because it is considered an abundant and widely distributed species. In Russia, the western Pacific harbor seal is listed in the Red Book of the Russian Federation, and its hunting is prohibited (Harvey 2016), while it is also cataloged in the IUCN's Red List as *Data Deficient* (Harvey 2016). In Mexico, it is also listed in the lowest category (*Under Special Protection*) in the Mexican *List of Species at Risk* (Official Mexican Standard NOM-059-ECOL-2010, SEMARNAT 2010), which may be due to a lack of information. Recent studies of harbor seals in Baja California, Mexico, conducted during the last 10 years have provided information which could lead to changing their status from *Under Special Protection* to *Threatened* based on criteria established in the mentioned Official Mexican Standard (Fernández-Martín 2018).

Acknowledgments The studies on the harbor seal in Mexico have been financed since 2006 by the National Science and Technology Council (CONACyT: Fondo SEMARNAT-CONACyT call S0010-2006-1, project 23702; Fondo SEP-CONACyT call CB-2012-01, project 179451). The following research permits were issued by the Ministry for Environment and National Resources: SGPA/DGVS/04007/08, SGPA/DGVS/04188/09, SGPA/DGVS/09252/09, SGPA/DGVS/03550/10, SGPA/DGVS/03551/10, SGPA/DGVS/12269/13, SGPA/DGVS/08370/14, and SGPA/DGVS/01140/16. Logistic support in the field was provided by the Mexican Navy, the Autonomous University of Baja California (UABC, for its Spanish acronym), the Ensenada Center for Scientific Research and Higher Education (CICESE), Island Conservation Group (GECI), Community and Biodiversity (COBI), and the Salt Exporting Company (Exportadora de Sal). Divers and fishermen at the cooperatives near the seal colonies transported us to the islands and were always good company: San Ignacio California Cooperative, Litoral Cooperative, Divers and Fishermen of Baja California, Ensenada Cooperative, and Pacifico Aquaculture. D. Lubinsky-Jinich, E.M. Fernández Martín, C. Tapia-Harris, A. Arias-Del Razo, E. Brassea-López, M. Juárez-Rodríguez, P. Durazo-Rodríguez, R. Alamán-de-Regules, M.G. Ruiz-Mar, and A.M. Pacheco-Sandoval finished their graduate theses on different subjects about the harbor seal in Mexico. M.C. García-Aguilar, M.E. Jiménez-López, A. Baez-Flores, and E. López-Reyes provided technical support. We thank more than 30 volunteers who collaborated in the field and lab. Finally, we thank A. Acevedo-Gutiérrez, Mark S. Lowry, and Fernando R. Elorriaga-Verplancken for reviewing the manuscript and their suggestions. Benjamin Stewart copyedited the final manuscript.

References

Alamán-De Regules R (2014) Hábitos alimentarios de la foca de puerto, *Phoca vitulina richardsi*, en la Bahía Todos Santos, Baja California, México (Foraging habits of the harbor seal, *Phoca vitulina richardsi*, in Todos Santos Bay, Baja California, Mexico). M.Sc. thesis, Universidad Autónoma de Baja California, Ensenada, Baja California, México

Allen BM, Angliss RP (2015) Alaska marine mammal stock assessments, 2014. NOAA Technical memorandum NMFS-AFSC-301. Alaska Fisheries Science Center, National Marine Fisheries Service, Seattle, Washington, USA

Andersen LW, Lydersen C, Frie AK et al (2011) A population on the edge: genetic diversity and population structure of the world's northernmost harbour seals (*Phoca vitulina*). Biol J Linn Soc Lond 102:420–439

Arias-Del-Razo A (2016) Factores que determinan la preferencia de hábitat de los pinnípedos en las islas del Pacífico de Baja California (Factors that determine pinniped habitat preferences on islands in the Pacific of Baja California). Ph.D. Dissertation, Centro de Investigación Científica y de Educación Superior de Ensenada, Ensenada, Baja California, México

Arias-Del-Razo A, Heckel G, Schramm Y et al (2016) Terrestrial habitat preferences and segregation of four pinniped species on the islands off the western coast of the Baja California Peninsula, Mexico. Mar Mamm Sci 32(4):1416–1432. https://doi.org/10.1111/mms.12339

Arias-Del-Razo A, Schramm Y, Heckel G et al (2017) Distribution of four pinnipeds (*Zalophus californianus, Arctocephalus philippii townsendi, Phoca vitulina richardii,* and *Mirounga angustirostris*) on islands off the west coast of the Baja California Peninsula, Mexico. Aquat Mamm 43(1):40–51. https://doi.org/10.1578/AM.43.1.2017.40

Barlow J, Baird RW, Heyning JE et al (1994) A review of cetacean and pinniped mortality in coastal fisheries along the west coast of the USA and Canada and the east coast of the Russian Federation. Rep Int Whal Commn 15:405–426

Bigg MA (1969) The harbour seal in British Columbia. Fish Res Board Can Bull 172:1–33

Bjørge A, Desportes G, Waring GT et al (2010) The harbour seal (*Phoca vitulina*) – a global perspective. NAMMCO Sci Publ 74:125–135

Bjorkland RH, Pearson SF, Jeffries SJ et al (2015) Stable isotope mixing models elucidate sex and size effects on the diet of a generalist marine predator. Mar Ecol Prog Ser 526:213–225. https://doi.org/10.3354/meps11230

Boness DJ, Bowen WD, Oftedal OT (1994) Evidence of a maternal foraging cycle resembling that of otariid seals in a small phocid, the harbor seal. Behav Ecol Sociobiol 34:95–104

Boness DJ, Bowen WD, Buhleier BM et al (2006) Mating tactics and mating system of an aquatic-mating pinniped: the harbor seal, *Phoca vitulina*. Behav Ecol Sociobiol 61:119–130. https://doi.org/10.1007/s00265-006-0242-9

Boulva J (1975) Temporal variations in birth period and characteristics of newborn harbour seals. Rapp P-V Réun Cons Int Explor Mer 169:405–408

Boulva J, McLaren IA (1979) Biology of the harbor seal, *Phoca vitulina*, in eastern Canada. Fish Res Board Can Bull 200:1–24

Bowen WD, Harrison GD (1996) Comparison of harbour seal diets in two inshore habitats of Atlantic Canada. Can J Zool 74:125–135

Bowen WD, Lidgard D (2013) Marine mammal culling programs: review of effects on predator and prey populations. Mamm Rev 43(3):207–220. https://doi.org/10.1111/j.1365-2907.2012.00217.x

Bowen WD, Oftedal OT, Boness DJ (1992) Mass and energy transfer during lactation in a small phocid, the harbor seal (*Phoca vitulina*). Physiol Zool 65:884–866

Bowen WD, Boness DJ, Iverson SJ (1999) Diving behaviour of lactating harbour seals and their pups during maternal foraging trips. Can J Zool 77:978–988

Bowen WD, Ellis SL, Iverson SJ et al (2003) Maternal and newborn life-history traits during periods of contrasting population trends: implications for explaining the decline of harbour seals (*Phoca vitulina*), in Sable Island. J Zool 261(2):155–163. https://doi.org/10.1017/S0952836903004047

Brassea-Pérez E, Schramm Y, Heckel G et al (2019) Metabarcoding analysis of the Pacific harbor seal diet in Mexico. Mar Biol 166(106):1–14. https://doi.org/10.1007/s00227-019-3555-8

Bravo E, Heckel G, Schramm Y et al (2005) Occurrence and distribution of marine mammal strandings in Todos Santos Bay, Baja California, Mexico, 1998-2001. Lat Am J Aquat Mamm 4(1):15–25

Bromaghin JF, Lance MM, Elliot EW et al (2013) New insights into the diets of harbor seals (*Phoca vitulina*) in the Salish Sea revealed by analysis of fatty acid signatures. Fish Bull 111:13–26. https://doi.org/10.7755/FB.111.1.2

Brown RF, Mate BR (1983) Abundance, movements, and feeding habits of harbor seals, *Phoca vitulina*, at Netarts and Tillamook Bays, Oregon. Fish Bull 81(2):291–301

Burg T, Trites AW, Smith M (1999) Mitochondrial and microsatellite DNA analyses of harbour seal population structure in the Northeast Pacific Ocean. Can J Zool 77:930–943

Burns JJ (1986) Ice seals. In: Haley D (ed) Marine mammals of eastern North Pacific and Arctic waters, 2nd edn. Pacific Search Press, Seattle, pp 216–229

Burns JJ (2009) Harbor seal and spotted seal *Phoca vitulina* and *P. largha*. In: Perrin WF, Würsig B, Thewissen JGM (eds) Encyclopedia of marine mammals, 2nd edn. Academic Press, San Diego, pp 533–542

Carretta JV, Oleson EM, Weller DW et al (2015) U.S. Pacific marine mammal stock assessments: 2014. Southwest Fisheries Science Center, National Marine Fisheries Service, La Jolla

Cavole LM, Demko AM, Diner RE et al (2016) Biological impacts of the 2013-2015 warm-water anomaly in the Northeast Pacific: winners, losers, and the future. Oceanography 29(2):273–285

Committee on Taxonomy (2019) List of marine mammal species and subspecies. Society for Marine Mammalogy. http://www.marinemammalscience.org. Accessed 30 Nov 2019

Cunningham L, Baxter JM, Boyd IL et al (2009) Harbour seal movements and haul-out patterns: implications for monitoring and management. Aquat Conserv 19(4):398–407

Daniel RG, Jemison LA, Pendleton GW et al (2003) Molting phenology of harbor seals on Tugidak Island, Alaska. Mar Mamm Sci 19(1):128–140. https://doi.org/10.1111/j.1748-7692.2003.tb01097.x

Davis MB, Renouf D (1987) Social behaviour of harbour seals, *Phoca vitulina*, on haulout grounds at Miquelon. Can Field-Nat 101:1–5

DeMaster DP, Miller D, Henderson J et al (1985) Conflicts between marine mammals and fisheries off the coast of California. In: Beddington JR, Beverton RJH, Lavigne DM (eds) Marine mammals and fisheries. George Allen & Unwin, London

DeMaster DP, Fowler CW, Perry SL et al (2001) Predation and competition: the impact of fisheries on marine mammal populations over the next one hundred years. J Mammal 82(3):641–651

Dierauf LA, Dougherty SA, Lowenstine LJ (1986) Survival versus non survival determinants for neonatal harbor seals. J Am Vet Med A 189:1024–1028

Dishman DL (2011) Genetic substructure of Pacific harbor seal (*Phoca vitulina richardsi*) populations along the coasts of Oregon and Washington M.Sc. thesis. Portland State University, Portland, Oregon, USA

Durazo-Rodríguez P (2015) Variación espacio-temporal de los hábitos alimentarios de la foca de puerto (*Phoca vitulina richardii*) en México (Spatial and temporal variation in foraging habits of the harbor seal (*Phoca vitulina richardii*) in Mexico). M.Sc. thesis. Universidad Autónoma de Baja California, Ensenada, Baja California, México

Eguchi T, Harvey JT (2005) Diving behavior of the Pacific harbor seal (*Phoca vitulina richardii*) in Monterey Bay, California. Mar Mamm Sci 21(2):283–295

Elorriaga-Verplancken FR, Hadrys LM-L, Moreno-Sánchez X et al (2013) Inferences on the diet of the eastern Pacific harbor seal (*Phoca vitulina richardii*) at the southern end of its distribution: stable isotopes and scats analyses. Aquat Mamm 39(4):415–421. https://doi.org/10.1578/AM.39.4.2013.415

Elorriaga-Verplancken FR, Morales-Luna L, Heckel G et al (2016) Foraging ecology of harbour seals (*Phoca vitulina*) and Northern elephant seals (*Mirounga angustirostris*) from

Baja California, Mexico: inferences from stable isotopes in pups. J Mar Biol Assoc UK 96(4):903–908. https://doi.org/10.1017/S0025315415002143

Fernández-Martín EM (2018) Fenología de los nacimientos, estado de salud de las crías, y estructura genética poblacional de *Phoca vitulina richardii* en México (Pupping phenology, pup health status, and population genetic structure of *Phoca vitulina richardii* in Mexico). Ph.D. Dissertation, Universidad Autónoma de Baja California, Ensenada, Baja California, México

Fernández-Martín EM, Heckel G, Schramm Y et al (2016) The timing of pupping and molting of the Pacific harbor seal, *Phoca vitulina richardii*, at Punta Banda Estuary, Baja California, Mexico. Cienc Mar 42(3):195–208. https://doi.org/10.7773/cm.v42i3.2615

Fisher HD (1954) Delayed implantation in the harbour seal, *Phoca vitulina*. Nature 173:879–880

Fraker M (1996) Interactions between salmon farms and marine mammals and other species. BC Salmon Farmers Association British Columbia, Canada

Frost KJ, Simpkins MA, Lowry LF (2001) Diving behavior of subadult and adult harbor seals in Prince William Sound, Alaska. Mar Mamm Sci 17(4):813–834

Frost KJ, Simpkins MA, Small RJ et al (2006) Development of diving by harbor seal pups in two regions of Alaska: use of the water column. Mar Mamm Sci 22(3):617–643. https://doi.org/10.1111/j.1748-7692.2006.00056.x

Gallo-Reynoso JP, Aurioles-Gamboa D (1984) Distribución y estado actual de la población de foca común (*Phoca vitulina richardsi* (Gray, 1864)), en la península de Baja California, México (Distribution and present status of the common seal (*Phoca vitulina richardsi* (Gray, 1864)) population, along the Baja California Peninsula, Mexico). An Inst Biol Univ Nal Autón México Ser Zoología 55(2):323–332

Gibble CM, Harvey JT (2015) Food habits of harbor seals (*Phoca vitulina richardii*) as an indicator of invasive species in San Francisco Bay, California. Mar Mamm Sci 31(3):1014–1034

Godsell J (1988) Herd formation and haul-out behavior in harbor seals (*Phoca vitulina*). J Zool 215:83–98. https://doi.org/10.1111/j.1469-7998.1988.tb04886.x

Goodman SJ (1998) Patterns of extensive genetic differentiation and variation among European harbour seals (*Phoca vitulina vitulina*) revealed using microsatellite DNA polymorphisms. Mol Biol Evol 15:104–118

Gray JE (1864) Notes on seals (Phocidae), including the description of a new seal (*Halicyon richardii*) from the west coast of North America. Proc Zool Soc London 1864:27–34

Hanggi EB, Schusterman RJ (1994) Underwater acoustic displays and individual variation in male harbour seals, *Phoca vitulina*. Anim Behav 48:1275–1283

Härkönen T, Dietz R, Reijnders P et al (2006) A review of the 1988 and 2002 phocine distemper virus epidemics in European harbour seals. Dis Aquat Org 68:115–130

Harvey JT (1987) Population dynamics, annual food consumption, movements and dive behavior of harbor seals, *Phoca vitulina richardsi,* in Oregon. Ph.D. Dissertation, Oregon State University, Corvallis, Oregon, USA

Harvey JT (2016) *Phoca vitulina* spp. *stejnegeri*. The IUCN Red list of threatened species 2016. https://doi.org/10.2305/IUCN.UK.2016-1.RLTS.T17014A66991317.en. Accessed 28 Feb 2020

Harvey JT, Torok ML (1994) Movements, dive behaviors, and food habits of harbor seals (*Phoca vitulina richardsi*) in San Francisco Bay, California. Final report. Department of Water Resources, Moss Landing Marine Laboratories, Moss Landing, California, USA

Hayes SA, Costa DP, Harvey JT et al (2004a) Aquatic mating strategies of the male Pacific harbor seal (*Phoca vitulina richardii*): are males defending the hotspot? Mar Mamm Sci 20(3):639–656. https://doi.org/10.1111/j.1748-7692.2004.tb01184.x

Hayes SA, Kumar A, Costa DP et al (2004b) Evaluating the function of the male harbour seal, *Phoca vitulina*, roar through playback experiments. Anim Behav 67:1133–1139. https://doi.org/10.1016/j.anbehav.2003.06.019

Hayes SA, Pearse DE, Costa DP et al (2006) Mating system and reproductive success in eastern Pacific harbour seals. Mol Ecol 15(10):3023–3034. https://doi.org/10.1111/j.1365-294X.2006.02984.x

Herreman JK, Blundell GM, McDonald DB et al (2009) Asymmetrical male mediated gene flow between harbor seal (*Phoca vitulina*) populations in Alaska. Can J Zool 87:498–507
Huber HR, Jeffries SJ, Lambourn DM et al (2010) Population substructure of harbor seals (*Phoca vitulina richardsi*) in Washington State using mtDNA. Can J Zool 88(3):280–288
Huber HR, Dickerson BR, Jeffries SJ et al (2012) Genetic analysis of Washington State harbor seals (*Phoca vitulina richardii*) using microsatellites. Can J Zool 90:1361–1369
Itoo T, Shukunobe T (1986) Number and present status of the Kuril seal (in Japanese with English abstract). In: Wada K, Itoo T, Niizuma A et al (eds) Ecology and protection of the Kuril seal. Tokai University Press, Tokyo
Jamieson GS, Olesiuk PF (2001) Salmon farm-pinniped interactions in British Columbia: an analysis of predator control, its justification and alternative approaches. Canadian Science Advisory Secretariat Research Document, 1480–4883, 2001/142. http://publications.gc.ca/site/eng/9.807175/publication.html. Accessed 28 Feb 2020
Jemison LA, Kelly BP (2001) Pupping phenology and demography of harbor seals (*Phoca vitulina richardsi*) on Tugidak Island, Alaska. Mar Mamm Sci 17(3):585–600. https://doi.org/10.1111/j.1748-7692.2001.tb01006.x
Jemison LA, Pendleton GW, Wilson CA et al (2006) Long-term trends in harbor seal numbers at Tugidak Island and Nanvak Bay, Alaska. Mar Mamm Sci 22:339–360. https://doi.org/10.1111/j.1748-7692.2006.00021.x
Juárez-Rodríguez M, Heckel G, Herguera-García JC et al (2020) Trophic ecology of Mexican Pacific harbor seal colonies using carbon and nitrogen stable isotopes. PLoS One 15(1):e0225889. https://doi.org/10.1371/journal.pone.0225889
Kappe AL, Bijlsma R, Osterhaus ADME et al (1997) Structure and amount of genetic variation at minisatellite loci within subspecies complex of *Phoca vitulina* (the harbour seal). Heredity 78:457–463
Kauppinen T, Siira A, Suuronen P (2005) Temporal and regional patterns in seal-induced catch and gear damage in the coastal trap-net fishery in the northern Baltic Sea: effect of netting material on damage. Fish Res 73:99–109
Kelly BP (1980) Pelage polymorphism in Pacific harbor seals. Can J Zool 59:1212–1219
King J (1983) Seals of the world, 2nd edn. Oxford University Press, Oxford
Kobayashi Y, Kariya T, Chishima J et al (2014) Population trends of the Kuril harbour seal *Phoca vitulina stejnegeri* from 1974 to 2010 in southeastern Hokkaido, Japan. Endanger Species Res 24:61–72. https://doi.org/10.3354/esr00553
Kovacs KM, Jonas KM, Welke SE (1990) Sex and age segregation by *Phoca vitulina concolor* at haul-out sites during the breeding season in the Passamaquoddy Bay region, New Brunswick. Mar Mamm Sci 6(3):204–214
Kovacs KM, Aguilar A, Aurioles-Gamboa D et al (2012) Global threats to pinnipeds. Mar Mamm Sci 28:414–436. https://doi.org/10.1111/j.1748-7692.2011.00479.x
Lamont M (2002) Genetic diversity and substructure of the Pacific Harbor seal, *Phoca vitulina richardsi*. In: Pfeiffer C (ed) Molecular and cell biology of marine mammals. Krieger Publishing, Malabar, pp 130–137
Lamont MM, Vida JT, Harvey JT et al (1996) Genetic substructure of the Pacific harbor seal (*Phoca vitulina richardsi*) off Washington, Oregon, and California. Mar Mamm Sci 12(3):402–413
Lance MM, Jeffries SJ (2007) Temporal and spatial variability of harbor seal diet in the San Juan Island archipelago. Contract report to SeaDoc society research agreement no. K004431–25. Washington Department of Fish and Wildlife, Olympia, Washington, USA
Lance MM, Chang WY, Jeffries SJ et al (2012) Harbor seal diet in northern Puget Sound: implications for the recovery of depressed fish stocks. Mar Ecol Prog Ser 464:257–271
Lawson J, Renouf D (1987) Bonding and weaning in harbor seals, *Phoca vitulina*. J Mammal 68(2):445–449. https://doi.org/10.2307/1381496
Lowry LF, Forst KJ, Ver-Hoef JM et al (2001) Movements of satellite-tagged subadult and adult harbor seals in Prince William Sound, Alaska. Mar Mamm Sci 17(4):835–861

Lubinsky-Jinich D (2019) Abundancia y movimientos de la foca de puerto (*Phoca vitulina richardii*), en Baja California, México (Abundance and movements of the harbor seal (*Phoca vitulina richardii*) in Baja California, Mexico). Ph.D. Dissertation, Centro de Investigación Científica y de Educación Superior de Ensenada, Ensenada, Baja California, México

Lubinsky-Jinich D, Schramm Y, Heckel G (2017) The Pacific harbor seal's (*Phoca vitulina richardii*) breeding colonies in Mexico: abundance and distribution. Aquat Mamm 43(1):73–81. https://doi.org/10.1578/AM.43.1.2017.73

Lubinsky-Jinich D, Pardo MA, Heckel G et al (2019) Temporal variations in the terrestrial counts of Pacific harbor seals in the southern California current system. Mar Mamm Sci 35(3):1173–1182. https://doi.org/10.1111/mms.12579

Lunneryd SG (2001) Fish preference by the harbour seal (*Phoca vitulina*), with implications for the control of damage to fishing gear. ICES J Mar Sci 58:824–829

Luxa K, Acevedo-Gutiérrez A (2013) Food habits of harbor seals (*Phoca vitulina*) in two estuaries in the Central Salish Sea. Aquat Mamm 39:10–22. https://doi.org/10.1578/AM.39.1.2013.10

Mathews EA, Pendleton GW (2006) Declines in harbor seal (*Phoca vitulina*) numbers in Glacier Bay National Park, Alaska, 1992-2002. Mar Mamm Sci 22(1):167–189. https://doi.org/10.1111/j.1748-7692.2006.00011.x

McLaren IA (1993) Growth in pinnipeds. Biol Rev Camb Philos Soc 68(1):1–79. https://doi.org/10.1111/j.1469-185X.1993.tb00731.x

Moore JE, Wallace BP, Lewison RL et al (2009) A review of marine mammal, sea turtle and seabird by catch in USA fisheries and the role of policy in shaping management. Mar Policy 33:435–451

Newby TC (1973a) Observations on the breeding behavior of the harbor seal in the state of Washington. J Mammal 54(2):540–543

Newby TC (1973b) Changes in the Washington state harbor seal population, 1942-1972. Murrelet 54(1):4–7

Nicholson T (1997) Underwater vocalization and subsequent social behavior of the harbor seal (*Phoca vitulina richardsi*) near Hopkins Marine Station, Pacific Grove. CA. Monterey Bay Aquarium Research Institute, Monterey

O'Corry-Crowe G, Martien KK, Taylor BL (2003) The analysis of population genetic structure in Alaskan harbor seals, *Phoca vitulina*, as a framework for the identification of management stocks. Administrative report no. LJ-03-08. Southwest Fisheries Science Center-National Marine Fisheries Service, La Jolla, California, USA

Olesiuk PF (2010) Assessment Pacific harbour seal (*Phoca vitulina richardsi*). Science Advisory Report 2009/011. Department of Fisheries and Oceans, Canadian Stock Assessment Secretariat, Ottawa, Canada

Pacheco-Sandoval A, Schramm Y, Heckel G et al (2019) The Pacific harbor seal gut microbiota in Mexico: its relationship with diet and functional inferences. PLoS One 14(8):e0221770. https://doi.org/10.1371/journal.pone.0221770

Pauly D, Trites AW, Capuli E et al (1998) Diet composition and trophic levels of marine mammals. ICES J Mar Sci 55(3):467–481

Payne PM, Selzer LA (1989) The distribution, abundance and selected prey of the harbor seal, *Phoca vitulina concolor*, in southern New England. Mar Mamm Sci 5(2):173–192. https://doi.org/10.1111/j.1748-7692.1989.tb00331.x

Peterson SM (1998) Genetic and behavioral evidence that harbor seal (*Phoca vitulina*) females may mate with multiple males. Mar Mamm Sci 14(1):178–182

Peterson SH, Lance MM, Jeffries SJ et al (2012) Long distance movements and disjunct spatial use of harbor seals (*Phoca vitulina*) in the inlands waters of the Pacific Northwest. PLoS One 7(6):e39046. https://doi.org/10.1371/journal.pone.0039046

Pitcher KW (1980) Stomach contents and feces as indicator of harbor seal (*Phoca vitulina*) foods in the Gulf of Alaska. Fish Bull 78:797–798

Pitcher KW (1990) Major decline in number of harbor seals, *Phoca vitulina richardsi*, on Tugidak Island, Gulf of Alaska. Mar Mamm Sci 6(2):121–134

Read AJ, Drinker P, Northridge S (2006) Bycatch of marine mammals in U.S. and global fisheries. Conserv Biol 20(1):163–169

Reeves RR, Stewart BS, Clapham PJ et al (2002) Guide to marine mammals of the world. Alfred A. Knopf, New York

Renouf D, Lawson J, Gaborko L (1983) Attachment between harbour seal (*Phoca vitulina*) mothers and pups. J Zool 199(2):179–187. https://doi.org/10.1111/j.1469-7998.1983.tb02089.x

Ruiz-Mar MG (2016) Determinación del periodo de lactancia y cuidado materno en la foca de puerto *Phoca vitulina richardii* y su relación con el disturbio humano en el estero de Punta Banda, Baja California, México (Lactation period and maternal care in the harbor seal *Phoca vitulina richardii* and its relation with human disturbance at Punta Banda Estuary, Baja California, Mexico). M.Sc. Thesis, Centro de Investigación Científica y de Educación Superior de Ensenada, Ensenada, Baja California, México

Scheffer VB, Slipp JW (1944) The harbor seal in Washington State. Am Midl Nat 32(2):373–416

Scheffer TH, Sperry CC (1931) Food habits of the Pacific harbor seal, *Phoca richardi*. J Mammal 12:214–226

Schwarz D, Spitzer SM, Thomas AC et al (2018) Large-scale molecular diet analysis in a generalist marine mammal reveals male preference for prey of conservation concern. Ecol Evol 8(19):9889–9905. https://doi.org/10.1002/ece3.4474

SEMARNAT (2010) Norma Oficial Mexicana NOM-059-ECOL-2010, Protección ambiental-Especies nativas de México de flora y fauna silvestres-Categorías de riesgo y especificaciones para su inclusión, exclusión o cambio-Lista de especies en riesgo (Mexican Official Norm NOM-059-ECOL-2010, Environmental protection – Mexican wildlife flora and fauna native to Mexico: Risk categories and specifications for their inclusion, exclusion or change – list of species at risk). Diario Oficial de la Federación (Federal Register), 30 Dec 2010

Shaughnessy PD, Fay FH (1977) A review of the taxonomy and nomenclature of North Pacific harbour seals. J Zool 182:385–419

Skinner JP (2006) Physical and behavioural development of nursing harbour seal (*Phoca vitulina*) pups in Maine. University of Maryland, Orono

Slater LM, Markowitz H (1983) Spring population trends in *Phoca vitulina richardi* in two Central California coastal populations. Calif Fish Game 69:217–226

Stanley HF, Casey S, Carnahan JM et al (1996) Worldwide patterns of mitochondrial DNA differentiation in the harbour seal (*Phoca vitulina*). Mol Biol Evol 13(2):368–382

Stewart BS, Yochem PK (1994) Ecology of harbor seals in the Southern California bight. In: Halvorson WL, Maender GJ (eds) The Fourth California Islands symposium: update on the status of resources, Santa Barbara, California, USA. Santa Barbara Museum of Natural History, pp 123–134

Sullivan RM (1981) Aquatic displays and interactions in harbor seals, *Phoca vitulina*, with comments on mating systems. J Mammal 62(4):825–831. https://doi.org/10.2307/1380606

Sullivan RM (1982) Agonistic behavior and dominance relationships in the harbor seal, *Phoca vitulina*. J Mammal 63:554–569

Tapia-Harris C, Heckel G, Schramm Y et al (2017) Molting phenology of the Pacific harbor seal (*Phoca vitulina richardii*) on two islands off the Baja California Peninsula, Mexico. Mar Mamm Sci 33(3):817–829. https://doi.org/10.1111/mms.12400

Temte JL (1991) Precise birth timing in captive harbor seals (*Phoca vitulina*) and California Sea lions (*Zalophus californianus*). Mar Mamm Sci 7(2):145–156

Temte JL (1994) Photoperiod control of birth timing in the harbour seal (*Phoca vitulina*). J Zool 233(3):369–384. https://doi.org/10.1111/j.1469-7998.1994.tb05271.x

Thompson PM, Wheeler H (2008) Photo-ID-based estimates of reproductive patterns in female harbor seals. Mar Mamm Sci 24(1):138–146. https://doi.org/10.1111/j.1748-7692.2007.00179.x

Thompson PM, Mackey B, Barton TR et al (2007) Assessing the potential impact of salmon fisheries management on the conservation status of harbour seals (*Phoca vitulina*) in north-East Scotland. Anim Conserv 10:48–56

Thompson D, Härkönen T, IUCN SSC Pinniped Specialist Group (2008) *Phoca vitulina*. The IUCN red list of threatened species 2008: eT17013A6723560. https://doi.org/10.2305/IUCN.UK.2008.RLTS.T17013A6723560.en. Accessed 12 Nov 2019
Tollit DJ, Greenstreet SPR, Thompson PM (1997) Prey selection by harbour seals, *Phoca vitulina*, in relation to variations in prey abundance. Can J Zool 75:1508–1518
Tollit DJ, Black AD, Thompson PM et al (1998) Variations in harbour seal *Phoca vitulina* diet and dive-depths in relation to foraging habitat. J Zool 244(2):209–222
Van-Parijs SM, Thompson PM, Tollit DJ et al (1997) Distribution and activity of male harbour seals during the mating season. Anim Behav 54(1):35–43. https://doi.org/10.1006/anbe.1996.0426
Venables UM, Venables LSV (1957) Mating behaviour of the seal, *Phoca vitulina*, in Shetland. Proc Zool Soc London 128:387–396. https://doi.org/10.1111/j.1096-3642.1957.tb00332.x
Venables UM, Venables LSV, Matthews L (1955) Observations on a breeding colony of the seal *Phoca vitulina* in Shetland. Proc Zool Soc London 125:521–532. https://doi.org/10.1111/j.1096-3642.1955.tb00614.x
Voelker M (2018) Cross-sectional scat sampling reveals intrapopulation feeding diversity in a marine predator. M.Sc. Thesis, Western Washington University, USA
Walker BG, Bowen WD (1993) Changes in body mass and feeding behaviour in male harbour seals (*Phoca vitulina*) in relation to female reproductive success. J Zool 231:423–436. https://doi.org/10.1111/j.1469-7998.1993.tb01929.x
Westlake RL, O'Corry-Crowe GM (2002) Macrogeographic structure and patterns of genetic diversity in harbor seals (*Phoca vitulina*) from Alaska to Japan. J Mammal 83(4):1111–1126
Wilson SC (1974) Mother-young interactions in the common seal, *Phoca vitulina vitulina*. Behaviour 48:23–36
Wilson SC (1978) Social organization and behavior of harbor seals, *Phoca vitulina concolor*, in Maine. Report to U.S. Marine Mammal Commission, Contract MM6AC013. Office of Zoological Research, National Zoological Park, Smithsonian Institution, Washington, DC, USA
Wilson SC (2001) Population growth, reproductive rate and neo-natal morbidity in a re-establishing harbour seal colony. Mammalia 65:319–334. https://doi.org/10.1515/mamm.2001.65.3.319
Wilson SC, Jones KA (2018) Behaviour of harbour seal (*Phoca vitulina vitulina*) mother-pup pairs in Irish Sea intertidal habitats. Biol Environ Proc R Irish Acad 118B(1):13–27. https://doi.org/10.3318/BIOE.2018.02
Wilson K, Lance MM, Jeffries S et al (2014) Fine-scale variability in harbor seal foraging behavior. PLoS One 9(4):e92838. https://doi.org/10.1371/journal.pone.0092838
Wright BE, Riemer SD, Brown RF et al (2007) Assessment of harbour seal predation on adult salmonids in a Pacific Northwest estuary. Ecol Appl 17:338–351

Index

G. Heckel, Y. Schramm (eds.), *Ecology and Conservation of Pinnipeds in Latin America*, https://doi.org/10.1007/978-3-030-63177-2

The manufacturer's authorised representative in the EU is Springer Nature Customer Service Centre GmbH, Europaplatz 3, 69115 Heidelberg, Germany. If you have any concerns regarding our products, please contact ProductSafety@springernature.com

Printed and bound by CPI Group (UK) Ltd, Croydon, CR0 4YY
30/09/2026
02225692-0003